Have you been to our website?

For code downloads, print and e-book bundles, extensive samples from all books, special deals, and our blog, please visit us at:

www.rheinwerk-computing.com

Rheinwerk Computing

The Rheinwerk Computing series offers new and established professionals comprehensive guidance to enrich their skillsets and enhance their career prospects. Our publications are written by the leading experts in their fields. Each book is detailed and hands-on to help readers develop essential, practical skills that they can apply to their daily work.

Explore more of the Rheinwerk Computing library!

www.rheinwerk-computing.com

Michael Kofler

Linux Command Reference

Editor Rachel Gibson
Acquisitions Editor Hareem Shafi
German Edition Editor Christoph Meister
Translation Winema Language Services, Inc.
Copyeditor Julie McNamee
Cover Design Graham Geary
Photo Credits iStockphoto: 1467484831/© burroblando; Shutterstock: 2210496609/© Dragon Claws
Layout Design Vera Brauner
Production Kelly O'Callaghan
Typesetting SatzPro, Germany
Printed and bound in Canada, on paper from sustainable sources

ISBN 978-1-4932-2749-5
1st edition 2025
6th German edition published 2024 by Rheinwerk Verlag, Bonn, Germany

© 2025 by:
Rheinwerk Publishing, Inc.
2 Heritage Drive, Suite 305
Quincy, MA 02171
USA
info@rheinwerk-publishing.com
+1.781.228.5070

Represented in the E.U. by:
Rheinwerk Verlag GmbH
Rheinwerkallee 4
53227 Bonn
Germany
service@rheinwerk-verlag.de
+49 (0) 228 42150-0

Library of Congress Cataloging-in-Publication Control Number: 2025939046

Contents at a Glance

Overview of Commands Sorted by Topic

Managing Files

Finding Files

Compressing and Archiving Files

Processing Text Files

Access Control Lists (ACLs) and Extended Attributes (EAs)

Converters

Managing Processes

Managing Users and Groups

Administrating the File System

Logical Volume Manager (LVM)

SELinux and AppArmor

Package Management

Network Administration

Hacking and Security

Printer, Database, and Server Administration

Audio Functions and Hardware Management

Bluetooth

Kernel

System Start and Stop, init System, Logging, and GRUB

Virtualization, Containers, Cloud

Terminal and Text Console

Online Help

Graphics System and Gnome

Miscellaneous

"bash" Programming

"bash" Variable Management

Additional "bash" Commands and Special Characters

Configuration Files

Keyboard Shortcuts

Introduction

This book contains short descriptions of the most important Linux commands for managing the file system, for starting and terminating processes, for editing text files, for other administrative tasks, and for bash programming. The book also summarizes the syntax of basic configuration files and contains a keyboard shortcut reference for the emacs, nano, and vi editors, as well as some other interactive commands such as less or info.

The aim of this book is to provide a compact reference work for using Linux in the terminal. The basic principle of this book is: *less is more*. The book cannot and is not intended to replace the man and info pages of complex commands! Thus, you will still have to look up or research exotic options yourself.

With this book, however, I'm trying to relieve you of the work of searching through the often dozens of pages of original documentation for options for everyday use. Numerous examples also show the basic use of a command at a glance.

Sometimes, you're looking for a command for a specific task, but don't know its name or have just forgotten it. The thematically organized table of contents is intended for these cases.

Depending on which distribution you are using, some commands aren't available by default and must be installed separately. There are also distribution-specific commands that are only available in certain distributions, such as the package management commands dpkg and apt (Debian, Ubuntu), rpm and dnf or yum (Fedora, Red Hat), and zypper (SUSE). This is pointed out in the respective command description.

What Is a Command?

Linux doesn't distinguish between commands (as described in this book) and programs such as Firefox, LibreOffice, or GIMP. Here, "command" refers to programs without a graphical user interface, which are usually executed in a terminal window.

In this book, I also describe some commands that aren't real programs at all, but only commands of the currently active shell. I assume that you're using the bash (Bourne Again Shell) or the largely compatible zsh. These shells are used in most Linux distributions for the interactive execution of commands. An example of a shell command is cd to change the current directory.

Options

Most of the commands described in this book are controlled by options. The options are specified *before* all other parameters. There are two ways of notation for many commands: -x for short options (one letter) and --xyz for long options (multiple letters).

Thus, the following two ls commands are equivalent and display all files and directories in the /usr directory:

```
user$ ls -l -A /usr
user$ ls --format=long --almost-all /usr
```

For some commands, multiple options can be specified as a group (i.e., -ab instead of -a -b). Some commands also work with options that are specified after the actual parameter(s). However, this shouldn't lead you to the conclusion that this applies to all commands!

```
user$ ls -lA /usr
user$ ls /usr -lA
```

For a few commands, the order of the parameters has an influence on how the command is executed. If options are specified that are logically mutually exclusive, the last option specified applies.

man, info, and help

To avoid making this book unnecessarily long, I'll only describe the most important options. A complete overview of all options for the majority of commands is provided by command name --help. More detailed information is usually contained in the manual pages, which you can access via man *name* or man 1 *name*. For some commands, the man pages only contain a reference to the info texts, which are displayed accordingly via info *name*.

For commands that are directly integrated into the bash (e.g., cd), man *name* leads to the man page of the bash. The command is actually described there, but searching through the very long documentation is tedious. In this context, it's more helpful to use help *name*.

The help texts from man and info sometimes overshoot the mark. If you're just looking for a few examples of how to use a command, you should use either this book or the tldr command. Incidentally, the command name stands for *Too long, didn't read*. :-)

Commands from A to Z

aa-complain program

The aa-complain command activates the complain mode for the AppArmor profile of the specified program. Rule violations are logged but not punished, that is, the program continues to run undisturbed.

Like all other aa-xxx commands, aa-complain is only available for distributions that use the AppArmor security system (Debian, SUSE, and Ubuntu). On Ubuntu systems, the command can be found in the apparmor-utils package, which must be installed separately.

Example

The dovecot program is still monitored, but rule violations are only recorded in a logging file:

```
root# aa-complain /usr/sbin/dovecot
Setting /usr/sbin/dovecot to complain mode.
```

aa-disable program

aa-disable deactivates the AppArmor profile for the specified program. The program is therefore no longer monitored by AppArmor. To reactivate this, you can use aa-enforce or aa-complain.

aa-enforce program

The aa-complain command activates the enforce mode for the AppArmor profile of the specified program. AppArmor thus prevents file or network access that is prohibited by AppArmor rules.

aa-status [option]

aa-status provides a summary of the status of the AppArmor system. By specifying exactly one option from --complaining, --enabled, --enforced, or --profiled, you can shorten the output to detailed information.

Example

AppArmor is active on the test computer. Although 50 rule profiles are loaded, only 5 of them actually monitor running programs.

```
root# aa-status
50 profiles are loaded.
48 profiles are in enforce mode.
   /snap/snapd/20674/usr/lib/snapd/snap-confine
   /snap/snapd/20674/.../mount-namespace-capture-helper
   /usr/bin/evince
   /usr/bin/evince-previewer
5 processes are in enforce mode.
   /usr/sbin/cups-browsed (856)
   /usr/sbin/cupsd (552)
   /usr/sbin/libvirtd (566) libvirtd
   /snap/snapd-desktop-.../snapd-desktop-integration (1902)
   /snap/snapd-desktop-.../snapd-desktop-integration (1975)
```

---

ack [options] search term

---

ack is a variant of grep optimized for programmers and administrators. Unlike the original, ack ignores GIT and SVN directories, backup files, and binary files. In many distributions, ack can be found in the package of the same name or in ack-grep. Alternatively, ack can also be installed as a Perl module (cpan App::Ack).

ack is compatible with grep, but searches all files in the current directory recursively by default (e.g., grep -r). The command is most frequently called in the forms ack *search term* or ack -i *search term* (not case-sensitive). A reference of ack-specific options is provided by ack --help and man ack.

An alternative to the ack variant is the ag command. Some distributions provide the command in the silversearcher-ag package. Alternatively, you can clone the program from GitHub and then compile it yourself:

https://github.com/ggreer/the_silver_searcher/wiki

Example

The following example searches for files in /etc that contain the search term localhost and displays the corresponding hits together with the line numbers:

```
root# cd /etc
root# ack localhost
security/pam_env.conf
   52:# to "localhost" rather than not being set at all
```

```
  53:#REMOTEHOST  DEFAULT=localhost OVERRIDE=@{PAM_RHOST
  64:#NNTPSERVER  DEFAULT=localhost
speech-dispatcher/speechd.conf
  38:# for connections coming from localhost. If LocalhostAccessOnly is set
        to 0 it disables this
 199:#AudioNASServer "tcp/localhost:5450"
speech-dispatcher/modules/festival.conf
   7:#FestivalServerHost  "localhost"
...
```

acme.sh [options]

acme.sh is a script that helps with the installation of certificates. acme.sh uses *certificate authority* (CA) *https://zerossl.com* by default. The --server option can be used to set alternative CAs, including Let's Encrypt (*https://letsencrypt.org*).

The script is a useful alternative to the certbot command, whose Python dependencies have often led to installation problems in the past.

acme.sh must also be installed. But this can be done with almost no effort:

```
user$ sudo apt/dnf/yum/zypper install curl socat
user$ curl https://get.acme.sh -o acme-setup
user$ less acme-setup     (briefly check the script code)
user$ sudo sh acme-setup email=name@myhostname.com

  ...
  Installed to /root/.acme.sh/acme.sh
  Installing alias to '/root/.bashrc'
  OK, Close and reopen your terminal to start using acme.sh
```

As part of the installation, /var/spool/cron/root is extended by one line. This makes Cron take care of calling acme.sh --cron on a regular basis and thus automatically renewing all certificates set up by using acme.sh.

▶ --cron

Renews all certificates installed via acme.sh that are older than 60 days. The additional --force option can be used to force a premature renewal.

▶ --install-cert -d *domain*

[--cert-file *path/to/certfile*]

[--key-file *path/to/keyfile*]

[--fullchain-file *path/to/fullchainfile*]

[--reloadcmd "*systemctl restart httpd*"]

Copies the certificates previously generated with acme.sh --issue to a location where the web server can access them and then restarts the web server. Instead of systemctl restart httpd, you must enter a command suitable for your distribution to restart the web server.

Note that it's not advisable to copy the certificates stored in /root/.acme.sh to another directory yourself. acme.sh remembers the paths and the reload command for automatic certificate renewal.

acme.sh doesn't change the configuration files of your web server! You must add the statements with the paths to the certificate files yourself.

► --issue -d *domain* -w *webrootdir*
Sets up a certificate for the specified domain. The -d option can be used multiple times to create a certificate for related domains (e.g., -d a-company.com -d www.a-company.com). If you want a wildcard certificate, you need to pass -d a-company.com -d *.a-company.com, for example. The certificates are saved in the /root/.acme.sh directory.

If no web server is running on the computer, you should specify the --standalone option instead of -w *webrootdir*. acme.sh itself will then temporarily act as a web server. This assumes that port 80 isn't blocked by any program.

► --remove -d *domain*
Deactivates the certificate renewal for the specified domain, which takes place automatically every 60 days. Although the name of the option suggests otherwise, the certificates themselves won't be deleted. However, you can then delete the corresponding /root/.acme.sh/domain directory yourself.

► --server letsencrypt|buypass|google|sslcom|zerossl
Specifies the CA to be used. If this option isn't used, zerossl applies, that is, free certificates are obtained from *http://zerossl.com*.

► --upgrade
Updates the acme.sh script.

Example

The following commands generate a certificate for the domains a-company.com and www.a-company.com and copy the certificates to the /etc/acme-letsencrypt directory set up for this purpose:

```
root# mkdir /etc/acme-letsencrypt
root# acme.sh --issue --server letsencrypt -d a-company.com -w /var/www/html
  Your cert is in        /root/.acme.sh/a-company.com/a-company.com.cer
  Your cert key is in  /root/.acme.sh/a-company.com/a-company.com.key
  The intermediate CA cert is in      /root/.acme.sh/a-company.com/ca.cer
  And the full chain certs is there:  /root/.acme.sh/a-company.com/
                                      fullchain.cer
```

```
root# acme.sh --install-cert -d a-company.com \
        --cert-file     /etc/acme-letsencrypt/a-company.com.cert \
        --key-file      /etc/acme-letsencrypt/a-company.com.key \
        --fullchain-file /etc/acme-letsencrypt/a-company.com.fullchain
        --reloadcmd 'systemctl restart httpd'
Installing cert to:     /etc/acme-letsencrypt/a-company.com.cert
Installing key to:      /etc/acme-letsencrypt/a-company.com.key
Installing full chain to: /etc/acme-letsencrypt/a-company.com.fullchain
Run reload cmd: systemctl restart httpd
Reload success
```

Alternatives

In the past, certbot was often used instead of acme.sh. This Python script is still in use, but often causes problems during installation. Another alternative is getssl (see *https://github.com/srvrco/getssl*).

acpi [options]

acpi provides information on the computer's power supply:

▶ -a

Indicates whether the computer is connected to the power supply.

▶ -b

Indicates the battery status.

▶ -t

Displays the temperature of the battery.

▶ -V

Displays all available information.

add-apt-repository ppa:name

The Ubuntu-specific add-apt-repository command from the software-properties-common package sets up a new package source for a *personal package archive* (PPA). PPAs are unofficial package sources from Ubuntu developers, which often contain the latest versions or developer versions of popular programs.

Example

The following command sets up a package source for the latest version of LibreOffice:

```
root# add-apt-repository ppa:libreoffice/ppa
```

---

addgroup [options] name

addgroup sets up a new group on Debian and Ubuntu systems, taking into account the settings from adduser. In Fedora and Red Hat, addgroup is a link to the groupadd command with its own syntax.

▶ --gid *n*
Assigns the specified GID (Group Identifier) to the group.

---

adduser [options] name
adduser name group

adduser sets up a new user on Debian and Ubuntu systems, taking into account the settings from /etc/adduser.conf. In particular, a group with the same name is created for each user and assigned to the new user. In the second syntax variant, adduser adds the existing user to an additional group.

In Fedora and Red Hat, adduser is a link to the useradd command, which means a different syntax applies there. On (open)SUSE, adduser isn't available, and you have to use useradd instead.

▶ --disabled-login
Dispenses with the password prompt. It's impossible to log in until the password has been set.

▶ --gecos '*full name*'
Doesn't ask for the full name, (office) room, phone numbers, and other information.

▶ --group
Only creates a new group (no user). adduser --group corresponds to addgroup.

▶ --home *directory*
Specifies the desired home directory if that isn't supposed to be /home/*name*.

▶ --system
Sets up a system user. The UID is selected from the range for system users (usually 100–999), and the login is blocked (/bin/false shell). No password can be entered. The system user is also not assigned a valid home directory. These precautionary measures are intended to prevent a misuse of the account.

▶ --uid *n*
Assigns the specified UID (*User Identifier*) to the user.

Example

The following command sets up the new user kofler on a Debian/Ubuntu system and then adds it to the libvirtd group:

```
root# adduser --gecos 'Michael Kofler' kofler
root# adduser kofler libvirtd
```

To set up a user that can't log in (e.g., for a mail account), you should proceed as follows:

```
root# adduser --shell /bin/false --gecos ,,, mailuser
```

alias abbreviation=command

The alias shell command defines a new abbreviation or displays an existing abbreviation. If alias is used without any other parameters, all defined abbreviations are displayed.

On Fedora and Red Hat systems, there are some predefined aliases for root. These are loaded from the /root/.bashrc file when you log in.

Example

The following command defines the ll abbreviation for the ls -l command:

```
user$ alias ll='ls -l'
```

alien [options] package file

alien converts the specified package file into a different format. The desired format is specified by the options --to-deb (Debian), --to-rpm (RPM), or --to-tgz (tar archive). alien must be run by root so that the owners and access rights of the new packages are set correctly.

alien only works without problems for simple packages. However, if a package uses installation scripts or other specific features of the respective package format, the installation of the previously converted package will most likely fail.

alsactl [options] [command] [card/id/device]

You can use alsactl to display, save, and restore the status of the audio system. The command also provides some advanced control options that aren't available in the simpler amixer command.

alsactl generally applies to all audio devices, unless you explicitly specify the number or the device name of the audio device when calling the command (see also the /proc/asound/cards file).

▶ -b

Runs alsactl in the background. This option helps the init system to start alsactl as a background service.

▶ -f *file*
Specifies which file should be read by asactl or used for saving (by default, this is /var/lib/alsa/asound.state).

▶ -s
Uses the syslog service to log messages, warnings, and errors.

The most important commands of alsactl are listed here. There are a few other commands that are specifically intended for use as a background service and are rarely of interest for interactive calls or scripts.

▶ init
Initializes the audio system with default settings.

▶ monitor
Tracks all changes to the audio system and displays them until the command is ended using ⌞Ctrl⌟+⌞C⌟.

▶ restore
Restores a previously saved state of the audio system. The data is read from /var/lib/alsa/asound.state or from the file specified by -f.

▶ store
Saves the state of the audio system in the text file /var/lib/alsa/asound.state or in another file specified by -f.

alsamixer [options]

alsamixer is an interactive command in which you can set the volume or recording level of the audio channels using the function and cursor keys. ⌞Esc⌟ terminates the program. alsamixer isn't suitable for script programming. For this purpose, you can use amixer.

▶ -c *n*
Selects the desired audio card if there is more than one. The audio cards are numbered consecutively, starting with 0. If necessary, you should take a look at the file /proc/asound/cards, which lists all audio cards.

▶ -D *device*
Selects the audio system. For example, you can use -D pulse to change the PulseAudio system settings. Without this option, alsamixer controls the ALSA system directly.

▶ -V playback|capture|all
Determines which controllers are initially displayed: for audio playback, for recording, or all of them. When alsamixer is running, you can switch between these display formats using ⌞F3⌟, ⌞F4⌟, and ⌞F5⌟.

`alternatives` [options] command

`alternatives` is the Red Hat or Fedora version of `update-alternatives`. The syntax of the two commands is largely identical and is described in the `update-alternatives` command.

`amixer` [options] [command]

You can use the `amixer` command to change the volume or level of the audio system channels by using options. If you want to adjust the volume or other settings interactively, you can use the `alsamixer` command with a simple, text-based user interface.

▶ `-c n`

Selects the desired audio card. The audio cards are numbered consecutively, starting with 0. If necessary, you can take a look at the `/proc/#asound/cards` file, which lists all audio cards. If the option is missing, `amixer` processes the default audio device.

▶ `-D device`

Selects the audio system (e.g., `-D pulse` for the PulseAudio system). Without this option, `amixer` controls the ALSA system directly.

▶ `-q`

Doesn't output the new status after the settings have been changed.

▶ `-s`

Reads the commands to be executed from the standard input. Only the `sset` and `cset` commands are taken into account.

The following list describes the most important `amixer` commands:

▶ `cget id`

Indicates the status of a switch. The switch can be designated by one or more parameters, for example, as `numid=3` or `iface=MIXER,name='Headphone Playback Switch'`. A list of all controls and switches, including the corresponding ID data, is created by `amixer controls`.

▶ `controls|scontents|scontrols`

Lists all the controls and switches in varying degrees of detail and indicates their current status.

▶ `cset id parameter`

Changes the status of a switch. For example, `amixer cset numid=3 1` switches on the audio output on Raspberry Pi.

▶ `info`

Displays a brief description of the audio card.

- sget '*name*',*n*

  Returns the status of the specified control. Controls are described by a combination of character string and number, such as 'Master',0. The number is usually 0, unless an audio card provides multiple controls with the same name, which are then numbered consecutively.

- sset '*name*',*n parameter*

  Changes the status of the control. In the simplest case, you can enter a percentage value as a parameter, such as 50%. If it's a stereo control, you can set the two channels separately using a pair of numbers, for example, 40%,55%. Alternatively, many controls that have the suffix dB can also be set in decibels, such as -12.5dB.

  You can also use trailing plus or minus signs to change controls in relative terms. 5%+ increases the volume or recording level by five percentage points, 5%- reduces the volume or level by five percentage points.

  You can use mute, unmute, and toggle to change the mute mode. cap activates the recording (*capture*), while nocap deactivates this function.

  If the control supports multiple channels, you can select them with the additional parameters front, rear, center, and woofer.

Example

On the test computer, the third audio device (-c 2, the numbering starts with 0) is a USB headset. There are only two *simple controls* for controlling the headphones and the microphone. amixer sset then sets the headphones to a medium volume.

```
user$ amixer -c 2 scontrols
Simple mixer control 'Headphone',0
Simple mixer control 'Mic',0
user$ amixer -c 2 sset 'Headphone',0 50% unmute
Simple mixer control 'Headphone',0
  Capabilities: pvolume pswitch pswitch-joined
  Playback channels: Front Left - Front Right
  Limits: Playback 0 - 41
  Front Left: Playback 21 [51%] [-20.00dB] [on]
  Front Right: Playback 21 [51%] [-20.00dB] [on]
```

apk command

The apk command is used for package management in Alpine Linux. This minimalist Linux distribution is particularly popular in the Docker context.

▶ add *name*
Installs the specified package. If the --update option is passed to the command, apk
updates the package cache beforehand (as with apk update). --no-cache has almost
the same effect, but the cache is deleted immediately afterwards to save space.

▶ del *name*
Removes the specified package.

▶ info
Lists all installed packages.

▶ search *search term*
Searches for packages whose package name matches the search term.

▶ status
Shows how many packages are currently installed.

▶ update
Updates the cache of all package sources listed in /etc/apk/repositories.

▶ upgrade
Updates all installed packages.

Example

The following command first updates the cache of the package sources and then
installs the two packages build-base and python-dev:

```
root# apk add --update build-base python-dev
```

apropos topic

apropos provides a list of all man texts that contain information on the specified topic. If
apropos doesn't work, the underlying databases that can be created via mandb or
makewhatis are probably missing.

Example

apropos editor provides a list of various installed edit commands:

```
user$ apropos editor
ed (1)     - line-oriented text editor
editor (1) - Vi IMproved, a programmer's text editor
ex (1)     - Vi IMproved, a programmer's text editor
jmacs (1) - Joe's Own Editor
joe (1)    - Joe's Own Editor
...
```

apt command

`apt` is the most popular command for package management on Debian-based distributions. This also applies to Linux Mint, Raspberry Pi OS, and Ubuntu. Possible alternatives are the previously very popular `apt-get` command (it's still preferable for automated installations and script programming) and `aptitude` with a text-based graphical user interface.

`apt` installs, updates, and removes packages, taking into account all packages in the package sources defined in `/etc/apt/sources.list`. The following list describes the most important commands:

- `autoclean`
  Only removes packages from the package cache for which a newer version is already available.

- `autoremove`
  Uninstalls all packages that are no longer required and have been installed due to dependencies.

- `full-upgrade`
  Works in a similar way to `upgrade`, but also carries out updates for which packages have to be removed.

- `install` *name1 name2* ...
  Searches for the packages `name1`, `name2`, and so on in all APT package sources; downloads them; and installs them. If necessary, additional packages are also loaded and installed or updated to fulfill package dependencies.

- `install` *./filename*.`deb`
  Installs a Debian package previously downloaded locally. All APT package sources are still used to resolve package dependencies. It's important that the path to the file is recognizable as such and begins with `./` for the local directory or `/` for an absolute path. Otherwise, `apt` thinks it's a package name, searches for it in the package sources, and finally returns an error message.

- `list`
  Lists all available packages. With the `--installed` option, the command only returns the currently installed packages.

- `remove` *name1 name2* ...
  Uninstalls the specified packages.

- `search` *search term*
  Lists all packages whose name or description contains the search term.

- `show` *name1*
  Displays information on the specified package.

▶ update
Updates the package lists of the archives specified in `sources.list`. Only the meta-data of the package sources is read and entered in APT cache files. However, no packages are downloaded or updated! The sole purpose of this command is to let apt know which packages are available on the internet. The command should be run before any other apt command. Note: If you want to update only one package (not all packages), you need to run apt install *name*, instead of apt update *name*!

▶ upgrade
Updates all installed packages if newer versions are available in the package sources. If necessary, additional packages are also installed for the update. However, apt upgrade doesn't perform updates for which it's necessary to remove packages.

Example

The following commands show how you can use apt:

```
root# apt update         (update package sources)
root# apt full-upgrade   (update all installed packages)
root# apt install apache2 (install Apache web server)
```

apt-cache command

The apt-cache command, which is specific to Debian and Ubuntu, provides information about available or previously installed packages.

▶ madison *package name*
Provides a one-line summary of the results of apt-cache policy (see below).

▶ policy
Displays a detailed list of all package sources.

▶ policy *package name*
Shows which package source the currently installed package comes from and which package sources may provide alternative versions of the package.

▶ search *pattern*
Returns a list of all available packages (regardless of whether they have already been installed or not) whose package name or package short description contains the search text *pattern*. The additional --names-only option restricts the search to the package name.

▶ show *package name*
Provides a short package description. This also works for packages that aren't installed, as long as the package description is in the cache.

▶ showpkg *package name*
Displays the dependency information for the package.

▶ stats
Provides statistics on the number of installed and available packages.

apt-cache can't provide detailed information about the specific contents of a package or establish an association between a file and the corresponding package. If you're interested in this information, you must use dpkg.

Examples

The following command returns a sorted list of all packages whose names contain gimp:

```
root# apt-cache search --names-only gimp | sort
  gimp-cbmplugs    - Extensions for GIMP to import/export
                     Commodore-64 files
  gimp-data        - Data for GIMP
  gimp-data-extras - Extra brushes and patterns for GIMP
  ...
```

The second example shows which version of the openjdk-17-jre package is installed and whether there are alternative versions. apt-cache only finds an older version (17.0.2 instead of 17.0.10):

```
user$ apt-cache policy openjdk-11-jre
openjdk-17-jre:
  Installed: 17.0.10+7-1~22.04.1
  Candidate: 17.0.10+7-1~22.04.1
  Version table:
  ** 17.0.10+7-1~22.04.1 500
        500 http://de.../ubuntu-ports jammy-updates/universe arm64 Packages
        500 http://ports... jammy-security/universe arm64 Packages
        100 /var/lib/dpkg/status
     17.0.2+8-1 500
        500 http://de../ubuntu-ports jammy/universe arm64 Packages
```

apt-file [options]

The apt-file command from the package of the same name makes it possible to search for a file that is contained in a Debian package—even if the package hasn't yet been installed.

Example

The following commands first create or update the local database and then search for packages with files containing main.cf:

```
root# apt-file update
user$ apt-file find main.cf
postfix:     /usr/share/postfix/main.cf.debian
postfix:     /usr/share/postfix/main.cf.dist
postfix:     /usr/share/postfix/main.cf.tls
postfix-doc: /usr/share/doc/postfix/examples/main.cf.default
postsrsd:    /usr/share/doc/postsrsd/main.cf.ex
...
```

apt-get [options] command

The apt-get command, which is specific to Debian and Ubuntu, installs, updates, and removes packages. For interactive use, the apt command is recommended today instead of apt-get. It has been optimized for ease of use and also provides subcommands for package search (apt show, apt search, and apt list), which are missing in apt-get.

However, apt-get isn't obsolete. It can still be used on all Debian- and Ubuntu-based distributions. Its use is particularly recommended if you want to automate package management in scripts. The following list describes the most important commands:

▶ autoclean
Removes only those packages from the package cache for which a newer version is already available.

▶ autoremove
Uninstalls all packages that are no longer required and have been installed due to dependencies.

▶ check
Updates the cache of all installed packages and detects any existing package conflicts and unresolved dependencies. This is only necessary if packages have been (de)installed without APT and conflicts have occurred as a result.

▶ clean
Removes all downloaded packages from the package cache.

▶ dist-upgrade
Works in a similar way to upgrade but also installs new packages if required.

▶ install *name1 name2* ...
Installs the packages name1, name2, and so on, and also takes package dependencies into account. If a package has already been installed, apt-get install *name* will update this package.

▶ remove *name1 name2* ...
Uninstalls the specified packages.

▶ source *name*
Installs the source code of the package in the current directory.

▶ update
Updates the package lists of the archives specified in sources.list. (Further details are given in apt update.)

▶ upgrade
Updates all installed packages if newer versions are available in the package sources.

The detailed behavior of the commands is controlled by options:

▶ -d or --download-only
Only downloads the packages to the /var/cache/apt/archives directory, but doesn't install them.

▶ --no-install-recommends
Dispenses with the installation of recommended packages.

▶ -s or --simulate
Simulates the installation, but doesn't make any actual changes.

▶ -y or --assume-yes
Answers all questions with *yes*, enabling the command to be used in a script.

Example

The following commands show how you can use apt-get:

```
root# apt-get update && apt-get dist-upgrade    (perform complete update)
root# apt-get remove -y emacs                   (remove emacs without asking)
```

apt-key command

Although the apt-key command is still mentioned in countless internet manuals, it's outdated. In the past, the command helped to manage the keys of package sources. It was usually executed in the form apt-key add keyfile.

The command is no longer available in current distributions. The key files must now be saved manually in the /etc/apt/trusted.gpg.d directory. In addition, when specifying the package source in /etc/apt/sources.list or in /etc/apt/sources.d/repo.list, the key file must be explicitly referenced using signed-by:

```
# /etc/apt/sources.d/myrepo.list
deb [signed-by=/etc/apt/trusted.gpg.d/myrepo.asc] https://apt.myrepo.com ...
```

aptitude [options] [command]

The Debian-specific `aptitude` command installs, updates, and removes packages, as well as accesses the APT infrastructure like `apt` and `apt-get`. The advantage compared to these two commands is that `aptitude` remembers dependent packages during the installation process and automatically removes them when uninstalling.

All package management functions are performed in the form of commands (e.g., `aptitude install` *package name*). The basic commands correspond to those in `apt-get` and are described there. However, the `check` and `autoremove` commands aren't available in `aptitude`. The powerful `search` command is available for this purpose.

Alternatively, you can also use the program with a text user interface in a console by simply starting the program without any additional parameters. Use the keyboard shortcut [Ctrl]+[T] to select the menu. Despite the menu, `aptitude` isn't really intuitive to use which is why most users use `aptitude`, like `apt-get`, simply to run individual commands.

`aptitude` isn't installed by default in most distributions and must be installed via `apt` or `apt-get` before it can be used.

aptitude search

`aptitude search` provides comprehensive options for searching the package database. In the simplest form, you can use `aptitude search` *search term* to search for packages whose name contains the search term. You can use ~ and a letter to formulate various search criteria. For example, ~i restricts the search to installed packages, and ~U displays packages for which updates are available. The -F option allows you to control which details of the search result are to be displayed in columns of a specific width.

A number of other search options are described in `/usr/share/doc/aptitude/README` and on the following page:

www.debian.org/doc/manuals/aptitude/ch02s04s05.en.html

Examples

The following two commands first install the `mysql-server` package and then remove it again. It's noteworthy that the second command removes almost all dependent packages installed together with `mysql-server`. (However, the client tools and libraries that are independent of the server are retained).

```
root# aptitude install mysql-server
The following NEW packages will be installed:
   libcgi-fast-perl{a} libcgi-pm-perl{a} libevent-core-2.1-7{a} ...
   mysql-server mysql-server-8.0{a} mysql-server-core-8.0{a}
```

```
0 packages upgraded, 19 newly installed
...
root# aptitude remove mysql-server
The following packages will be REMOVED:
  libcgi-fast-perl{u} ... mysql-server-core-8.0{u}
0 packages upgraded, 0 newly installed, 19 to remove
```

The following command provides a list of all installed packages. The package group or package source is specified in the first column (%20s, *Source*, with a column width of 20 characters), and the package name is specified in the second column.

```
user$ aptitude search ~i -F "%20s  %p"
utils           acl
admin           adduser
gnome           adwaita-icon-theme
games           aisleriot
...
```

You can use egrep to filter those packages from the result that originate from the Ubuntu package sources universe or multiverse.

```
user$ aptitude search ~i -F "%20s  %p" | egrep "universe|multiverse"
universe/admin          aptitude
universe/admin          aptitude-common
universe/introspection  gir1.2-gtk-vnc-2.0
universe/editors        joe
universe/libs           libburn4
...
```

arp [options] [hosts/ipaddresses/network]

arp from the net-tools package helps to manage the cache for the *address resolution protocol* (ARP). This cache contains a list of local network devices and allows the assignment to their MAC addresses. Without any further parameters, the command returns the current cache content. With the -n option, the corresponding IP addresses are displayed instead of host names.

```
user$ arp
Address                   HWtype  HWaddress          Flags Mask  Iface
fritz.box                 ether   3c:37:12:b7:a1:cb  C           enp0s1
Archer-C6.fritz.box       ether   d8:07:b6:4f:19:af  C           enp0s1
m3.fritz.box              ether   80:a9:97:32:c5:d7  C           enp0s1
BRN001BA99C5DA4.fritz.b   ether   00:1b:a9:9c:5d:a4  C           enp0s1
```

```
user$ arp -n
Address              HWtype   HWaddress          Flags Mask  Iface
192.168.178.1        ether    3c:37:12:b7:a1:cb  C           enp0s1
192.168.178.21       ether    d8:07:b6:4f:19:af  C           enp0s1
192.168.178.146      ether    80:a9:97:32:c5:d7  C           enp0s1
192.168.178.20       ether    00:1b:a9:9c:5d:a4  C           enp0s1
```

arp-scan [options] [hosts/ipadresses/network]

arp-scan sends address resolution protocol (ARP) packets to the specified hosts or IP addresses and displays the resulting responses. arp-scan thus provides an extremely fast way of finding out which hosts or IP addresses are running active devices and which MAC addresses they have. As usual, the address range can also be specified in the form 10.17.0.0/16.

The nmap command provides similar functions, albeit on a completely different technical basis. In short, nmap is slower but more thorough.

▶ -f *filename* or --file=*filename*
  Loads the hosts or IP addresses from the specified file.

▶ -I *interface* or --interface=*interface*
  Explicitly selects the network interface via which the ARP packets are to be sent. If this option isn't specified, arp-scan simply uses the first interface, ignoring the loopback interface.

▶ -l or --localnet
  Searches the entire local network. There is then no need to specify hosts or IP addresses.

Example

The following arp-scan command is used to perform a network scan for the IP addresses 10.0.0.*:

```
root# arp-scan --interface=enp0s3 10.0.0.0/24
Interface: enp0s3, datalink type: EN10MB (Ethernet)
Starting arp-scan 1.9 with 256 hosts
10.0.0.9        00:16:b6:9d:ff:4b       Cisco-Linksys
10.0.0.22       b8:27:eb:11:44:2e       Raspberry Pi Foundation
10.0.0.39       ac:87:a3:1e:4a:87       (Unknown)
...
7 packets received by filter, 0 packets dropped by kernel
Ending arp-scan 1.9: 256 hosts scanned in 1.825 seconds
  (140.27 hosts/sec). 7 responded
```

```
at [options] time
```

You can use at to specify that one or more commands are to be run at a later time. The commands are entered interactively after the at command; the input ends with `Ctrl`+`D`.

If the job produces a standard output, the output is sent by email to the user who executed at once it has been completed. This requires /usr/sbin/sendmail to be installed on the computer.

Time Specification

There are a number of variants for the time specification, the most important of which are presented here using examples:

▶ 3:45 [tomorrow]
Runs the commands at 3:45 am—today, if it's earlier, or tomorrow, if 3:45 am has already passed. You can explicitly specify a time on the next day with a subsequent tomorrow.

▶ [16:30] [2025-]12-31
Runs the commands on the specified day and at the specified time. The year and time are optional. If this data is missing, the current year and the current time are used. at 11-30 therefore means the following: on November 30 of this year at the same time at which the at command was executed.

▶ noon | teatime | midnight
Corresponds to the times 12:00, 16:00, and 0:00.

▶ now
Runs the commands immediately. now is often combined with + xxx.

▶ ... + 2 days
Adds two days to the preceding time specification. Permissible time units are minutes, hours, days, and weeks. The grammatically questionable statement at now + 1 hours therefore means the following: in one hour. (Time units should always be given in the plural.)

Options

▶ -b
Runs the command as soon as the system has a low workload. The time specification is omitted with this option. at -b corresponds to the batch command.

▶ -c n
Specifies which commands are run by the job with the specified number. You can determine the job numbers via atq.

- -d or -r *n*

  Removes the job specified by its number from the list of marked jobs. at -d or at -r corresponds to atrm.

- -f *file*

  Loads the commands to be executed from the specified file, not from the standard input.

- -l

  Lists all jobs. at -l corresponds to the atq command.

- -m

  Always sends a mail, even if the job doesn't return any output.

- -M

  Doesn't send an email after completing the job.

- -q*x*

  Assigns the job to queue *x*, where *x* can be a lowercase or uppercase letter. The letters a and b are used for ordinary at jobs and for batch jobs. The higher the selected queue letter, the higher the nice value with which the job is executed. As a result, -qz is suitable for jobs with very low urgency.

- -t *YYYYMMDDhhmm*

  Runs the job at the specified time. The -t option is an alternative to the other forms of time specification.

Example

Changes have been made to the configuration of the web server, which will only take effect after a restart. However, to prevent active users of the website from losing session data, the web server shouldn't be restarted immediately, but at 1:00 am the following day. The service name apache2 applies to Debian and Ubuntu. On Fedora and RHEL, you must enter httpd instead.

```
root# at 1:00
service apache2 restart
<Ctrl>+<D>
```

atq

atq lists all jobs that have been set up using at for execution at a future point in time. If atq is run by root, it displays the jobs of all users; otherwise, only its own jobs are displayed.

---

atrm n

atrm deletes the job specified by its number. You can use atq to determine the numbers of all jobs that have been set up for future execution.

---

attr [options] files

attr determines or changes the extended access attributes of the specified files or directories. This only works if the file system supports *extended attributes* (EAs). For ext3/ext4 file systems, the mount option user_xattr must be used.

Instead of attr, you should prefer the getfattr or setfattr commands. attr is only available for reasons of compatibility with IRIX SGI.

---

awk [options] 'code' text files
awk [options] -f code file text files

awk isn't a simple command, but a separate programming language that helps with the processing and analysis of structured texts. For example, you can use it to search a text file for keywords and extract a text column from the subsequent text section. You can either enter the program code directly or read it from a code file using the -f option. awk then processes all text files passed as parameters line by line and writes the result to the standard output.

The program code to be processed by awk follows this simplified syntax:

```
/searchpattern1/ {actions}
/searchpattern2/ {more actions}
condition      {even more actions}
```

The same rules that are described in the grep command apply to the structure of the search pattern. Multiple actions are separated by semicolons. If the search pattern is missing, the action applies to every line of the text file. Conversely, if the action is missing, the text captured by the pattern simply gets output.

A comprehensive description of awk and its capabilities isn't possible here for reasons of space. Instead, I will restrict myself to a few examples. If you want to familiarize yourself more deeply with awk, the following tutorial is a good introduction:

www.grymoire.com/Unix/Awk.html

The official *User's Guide* goes into more detail, but with almost 500 pages, it may put off beginners:

www.gnu.org/software/gawk/manual/gawk.pdf

Examples

This excerpt from /etc/passwd serves as the starting point for the following examples:

```
root:x:0:0:root:/root:/bin/bash
bin:x:1:1:bin:/bin:/sbin/nologin
daemon:x:2:2:daemon:/sbin:/sbin/nologin
adm:x:3:4:adm:/var/adm:/sbin/nologin
lp:x:4:7:lp:/var/spool/lpd:/sbin/nologin
...
```

The first awk example only returns the first and seventh column from /etc/passwd, that is, the login name and the assigned shell. The -F option specifies that the columns are separated by colons and not by spaces and tab characters:

```
user$ awk -F':' '{print $1,$7}' /etc/passwd
root /bin/bash
bin /sbin/nologin
daemon /sbin/nologin ...
```

The second example determines the home directory and the shell of kofler:

```
user$ awk -F':' '/kofler/ {print $6; print $7;}' /etc/passwd
/home/kofler
/bin/bash
```

The third example displays the login names of all users with a UID greater than or equal to 1000:

```
user$ awk -F':' '$3>=1000 {print $1}' /etc/passwd
nfsnobody
kofler
test
```

The last example extracts the file size and the file name from the output of ls -l. Note, however, that this only works for file names that don't contain spaces.

```
user$ ls -l *.txt | awk '{print $5,$9}'
123 file1.txt
213231 file2.txt
...
```

```
aws [options] main command [subcommand] [parameter]
```

The aws command helps with the administration of various Amazon cloud services. As these services are often used from Linux computers, many administrators come into contact with aws.

This section refers to aws version 2 recommended by Amazon. You can use aws --version to make sure that you haven't installed the still widely used version 1! Both aws versions use the same command name and are largely compatible with each other, but there are also some differences:

https://docs.aws.amazon.com/cli/latest/userguide/cliv2-migration.html

Installation and Commissioning

aws isn't installed by default and usually isn't available as a package in your distribution. Instead, you must run the following commands for installation. On computers with ARM CPUs, you need to replace x86_64 with aarch64.

```
user$ sudo apt/dnf/yum/zypper install curl
user$ curl "https://awscli.amazonaws.com/awscli-exe-linux-x86_64.zip" \
      -o "awscliv2.zip"
user$ unzip awscliv2.zip
user$ sudo ./aws/install
```

The ZIP file contains a minimal Python 3 distribution with all the required packages. This means that there are neither dependencies on Python packages in your distribution nor conflicts due to differing version numbers. aws is installed in the /usr/local/bin directory. If this directory isn't part of PATH, you must specify the path explicitly.

```
root# aws --version
aws-cli/2.15.28 Python/3.11.8 Linux/6.1.0-18-arm64 exe/aarch64.debian.12
```

The connection to the Amazon cloud services is encrypted. Before you can configure the aws command for further use, you must set up a user and group on the AWS **Identity and Access Management** (IAM) page. The group should be linked to the AmazonS3Full Access policy and to the new user. You can find the AWS page at:

https://console.aws.amazon.com/iam

Now run aws configure in the terminal, and enter the *Access Key ID* and the *Secret Access Key* you received when setting up the user. aws saves this data and the other options in .aws/credentials and .aws/config.

```
root# aws configure
AWS Access Key ID [None]:     AKxxxxxxxx
AWS Secret Access Key [None]: ZRxxxxxxxxxxxxxxxxx
Default region name [None]:   eu-central-1
Default output format [None]:
```

Syntax and Online Help

You usually pass a main command to the aws command, which selects the desired AWS service, as well as a subcommand that applies specifically to the service in question. For example, aws s3 ls lists your buckets in the S3 service. For reasons of space, I'll restrict this to a few commands for the EC2 (*Elastic Compute Cloud*) and S3 (*Simple Storage Service*) services.

There is no man page available for aws. Instead, aws help provides an overview of the syntax. An overview of the commands for a specific service is provided by aws <service> help, such as aws ec2 help. Details of a subcommand can be found in aws <service> <subcommand> help, for example, aws s3 cp help.

S3 Administration

▶ aws s3 ls [*path*]
Lists all S3 buckets without further parameters or displays the content of a bucket directory or a file located there. Paths are entered in the form s3://<bucketname>/<directory>/<file> (e.g., s3://my-bucket/my-dir/my-file).

▶ aws s3 cp *path1 path2*
Copies a local file to an S3 bucket and from there back to the local file system. You can also copy an S3 file to another S3 location. You can use the --recursive option to copy directories recursively. For testing purposes, you need to pass --dryrun: the command then shows what it would do, but doesn't change anything.

If you use S3 for data backup, it's recommended that you only upload files to S3 that you've previously encrypted yourself. In the past, configuration errors have repeatedly occurred so that supposedly private buckets were publicly accessible after all. In addition, the possibility of intelligence services gaining access can't be ruled out.

▶ aws s3 mv *path1 path2*
As in the previous entry, but moves the file or renames a file that is already in the S3 memory.

▶ aws s3 sync *path1 path2*
Synchronizes two directories: one local and one in the S3 storage. This is faster than aws s3 cp because unchanged files won't be transferred again. Files deleted in the source path are retained in the target path. If you don't want this, you must pass the --delete option.

▶ aws s3 mb *newbucket*
Creates a new bucket, specifying the name in the usual manner (i.e., s3://<bucketname>). The name must be unique worldwide.

With the --region option, you can specify the data center in which the bucket is to be set up, which differs from the region preset in aws configure.

The new bucket isn't publicly accessible by default. Access rights and access rules (ACLs, i.e., *Access Control Lists*) must be set via the AWS web console.

▶ aws s3 rb *bucket*
Deletes the specified bucket. The operation is only successful if the bucket is empty. Otherwise, you can force the deletion using --force.

EC2 Administration

While the aws commands for S3 administration are easy to use, this unfortunately doesn't apply to EC2: The number of subcommands alone is daunting—there are more than 400 of them. Many commands also provide or expect JSON data structures and are therefore only intended for script programming, but not for use by humans. You can try out aws ec2 describe instances as a starting point. The command provides a list of all your instances already set up as a JSON document:

```
root# aws ec2 describe-instances
{
    "Reservations": [
        {
            "Groups": [],
            "Instances": [
                {
                    "AmiLaunchIndex": 0,
                    "ImageId": "ami-02fc41eea1852",
                    ...
```

Example

The following two commands set up a new bucket and copy all files from the /etc directory to it:

```
root# aws s3 mb s3://info.kofler.etcbak
root# aws s3 sync /etc s3://info.kofler.etcbak
```

badblocks device

badblocks performs a low-level check of the data medium. This is only useful if you suspect that your hard disk or SSD may be defective. If you intend to set up an ext file system on the hard disk (partition), it's better to perform the check using mkfs.ext2 -c as part of the formatting.

Usually, badblocks doesn't destroy any data and only performs a read-only test. You can carry out more thorough tests by adding the -n or -w options:

▶ -n

Performs read and write tests, but restores the original content of the data blocks. The option can't be used if the data medium or partition is currently being used, that is, is being included in the directory tree via mount.

▶ -w

Performs write tests with bit patterns. Caution: The contents of the data medium or partition will be deleted!

basename string [ending]

basename returns the file name of the transferred path. basename /etc/X11/Xmodmap thus leads to the result Xmodmap. If a file extension is specified as an additional parameter, this file extension (if present) will be removed from the file name.

Example

basename eliminates the path and the identifier .jpg. What remains is white.

```
user$ basename /home/kofler/Bilder/Wallpapers/white.jpg .jpg
 white
```

bat text file

The bat command from the package of the same name is a variant of the cat command. In contrast to cat, bat attempts to recognize the content of the text and then performs colored syntax highlighting when it's displayed. In some distributions (e.g., Debian), bat must be called under the name batcat to avoid a name conflict.

bat relies on your terminal to use a dark background by default. If you use a light background, you must pass a corresponding option to bat (e.g., --theme gruvbox-light) or permanently set the color theme in configuration file .config/bat/config.

batch

batch marks a job order for later execution. As with at, the commands to be executed are then entered interactively. The entry must be completed using Ctrl+D. Unlike at, there's no time specification for when the job is supposed to be run. Instead, the background daemon atd waits until the system has a low load. The threshold is defined as a load average of 1.5. If the job provides an output, this is sent to the user via email, which requires /usr/sbin/sendmail to be available on the computer.

batch processes neither options nor parameters. If you want to read the job order from a file, you must run at -qb -f file now.

Example

The execution of the backup script is started as soon as the system load is low:

```
user$ batch
./backup-script
<Ctrl>+<D>
```

bc [file]

bc stands for *basic calculator* and is a command-oriented pocket calculator. It can be used interactively to perform simple calculations. bc is also useful in scripts if the limited calculation functions of bash are insufficient. bc expects the calculation statements from the standard input or from a file.

Examples

The following examples show both the interactive use of bc and its use in commands. The `scale=3` statement controls the number of decimal places in the output.

```
user bc
> 2+3
5
> <Ctrl>+<D>
user$ echo "5 + 3 * 7" | bc
26
user$ echo "scale=3; sqrt(2)" | bc
1.414
```

Alternatives

bc only knows a few advanced mathematical functions. For more complex calculations, it's better to use qalc. However, qalc has the disadvantage that it has to be installed separately.

bg [process]

The bg shell command continues an interrupted process in the background. If no process number is specified, bg applies to the last process that was interrupted using Ctrl+Z. Otherwise, the process must be specified by its name or by the bash-internal job number (not by the PID!).

blkid [options] [device]

blkid provides information about the block device specified by the device name, for example, about hard disk partitions, logical volumes, or RAID devices. blkid specifies the file system type, the UUID, and the name (label) of the file system. If the device specification is missing when blkid is called, blkid provides this information for all active partitions and logical volumes. By specifying options, you can search for specific file systems:

▶ -k

Lists all file system types supported by the kernel.

▶ -L *label*

Searches for file systems with the specified name.

▶ -n *fstyp1,fstyp2,fstyp3*

Searches for file systems that correspond to the specified types.

▶ -t *name=value*

Searches for file systems that match the search criterion. Permitted criteria are TYPE (file system type), LABEL (name of the file system), PARTLABEL (name of the partition), UUID (UUID of the file system), and PARTUUID (UUID of the partition). blkid -t TYPE= ext4 corresponds to blkid -n ext4 and returns a list of all ext4 file systems.

▶ -U *uuid*

Searches for file systems with the specified UUID.

Examples

The /dev/sda1 partition contains an ext4 file system with the UUID 2716...19e4:

```
root# blkid /dev/sda1
/dev/sda1: UUID="27162884-8fe9-4fa9-8b5e-712ab82d19e4" TYPE="ext4"
```

The search for a partition with the label lvm2 returns /dev/sdb2:

```
root# blkid -t PARTLABEL=lvm2
/dev/sdb2: PARTLABEL="lvm2" PARTUUID="371cc374-847f-407a-bb5b-7ff015722383"
```

bluetoothctl

On desktop systems, a Bluetooth configuration is usually done via graphical user interfaces. If that's not available (e.g., on Raspberry Pi to which no monitor is connected) or doesn't work, the configuration with bluetoothctl can also be performed in text mode. For this purpose, you need to start an interactive session with bluetoothctl and then run commands to find and pair the Bluetooth device.

You can run the commands described in the following list within a `bluetoothctl` session. It's important to keep pressing the Bluetooth button (pairing button) on Bluetooth devices that you want to reconfigure, so that the device signals to the environment that it's ready to connect. On many devices, a blue LED then flashes. If there's no such button, you can also try turning the device off and on again.

► `agent on/off`

(De)activates the Bluetooth agent. This software component takes care of authorizing new devices and allows, for example, the keypad entry of a pairing code.

► `connect` *xx:xx:xx*

Establishes the connection. (This only works if `pair` and `trust` have been executed beforehand!) If everything works out, the response is *connection successful*. The device can now be used.

► `exit` or `quit`

Ends the `bluetoothctl` session.

► `info` *xx:xx:xx*

Shows the connection status and various other information about the device.

► `pair` *xx:xx:xx*

Initiates the connection to a device. If you use a keypad, you'll be prompted to enter a multi-digit code. Don't forget to finish the input by pressing ⏎ Enter ! You can recognize successful pairing by the *pairing successful* message. Devices without an input option (mouse, headset, loudspeaker, etc.) can be paired without entering a code.

► `pairable on/off`

Activates/deactivates the coupling mode.

► `scan on/off`

(De)activates the scan mode. In scan mode, the program lists all devices. This process can take some time and individual devices are displayed again and again. When you've found the desired device, you should switch the mode off again via `scan off`.

► `trust` *xx:xx:xx*

Trusts the specified device.

`bluetoothctl` saves the Bluetooth configuration for a specific device permanently in `/var/lib/bluetooth/id1/id2/info`. There, `id1` is the ID of the Bluetooth controller, and `id2` is the ID of the Bluetooth device.

`bluetoothctl` is actually intended for interactive use. To run a single command immediately, you can proceed as follows:

```
root# echo -e "connect FC:58:FA:A0:4D:E7\nquit" | bluetoothctl
```

Example

The following lines show in slightly abbreviated form the inputs and outputs for configuring a Bluetooth keyboard:

```
user$ bluetoothctl
[bluetooth]# agent on
[bluetooth]# pairable on
[bluetooth]# scan on
Discovery started
[CHG] Controller 00:1A:7D:DA:71:13 Discovering: yes
[NEW] Device 70:10:00:1A:92:20 70-10-00-1A-92-20
[CHG] Device 70:10:00:1A:92:20 Name: Bluetooth 3.0 Keyboard
...
[bluetooth]# scan off
[bluetooth]# pair 70:10:00:1A:92:20
Attempting to pair with 70:10:00:1A:92:20
[CHG] Device 70:10:00:1A:92:20 Connected: yes
[agent] PIN code: 963064
[CHG] Device 70:10:00:1A:92:20 Paired: yes
Pairing successful
[bluetooth]# trust 70:10:00:1A:92:20
[bluetooth]# connect 70:10:00:1A:92:20
[bluetooth]# info 70:10:00:1A:92:20
Device 70:10:00:1A:92:20
        Name: Bluetooth 3.0 Keyboard
        Paired: yes
        Trusted: yes
        Connected: yes
    ...
[bluetooth]# exit
```

boltctl command

boltctl controls devices connected via the Thunderbolt interface. Thunderbolt is currently mainly used in notebooks with Intel CPUs. It's likely to become more widespread in the future because Thunderbolt is part of the new USB4 standard.

▶ authorize *device*
Allows the use of a Thunderbolt device once. You can determine a list of all connected devices via boltctl list. To give permanent permission, you must use boltctl enroll.

▶ config *key*
config *key newvalue*

57

Loads or changes configuration options (requires `root` permissions). A list of all options is provided by `boltctl config --describe`.

▶ enroll [`--policy auto|manual|default`] *device*
Allows the device to communicate permanently via the Thunderbolt interface. Explicit permission is only required for devices that use the PCI bus via Thunderbolt (e.g., for a docking station) and only if strict Thunderbolt security guidelines are set in the BIOS/EFI.

auto means that the device can be used in the future without further queries. manual requires confirmation using `boltctl authorize` each time it's reconnected. default accepts the Thunderbolt daemon's specifications (usually this is auto).

▶ forget *device*
Removes the ID of a device from the enroll database and thus reverses `boltctl enroll`.

▶ info *device*
Displays detailed information on the specified device.

▶ list `-a`
Lists all connected devices.

▶ monitor
Tracks and logs which devices are connected or removed until the command is ended using Ctrl+C.

▶ power
Activates a Thunderbolt controller that is in sleep mode.

break [n]

The `bash` command `break` prematurely terminates a `for`, `while`, or `until` loop in shell scripts. The shell program then continues with the next command after the end of the loop. By specifying an optional numerical value, *n* loop levels can be canceled.

btrfs command

`btrfs` is the central administration command for Btrfs file systems. Provided there are no ambiguities, you can abbreviate the commands described in this section and save yourself a lot of typing. `btrfs fi sh` corresponds to `btrfs filesystem show`.

The administration of Btrfs file systems is exceptionally complex. For this reason, only the most important commands are summarized here. There's just not enough space here to describe the `btrfs` basics in detail.

▶ balance `start|pause|resume|cancel|status` *mount directory*
Distributes the data evenly across all devices in the file system. This process takes a

very long time and is carried out in the background. It can be interrupted via btrfs balance pause, resumed with btrfs balance resume, or stopped completely using btrfs balance cancel. btrfs balance status provides information on how far the balancing process has progressed.

▶ device add *devicename mount directory*
Adds another device to an active Btrfs file system and thus increases its data pool.

▶ device delete *devicename mount directory*
Removes a device from the data pool of a Btrfs file system. The data contained on the device is transferred to other devices in the file system, which, of course, takes a long time with large file systems.

If you want to remove a defective (no longer available) device, you should enter missing as the device name. For RAID-1 systems, you must first add a replacement device and run btrfs filesystem balance!

▶ filesystem df *directory*
Displays detailed information about how much space is reserved for data, metadata, and system data and how much of this space is already in use. However, the command doesn't provide any information about how much unreserved memory space is still available, which makes it difficult to interpret the data correctly.

filesystem resize [+/-]size *mount directory*
Enlarges or reduces the file system during operation. The new size can be specified either in absolute or relative terms. The abbreviations k, m, and g for KiB, MiB, and GiB are allowed. Max is also permitted as a size specification—the file system then uses the entire size of the underlying device in future. Remember that you may have to adjust the size of the underlying device *before* enlarging the file system, but *after* reducing it!

Instead of resize, it's easier to add an additional device to Btrfs file systems (btrfs add *device*).

▶ filesystem show [*directory/device*]
Displays basic information on Btrfs devices. Among other things, the command provides information on how large the devices are and how much space is available on the devices that hasn't yet been reserved for data, metadata, or system data.

▶ filesystem sync *directory*
Synchronizes the content of the file system or one of its directories.

▶ scrub start|cancel|resume|status *directory/device*
Controls the scrub process. scrub start loads all files in the file system and checks their integrity using checksums. This process runs in the background. Its progress can be checked using scrub status. scrub cancel interrupts the scrub process, while scrub resume resumes it.

- ▶ subvolume create *directory/name*
  Creates the subvolume name. The mount directory or a directory within the Btrfs file system must be specified as the directory. If no directory is specified, btrfs uses the current directory.

- ▶ subvolume delete *directory/name*
  Deletes the specified subvolume or the specified snapshot. However, the occupied memory is only gradually released by a background process.

- ▶ subvolume list *mount directory*
  Lists all subvolumes and snapshots. The mount directory of the file system is specified as the directory. The output can be influenced by several options. The combination -a -p -t has proven to be recommendable. It causes the location of the subvolume within the directory hierarchy to be specified exactly, the parent subvolume to also be specified for each subvolume or snapshot, and the result to be formatted as a table.

- ▶ subvolume set-default *id mount directory*
  Determines which subvolume or snapshot will be used by mount by default in the future. You can determine the volume ID via btrfs subvolume list.

- ▶ subvolume snapshot [-r] *source directory target directory/name*
  Creates the snapshot name. The mount directory of the file system or the directory of an existing subvolume or snapshot must be specified as the source directory. The target directory must be located within the Btrfs file system.

  Usually, the snapshot is mutable (i.e., the file system is split into two branches, both of which are separately mutable). If you want a read-only snapshot for backups, you must use the -r option.

Example

btrfs subvolume list shows the subvolumes set up by default in an (open)SUSE system as well as all snapshots:

```
root#  btrfs subvolume list / -p -a -t
ID     gen     parent  top level  path
--     ---     ------  ---------  ----
257    106     5       5          <FS_TREE>/@
260    106     257     257        <FS_TREE>/@/boot/grub2/i386-pc
261    106     257     257        <FS_TREE>/@/boot/grub2/x86_64-efi
262    1305    257     257        <FS_TREE>/@/home
...
268    106     257     257        <FS_TREE>/@/var/lib/libvirt/images
269    106     257     257        <FS_TREE>/@/var/lib/mailman
270    106     257     257        <FS_TREE>/@/var/lib/mariadb
271    106     257     257        <FS_TREE>/@/var/lib/mysql
```

```
...
258    1158    257    257         <FS_TREE>/@/.snapshots
259    1310    258    258         <FS_TREE>/@/.snapshots/1/snapshot
283    106     258    258         <FS_TREE>/@/.snapshots/2/snapshot
304    669     258    258         <FS_TREE>/@/.snapshots/20/snapshot
...
```

To create a backup of the MariaDB database files during operation, a read-only snapshot is temporarily created for subvolume /var/lib/mariadb:

```
root# btrfs subvolume snapshot -r /var/lib/mariadb/ /var/lib/mariadb-backup
Create a readonly snapshot of '/var/lib/mariadb/' in '/var/lib/mariadb-backup'
root# cd /var/lib/mariadb-backup
root# tar czf /home/kofler/mariadb.tgz .   (create backup)
root# cd                                    (leave directory)
root# btrfs subvolume delete /var/lib/mariadb-backup
Delete subvolume (no-commit): '/var/lib/mariadb-backup'
```

Documentation

The man pages for btrfs are spread over several pages, each with a name consisting of btrfs, a hyphen, and a main command. Details on btrfs scan can therefore be read via man btrfs-scan.

bunzip2 file.bz2

bunzip2 decompresses a file that has previously been compressed using bzip2. The .bz2 identifier is automatically removed from the file name. bunzip is a link to bzip2, whereby the -d option is automatically activated.

bzip2 file

bzip2 compresses the specified files and adds the extension .bz2 to them. The command usually produces files that are 20% to 30% smaller than files compressed using gzip. However, the computing time required for compression is significantly longer with bzip2.

▶ -1 to -9

Specifies how much memory (RAM) the compression algorithm is allowed to use. The default setting is -9 and provides the best results.

If only a small amount of RAM is available, you should select a smaller value; however, compression will then also be slightly worse.

▶ -c or --stdout or --to-stdout

Leaves the file to be (de)compressed unchanged and redirects the result to the standard output. From there, it can be redirected to any file by using >.

▶ -d or --decompress or --uncompress

Decompresses the specified file instead of compressing it (corresponds to bunzip2).

Example

The first command compresses file.eps. The new file name is now file.eps.bz2. The second command restores the original file.

```
user$ bzip2 file.eps
user$ bunzip2 file.eps.bz2
```

cadaver [options] [files]

The command with the strange name of cadaver from the package of the same name is a WebDAV client. You can use it to download or upload files in a similar way to ftp.

Most of the commands within cadaver correspond to FTP commands, that is, get, mget, put, mput, and so on (see ftp). The decisive difference is that the WebDAV protocol based on HTTP(S) is used.

Example

In the following command, a WebDAV connection is first established to the HTTP server with the IP address 172.28.128.5 on port 8585. The test.php file is then uploaded to the uploads directory.

```
user$ cadaver http://172.28.128.5:8585/uploads
dav:/uploads/> put test.php
  Uploading test.php to /uploads/test.php:
  Progress: ... 100,0% of 20 bytes succeeded.
dav:/uploads/> <Strg>+<D>
  Connection to 172.28.128.5 closed.
```

canonical-livepatch [command]

If the snap package of the same name is installed, canonical-livepatch controls the Ubuntu-specific kernel live patch functions.

▶ config

Sets up the live patch function or changes its configuration. This command requires an Ubuntu One account.

▶ enable/disable
Activates or deactivates the live patch system.

▶ refresh
Checks whether updates are available, downloads them if necessary, and applies them to the running kernel. This is done automatically at regular intervals.

▶ status
Shows the status of the live patch system.

Example

All available kernel patches are already applied to this server:

```
root# canonical-livepatch status
last check:       52 minutes ago
kernel:           5.4.0-173.191-generic
server check-in:  succeeded
kernel state:     kernel series 5.4 is covered by Livepatch
patch state:      no livepatches available for kernel 5.4.0-173.191-generic
```

```
case expression in
   pattern1 ) commands;;
   pattern2 ) commands;;
   ...
esac
```

The case construct of the bash shell forms multiple branches in scripts, whereby a character string is specified as a criterion for the branch—often a variable or a parameter that is passed to the shell program. This character string is compared in sequence with the patterns, whereby the wildcard characters for file names (*?[]) can be used in these patterns. Multiple patterns separated by the pipe operator (|) can also be specified in a case branch. As soon as a pattern matches, the commands that follow between the closing parenthesis ()) and the two semicolons (;;) are executed. The program then continues according to esac.

Example

In the following listing, case is used to classify the transferred parameters into file names and options. Within a loop, each individual parameter is analyzed using case. If the parameter begins with a hyphen, the parameter is added to the end of the opt variable; otherwise, it's added to the end of dat.

```
#!/bin/bash
for i do    # Loop for all transferred parameters
  case "$i" in
    -* ) opt="$opt $i";;
    *  ) dat="$dat $i";;
  esac
done        # End of the loop
echo "Options: $opt"
echo "Files:  $dat"
```

cat [options] [files]

cat displays the content of the specified text file. cat is also often used to combine multiple files into one larger file. For this purpose, the standard output must be redirected to a file by using > (see example). If cat is to read from standard input, you shouldn't enter a file name or enter the - character.

For longer texts, you should use the less command instead of cat because this allows you to scroll through the text line by line or page by page.

▶ -s

Reduces multiple empty lines to a single empty line.

▶ -T

Displays tab characters as ^I.

▶ -v

Displays nonprintable characters in ^xxx notation.

Examples

The following command combines the individual part1.tex file, part2.tex file, and so on into a total file called total.tex. The individual files are processed in alphabetical order.

```
user$ cat part*.tex > total.tex
```

In the second example, the standard input is redirected to a new file.

Once you've confirmed the command by pressing Enter, all further entries will be written to the new file. Ctrl+D terminates the input, so cat can be used to create a new text file without an editor.

```
root# cat > newfile
line 1
line 2
<Ctrl>+<D>
```

Variants

There are countless variants of cat:

▶ The bat and ccat commands display code files in common programming languages in color with syntax highlighting.

▶ The more and less commands allow you to scroll through the text line by line and page by page.

▶ The tac command outputs the text in reverse order, that is, the last line first.

▶ The zcat and zless commands make it possible to display compressed text files directly.

cd [directory]

The cd shell command changes to the specified directory. If no directory is specified, cd changes to the home directory. cd - activates the last valid directory.

pwd displays the path of the current directory. Modern alternatives to cd are j and z. Both commands memorize frequently used directories and allow you to change them with a minimum of typing work.

cfdisk [options] device

cfdisk helps with the formatting of hard disks and SSDs. It supports partition tables in MBR and GPT formats. Its operation is somewhat more convenient compared to the more familiar fdisk and parted commands because the current partition table is displayed as a menu. You can then use the cursor keys to select the partition to be edited. cfdisk is only intended for interactive use and isn't suitable for script programming.

chacl [options] files

chacl determines or changes the extended access rights of the specified files or directories. This only works if the file system supports ACLs (*Access Control Lists*). For ext3/ext4 file systems, you must use mount option acl.

If possible, you should use getfacl or setfacl instead of chacl. chacl is only available for reasons of compatibility with IRIX SGI.

chage [options] loginname

You can use chage to set how long a user account may be used, when its password expires and must be changed, and how often the password may or must be changed.

► -d *date*

Specifies the date on which the password was last changed. The date is entered in the notation YYYY-MM-DD (e.g., 2020-12-31). -d 0 means that the user must change their password immediately after the first login. The user can't run a command until the password has been changed.

► -E *date*

Determines the date on which the account expires. After this date, it will no longer be possible to log in. -E -1 means that no expiration date has been set. Expired accounts can be reactivated using passwd -u.

► -I *n*

Specifies how many days after the password has been expired an account will be deactivated. -I -1 prevents deactivation.

► -l

Displays the current settings.

► -m *n*

Specifies the minimum number of days that must elapse before a changed password can be changed again. The default setting 0 means that users can change their password at any time.

► -M *n*

Specifies the maximum number of days after which the password must be changed. -1 means that the password remains valid indefinitely. 183 means that the password must be changed twice a year.

► -W *n*

Specifies how many days prior to the password expiry a warning will be displayed (the default setting is 7).

Examples

You can use the following commands to set up a new user, define their initial password, and force them to change their password immediately when they log in for the first time and at least once a year in future. The password expiry warning appears one month in advance.

```
root# useradd peter
root# passwd peter
New password: ********
Retype new password: ********
root# chage -d 0 -M 365 -W 31 peter
```

The expiration date of an account (option -E) must be specified in ISO format. Relative times aren't possible. However, you can calculate the date using date. The following command causes the account to be deactivated in 100 days:

```
root# chage -E $(date -d +180days +%Y-%m-%d) peter
```

chattr [options] +-=[aAcCdDeijsStTu] files

In Linux file systems (ext2 to ext4, Btrfs, XFS, etc.), some additional attributes can be saved with each file in addition to the user information (see chmod and chown). For example, these attributes with the letter codes aAcCdDeijsStTu control whether files are to be compressed or whether changes to files are to be made using the copy-on-write procedure.

A short description of the attributes is provided in man chattr. However, only a few of the planned attributes are actually being used at the moment. To load the attributes, you can use lsattr.

▶ -R

Changes the attributes recursively in all files of the directory tree.

Example

An important use of chattr is in Btrfs file systems where you can use chattr +C to deactivate the automatically active copy-on-write (COW) procedure and chattr -C to reactivate it. The deactivation of COW is required to efficiently store image files from virtualization programs or large database files in Btrfs file systems.

You can pass individual files or entire directories to the command. With -R, the command changes the attribute recursively in directory trees:

```
root# chattr +C -R /var/lib/mysql
```

However, the attribute only applies to new files. To disable COW on existing files with a size not equal to 0 bytes, you must copy the files:

```
root# mv dir backup
root# mkdir dir
root# chattr +C dir
root# cp -a backup/* dir
root# rm -rf backup
```

```
chcon [options] context files
chcon [options] --reference=reference file files
chcon [options] [-u user] [-r role] [-t type] [-l level] files
```

chcon changes the SELinux context of the specified files. The new context can be specified in three ways: in the notation user:role:type:level, by specifying a reference file, or by using the options -u, -r, -l, and -t. The third variant has the advantage that only part of the entire context can be changed.

A manual setting of the SELinux context is normally required if you save files or directories in locations other than those specified by the SELinux rules—for example, Apache files in /disk2/var/www/html instead of /var/www/html.

If you're only interested in correcting the SELinux context of files that are already located in the intended directory, restorecon is less complicated. To determine the SELinux context of existing files, you should use the ls command with the -Z option.

▶ -l *level*
Changes the level of the SELinux context. Levels aren't used in the rules that usually apply on RHEL and Fedora (*targeted security*). The context level is generally s0 (see also the file /etc/selinux/targeted/setrans.conf).

▶ -r *role*
Changes the role of the SELinux context. SELinux roles determine which "domain" the file is assigned to. SELinux domains in turn determine which rights a process has.

▶ --reference=*reference file*
Reads the SELinux context of the reference file and transfers it to the other specified files.

▶ -R
Also takes into account all subdirectories (*recursive*).

▶ -t *type*
Changes the type of the SELinux context. SELinux types determine the rights of a file, just as SELinux domains control the rights of a process.

▶ -u *user*
Changes the user information of the SELinux context. The SELinux user is an identity known to the SELinux rules. Ordinary Linux users are connected ("mapped") to SELinux identities by SELinux rules.

▶ -v
Displays the changes made (*verbose*).

Examples

By default, Apache on RHEL has read-only rights in the /var/www/html directory. To allow the web server to change files in a specific directory, you must run the following command:

```
root# chcon -R system_u:object_r:httpd_sys_content_rw_t:s0 \
      /var/www/html/wordpress
```

If you want to enable Apache to access web files (read-only this time) that are located in a directory on another data medium, you should proceed as follows:

```
root# chcon -R system_u:object_r:httpd_sys_content_t:s0 /disk2/var/www/html/
```

chgrp [options] group files

chgrp changes the group membership of files. The owner of a file can assign this file only to his or her own groups, whereas root can make any assignments.

▶ -R or --recursive
  Also changes the group assignment of files in all subdirectories.

Example

The following command assigns all files in the /var/www directory to the www-data group. The Apache web server runs under this account on Debian and Ubuntu.

```
root# chgrp -R www-data /var/www/
```

chkrootkit [options]

The chkrootkit command from the package of the same name searches the system for known rootkits. The procedure is pretty simple: the command attempts to detect the malware by using simple criteria or signatures (e.g., the presence of certain character strings in binary files).

▶ -n
  Ignores NFS directories.

▶ -r directory
  Uses the specified directory as the starting point (as the root directory).

Although the project is still active, further development is only taking place in small steps. Both *false positives* and the failure to detect genuine threats are possible.

Example

```
root# chkrootkit
ROOTDIR is /
Checking amd...                                  not found
Checking basename...                             not infected
Checking biff...                                 not found
Checking chfn...                                 not infected
Searching for sniffer's logs ...                 nothing found
Searching for rootkit HiDrootkit's files...      nothing found
Searching for rootkit t0rn's default files...    nothing found
Checking asp...                                  not infected
Checking bindshell...                            not infected
Checking lkm...                                  chkproc: nothing detected
...
```

Variants

chkrootkit can be the starting point of a rootkit search due to its ease of use. Alternative or supplementary tools are, for example, rkhunter and the commercial tools Lynis and Snort. The latter two programs each have an open-source kernel that can be used free of charge.

chmod [options] changes files

chmod changes the nine access bits of files. Together with each file, it's stored whether the owner (*user*), the group members (*group*), and other users (*others*) are allowed to read, write, and execute the file.

The access bits can be changed using the character combination *group +/- access type*, for example, g+w, to give all group members write permission. You specify the group with u (*user*), g (*group*), o (*others*), or a (*all*); you specify the access type with r (*read*), w (*write*), or x (*execute*).

setuid, setgid, and sticky Bits

The setuid bit, which is often also referred to as the suid bit, causes programs to be executed as if the owner themselves had started the program. If the owner of a program is root, then anyone can run the program as if they were root.

The setgid bit has the same function for programs as setuid, but for group membership. For directories, the setgid bit causes newly created files in this directory to belong to the group of the directory and not, as usual, to the group of the user who created the file.

For directories in which everyone is allowed to change the files, the sticky bit makes sure that everyone is only allowed to delete their own files and not those of other users. For example, the bit is set for the /tmp directory. Every user is allowed to create temporary files in this directory. However, it must be avoided that every user can also rename or delete external files.

To set the special bits setuid, setgid, and sticky with chmod, the following character combinations are provided:

setuid: u+s setgid: g+s sticky: +t

For setuid to work, the x-bit for the owner must also be set (u+x).

For setgid to work, the x bit for the group must also be set (g+x).

Octal Notation

Instead of using letters, the access type can also be specified using an octal number that consists of a maximum of four digits. For the access bits, u, g, and o are each assigned a digit. Each digit is composed of the values 4, 2, and 1 for r, w, and x. 660 thus means rw-rw----, while 777 stands for rwxrwxrwx.

The three special bits setuid, setgid, and sticky have the octal values 4000, 2000, and 1000. While setting these bits in octal notation is simple, a security mechanism prevents the bits from being inadvertently deactivated for directories. After chmod 2770 dir and chmod 770 dir, the setgid bit remains set! To remove it, you must run chmod g-s dir or chmod 00770 dir (with two leading zeros).

To determine the access bits of an existing file or directory in octal notation, you must use the stat command.

Options

▶ -f or --silent or --quiet
Doesn't display any error messages.

▶ -R or --recursive
Additionally changes the access rights of files in all subdirectories.

Examples

The secure file can now be executed by all users. secure can be a shell script for creating a backup, for example.

```
user$ chmod a+rx secure
```

The following command revokes the read and write permission for all *.odt files in the current directory for all users outside of your own group:

```
user$ chmod o-rw *.odt
```

Only root is allowed to read the server.key file, nobody can change it:

```
root# chown root:root server.key
root# chmod 400 server.key
```

chown [options] user[:group] files

chown changes the owner and (optionally) also the group membership of a file. The owner of a file can only be changed by root, while the group can also be set by other users (see chgrp).

▶ -R or --recursive
Also changes the group assignment of files in all subdirectories.

Example

The following commands ensure that all files within /var/lib/mysql are assigned to the user and the mysql group so that only the MySQL server process can read and modify the files:

```
root# chown -R mysql:mysql /var/lib/mysql
root# chmod -R o-rwx /var/lib/mysql
```

chpasswd [options]

chpasswd makes it possible to change passwords in scripts, that is, in contrast to passwd, without user interaction. The command reads the usernames and passwords from the standard input, whereby the information must be entered line by line in the form name:password. The command is often called in the form chpasswd < pwfile.txt to process the combinations of names and passwords stored in a file.

You can use options to control how the passwords are encrypted (see man chpasswd). However, usually it's not advisable to use these options. chpasswd then uses the default settings specified by the distribution in the /etc/logins.def file.

Example

The following script uses useradd to set up 30 accounts: user01 to user30. For each account, a password generated via mkpasswd is set. For this purpose, the name and password are output using echo, and the output is passed directly to chpasswd using the pipe operator. All account names and the corresponding passwords are also saved in the text file, students.txt. Finally, chage is used to define that every user must change their password immediately after the first login.

```
#!/bin/bash
for i in {01..30}; do
  user="user$i"
  pw=$(mkpasswd -l 8 -d 0 -C 0 -s 0)
  echo "Account: $user  Password: $pw" >> students.txt
  useradd $user
  echo "$user:$pw" | chpasswd
  chage -d 0 $user
done
```

chroot directory [command]

Without further parameters, chroot starts a new shell that uses the specified directory as the root directory (/). You can work interactively in this shell. exit takes you back to the original shell.

If you specify an optional command, this command is started instead of the shell. During the execution of the command, the specified directory is again used as the root directory.

Example

chroot is useful for performing administrative tasks in a file system that wasn't activated at boot time. The following commands assume that you've started a computer from a rescue system or a USB flash drive. Then mount the system partition (in the example, it's /dev/sda4) in the /mnt/rescue directory, make this directory the new root directory, and then reset the forgotten root password using passwd:

```
root# mkdir /mnt/rescue
root# mount /dev/sda4 /mnt/rescue
root# chroot /mnt/rescue
root# passwd
Change password for root user.
Enter a new password: *********
Enter the new password again: *********
root# exit
root# reboot
```

`chsh` -s shell [user]

chsh changes the default shell that is automatically called after logging in. All shells entered in /etc/shells are available for selection—usually /bin/bash, /bin/csh, and /bin/zsh. The chsh command changes the passwd file and enters the new shell there.

Every user can change their own default shell as they wish. The shell of another user can only be changed by the root user. The new shell must be specified with the complete directory. The change will take effect with the next login.

Example

The following command sets up the nologin program instead of a shell. It prevents a direct login and is useful if you've set up a user only for automated (backup) scripts or only to access an email account:

```
root# chsh -s /usr/sbin/nologin mailuser
```

`chvt` n

chvt activates the console n. In older Linux distributions, the numbers 1 to 6 were reserved for text consoles, while the graphics system of the first user was executed on console 7. Consoles 8, 9, and so on were used for further parallel logins.

In newer distributions, the display manager usually runs in graphics mode on console 1. Console 2 is intended for the graphics system of the first user that has logged in first. Consoles 3 to 6 are operated in text mode.

`cksum` file

cksum determines the checksum and the length of the file in bytes. The checksum can be used to quickly determine whether two files are identical. cksum provides more reliable results than the related sum command. The sha command is mathematically and cryptographically much more secure.

`clear`

clear or [Ctrl]+[L] deletes the content of the console.

cmp [options] file1 file2

cmp compares two files byte by byte and returns the position of the first deviation. If the files are identical, the command doesn't display any message at all. diff is generally more suitable than cmp for comparing text files.

▶ -c or --show-chars
Shows the first text character where the files differ from each other.

▶ -l or --verbose
Provides a list of all deviations.

cnf command

cnf stands for *command not found* and is a SUSE-specific command that reveals which package doesn't yet contain the command you're looking for.

On Red Hat and Fedora distributions, the /usr/libexec/pk-command-not-found command executed by the bash contains a comparable function; on Ubuntu, this is /usr/lib/command-not-found from the command-not-found package. This package is also available on Debian. However, it must be installed there manually and set up once using the update-command-not-found command.

Example

```
user$ cnf jmacs
The 'jmacs' program can be found in the following package:
  * joe [ path: /usr/bin/jmacs, Repository: zypp (repo-oss) ]
Try installing with:
    sudo zypper install joe
```

column [options] file

column reads the specified file, splits each line into several parts, and then outputs all parts in indented columns.

▶ -o '*outseparator*'
Specifies how the columns are to be separated from each other in the output (by default, two spaces are used).

▶ -s '*inseparator*'
Specifies how the data to be imported is separated from each other. By default, *whitespace* is used as a separator, that is, space and tab characters.

▶ -t
Formats the output as a table.

compress [options] file

compress compresses or decompresses the specified file. For compressed files, the .Z identifier is appended to the file name. compress only exists for compatibility reasons. The alternatives bzip2, gzip, and xz are considerably more powerful.

continue [n]

The bash construct continue skips the body of a for, while, or until loop in shell scripts and continues the loop with the next run. This process can be carried out for several loop levels using the optional numerical value.

convert or magick [options] image old image new

convert from the ImageMagick package converts image files from one format to another. In its simplest form, it uses the syntax convert name.png name.jpg to convert the specified PNG file into a JPEG file. The original file is retained.

As of version 7, ImageMagick provides the magick command. It will replace convert in the future. For compatibility reasons, convert can still be used with distributions that already provide version 7. Other distributions haven't yet made the leap to Image-Magick 7. That's why I treat the command in this book under the established name convert.

Various image parameters can be changed simultaneously with more than 100 options. The following list is therefore only a selection:

▶ -blur *radius*
Blurs the image.

▶ -colors *n*
Reduces the number of RGB colors to *n*.

▶ -colorspace CMYK|GRAY|RGB|Transparent|YUV
Indicates the desired color model (with numerous other models to choose from).

▶ -compress None|BZip|Fax|Group4|JPEG|JPEG2000|Lossless|LZW|RLE|Zip
Specifies the desired compression format. However, the actual formats available for selection depend on the image format.

▶ -contrast or +contrast
Reduces or increases the contrast of the image.

▶ -crop *geometry*
Cuts out the desired part of the image. For example, -crop 50x50+100+100 describes a 50×50 pixel area that starts at the coordinate position (100, 100).

▶ `-filter Point|Box|Triangle|Hermite|Hanning|Hamming ...`
Applies the desired filter to the image.

▶ `-gaussian` *radius*
Blurs the filter with the Gaussian operator.

▶ `-normalize`
Normalizes the color distribution in the image.

▶ `-quality` *n*
Specifies the desired compression quality. The permissible values for *n* depend on the image format (e.g., 0 to 100 for JPEG).

▶ `-resize` *NxN* or `-resize` *n%*
Changes the resolution of the image.

▶ `-rotate` *angle*
Rotates the image clockwise by the specified angle in degrees.

▶ `-trim`
Cuts off monochrome image edges if they have the same color as the corner points of the image.

Further options are described in detail on the following website:

https://imagemagick.org/script/command-line-processing.php

For security reasons, `convert` refuses to process some file types by default (e.g., PDF files). This can be remedied by removing the corresponding line from `policy.xml`. I've commented out the PDF line in the following listing:

```
# file /etc/ImageMagick-6/policy.xml or /etc/ImageMagick-7/policy.xml
<policymap>
  ...
  <!-- <policy domain="coder" rights="none" pattern="PDF" /> - >
  ...
</policymap>
```

Examples

The following three examples show simple applications of the command:

```
user$ convert -resize 100x100 in.jpg out.png
user$ convert -type Grayscale in.jpg out.eps
user$ convert -quality 80     in.bmp out.jpg
```

The following command converts a PNG file into an EPS file according to the PostScript Level 2 specification (`convert` usually creates files for PostScript Level 3. These are then often significantly larger without any advantages). All image layers are placed on top of each other. Transparent areas of the PNG file are displayed in white, not black.

```
user$ convert -flatten -alpha remove -background white in.png EPS2:out.eps
```

The next command creates a PDF file containing all JPEG images from the current directory (one image per page). You can use the `--quality` and `--density` options to control the quality and size of the resulting PDFs.

```
user$ convert -density 200x200 -quality 80 *.jpg allebilder.pdf
```

```
cp [optionen] source  target
cp [options] files target directory
```

cp copies files and directories. Individual files can be renamed during copying. When editing multiple files (e.g., by specifying wildcards), these can only be copied to another directory, but not renamed. Statements of the type `cp *.tex *.bak` aren't permitted. Commands comparable to cp are mv for moving and renaming files, and ln for creating links. cp supports the following options, among others:

▶ -a or --archive
  Retains as many attributes of the files as possible. -a is an abbreviation for -dpR.

▶ -b or --backup
  Renames existing files to backup files (file name plus ~) instead of overwriting them.

▶ -d or --dereference
  Only copies the reference for links, but not the file that's referenced by the link.

▶ -i or --interactive
  Prompts before existing files are overwritten.

▶ -l or --link
  Creates *hard links* instead of copying the files. If cp is used with this option, it has the same functionality as ln.

▶ -p or --preserve
  Leaves the information about the owner, the group membership, the access rights and the time of the last modification. Without this option, the copy belongs to those who execute cp (user and group), and the time is set to the current time.

▶ -r or -R or --recursive
  Additionally copies subdirectories and the files they contain.

▶ --reflink[=auto]
  Initially copies only the metadata of the file, that is, creates a *shallow copy* of the file. The copy therefore takes up almost no space in the file system. Nevertheless, it's independent of the original: as soon as data blocks of the original or the copy are changed, these blocks are reallocated.

Shallow copies only work if this function is supported by the file system. This is the case with Btrfs and current versions of XFS, but not with ext4. If the file system isn't compatible, this option will result in an error unless you specify the auto mode in which cp itself decides which type of copy is possible.

▶ -s or --symbolic-link
Creates symbolic links instead of copying the files or directories. cp therefore has the functionality of ln -s.

▶ -u or --update
Copies files only if no file of the same name with a newer date will get overwritten.

Copying Directories

If you want to copy an entire directory with all the files and subdirectories it contains, you must run cp -r *source directory target directory*. This also copies hidden files and subdirectories. If you want access rights and times to be preserved when copying, you need to use the -a option instead of -r.

The question of whether the source directory itself or only its content is copied is somewhat difficult. If the target directory already exists, the new subdirectory named *sourcedirectory* will be created in it, and the entire contents of the source directory will be copied there. However, if the target directory doesn't yet exist, it will be created; in this case, only the content of the source directory is copied to the newly created target directory, but not the source directory itself. cpio, rsync, and tar are also suitable for copying entire directory trees.

Examples

The following command copies all *.tex files from the book subdirectory to the current directory. The dot indicates the current directory as the target directory.

```
user$ cp book/*.tex .
```

The second command creates a backup copy of the entire book directory:

```
user$ cp -a book bak-book
```

cp can't be used to rename multiple files during the copy process. cp *.xxx *.yyy therefore doesn't copy all *.xxx files to *.yyy files. To perform such operations, you must use for or sed. In the following command, for creates a loop over all *.xxx files. The expression ${i%.xxx}.yyy removes the ending *.xxx and replaces it with .yyy. If you replace cp with mv, the files won't get copied but renamed.

```
user$ for i in *.xxx; do cp $i ${i%.xxx}.yyy; done
```

The procedure with sed is somewhat more complicated: ls provides the list of files to be copied and passes it on to sed. sed uses the s command (*regular find and replace*) to create a list of cp commands and passes these on to a new sh shell, which then executes the commands.

```
user$ ls *.xxx | sed 's/\(.*\)\.xxx$/cp & \1.yyy/' | sh
```

cpio command [options]

cpio combines multiple files into one archive and copies them to another data medium, a streamer, or another directory. Similarly, the command can also be used to reload such data. On Linux, cpio is rather uncommon; tar is usually used instead. The major cpio commands are as follows:

- -i

  (*input*) Reads an archive from the standard input and extracts the files it contains.

- -o

  (*output*) Combines files into an archive and transfers the archive to the standard output. The files to be processed are often determined by find and then passed on to cpio via a pipe.

- -p

  (*pass through*) Transfers archives between different directories.

- -t

  Displays the contents of an archive from the standard input.

Details of these actions can be controlled by additional options.

Example

The following commands first archive all files from the /etc directory. Then, cpio -t displays the list of backed up files as a check.

```
root# cd /etc
root# find . | cpio -o > /tmp/etc-backup.cpio
root# cpio -t < /tmp/etc-backup.cpio
```

crontab [options]

The crontab command helps you administrate custom cron jobs in the */var/spool/cron/username* file (see crontab). Cron jobs are executed automatically in the background at the scheduled times, such as once an hour or every Sunday evening.

▶ `-e`
Opens an editor for editing the crontab file. Usually, `vi` is used as the editor. If you want to use a different editor, you must set the `EDITOR` environment variable accordingly. `crontab` checks compliance with the crontab syntax rules when saving.

▶ `-l`
Lists the current crontab entries.

▶ `-r`
Deletes all crontab entries.

▶ `-u` *username*
Edits the crontab file of the specified user. The `-u` option is only available to `root`.

cryptsetup `[options] command`

`cryptsetup` from the package of the same name uses functions of the `dm_crypt` kernel module. It sets up, activates, and deactivates crypto devices. A sufficiently large image file can also be used instead of a device. This can be created in advance, for example, via `dd if=/dev/zero of=cryptoImage bs=1M count=n`, where n is the desired size in mebibytes (MiB).

At this point, I'll only describe the most important LUKS-specific commands of `cryptsetup`:

▶ `luksAddKey` *device*
Sets up an additional password that provides access to the crypto container. To run the command, an existing password must be entered (use any password you like). A maximum of eight passwords are allowed.

▶ `luksClose` *device mappingname*
Deactivates a crypto device.

▶ `luksDump` *device*
Provides meta information about the crypto container (e.g., the encryption algorithm).

▶ `luksFormat --type luks2` *device*
Sets up a crypto container in the specified device in the current LUKS2 format. Without the `--type` option, the command uses the still supported but obsolete LUKS1 format.

You must then enter the *passphrase* twice, which should be at least 20 characters long for security reasons. You can select the desired encryption algorithm with `-c` and set a different key length with `-s`. The algorithms available for selection are listed in the pseudo file; `/proc/crypto`. The output of the `cryptsetup --help` command reveals in the last lines which algorithm and which key length are used by default. This is currently the `aes-xts-plain64` algorithm with a 256-bit key.

▶ luksOpen *device mappingname*

Activates the crypto container in the specified device and assigns it a name. The resulting crypto device can then be used via */dev/mapper/mappingname*.

Example

The following commands show how to first format a USB flash drive (/dev/sdh1) as a crypto device (luksFormat) and then activate the device under the arbitrarily chosen name of mycontainer (luksOpen). Needless to say, your data is only as secure as your password or your passphrase, which consists of several words. A password length of at least 20 characters is recommended.

You can then use /dev/mapper/mycontainer like a hard disk partition or a logical volume—that is, set up a file system, integrate it into the directory tree, and so on. After umount, you must remember to deactivate the crypto device again (luksClose) to release /dev/sdh1. Only at this point can you remove the USB flash drive.

```
root# cryptsetup luksFormat --type luks2 /dev/sdh1
This will overwrite data on /dev/sdh1 irrevocably.
Are you sure? (Type uppercase yes): YES
Enter LUKS passphrase: ********************
Verify passphrase: ********************
Command successful.
root# cryptsetup luksOpen /dev/sdh1 mycontainer
Enter LUKS passphrase: ********************
root# mkfs.ext4 /dev/mapper/mycontainer
root# mount /dev/mapper/mycontainer /test
root# ... use the encrypted file system ...
root# umount /test/
root# cryptsetup luksClose mycontainer
```

csplit [options] file splitting position

csplit splits a text file into several individual files at predefined points. The splitting position can be specified either by a direct line specification or by a search pattern. As a result, the command returns the files xx00, xx01, and so on and then displays their lengths on the screen. By specifying the appropriate options, "nicer" file names are of course also possible.

cat can reassemble the original file from these individual files. Instead of csplit, the split command can be used to split binary files.

Specification of the Splitting Positions

The splitting positions are specified either by a number of lines or by a search pattern. In the former case, the file is split after *n* lines, whereas in the latter case, it's split before or after the occurrence of the search pattern. If csplit is to split the file multiple times, which is usually the case, the number of lines or the split pattern must be followed by the number of times the operation is to be repeated.

▶ */pattern/*
  Splits the file in the line before the pattern occurs. (The line with the pattern found becomes the first line of the subsequent file.)

▶ */pattern/+n*
  /pattern/-n
  Splits the file *n* lines after (+) or before (-) the occurrence of the pattern.

▶ n
  Splits the file after *n* lines.

▶ {*n*}
  Splits the file into *n*+1 individual files (and not just two files).

▶ {*}
  Splits the file into the corresponding number of individual files each time the search pattern occurs or each time after *n* lines.

Options

▶ -f *file* or --prefix=*file*
  Uses the specified file name to name the output files.

▶ -k or --keep-files
  Prevents files that have already been created from being deleted again if an error occurs. The option must be used in particular for pattern specifications in the form *n* {*}. The pattern is specified as in grep.

▶ -z or --elide-empty-files
  Prevents the creation of empty files. Without this option, empty files can occur in particular if the first line of the source file already matches the search pattern.

Example

csplit splits total.txt into the files part.00, part.01, and so on. The individual files are each 100 lines long. cat then creates a copy of the original file.

```
user$ csplit -k -f part. total.txt 100 {*}
user$ cat part.* > copy.txt
```

In the second example, `csplit` splits the `total.txt` file into smaller files, whereby the split always occurs when a line begins with the text `% ===`.

`user$ csplit -k -f part. total.txt '/^% ===/' {*}`

curl [options] [url]

`curl` helps with the transfer of files from or to a server, whereby all conceivable protocols are supported (HTTP, HTTPS, FTP, SFTP, SCP, etc.).

The external file or directory is specified by a URL string (*Uniform Resource Locator*) that starts with the protocol name (e.g., `http://server.com/file`).

▶ `-d` or `--data 'data'`
Sends the specified data with the request. For `--data`, the following alternatives are available: `--data-binary`, `--data-raw`, and `--data-urlencode`. With all variants, the data can also be read from the specified file via `@file`.

▶ `--limit-rate n`
Limits the transfer rate to the specified number of bytes per second. *n* can be followed by the letter `k` or `m` to limit the transfer rate to *n* KiB or MiB per second.

▶ `-n` or `--netrc`
Instructs the command to read the login data from the `.netrc` file. Using `--netrc-file file.txt`, you can alternatively specify a different location for this file. Both variants are more secure than the password transfer via `-u`.

▶ `-o file`
Saves the downloaded data in the specified file instead of forwarding it to the standard output.

▶ `--oauth2-bearer 'token'`
Sends the specified bearer token with the request.

▶ `-r n1-n2`
Transfers the specified byte range of the file.

▶ `-s` or `--silent`
Dispenses with status outputs.

▶ `-T file`
Uploads the specified file to the server, that is, in the opposite direction of the usual use of `curl` (*transfer*). Instead of the file name, the hyphen character (-) can also be entered to process data from the standard input.

▶ `-u name:password`
Specifies the login name and password.

▶ `-X` or `--request GET|POST|PUT|...`
Specifies the desired request type (by default, this is `GET`).

Examples

The following command transfers the specified file to the FTP server backupserver and saves it in the dir directory:

```
user$ curl -T file -u username:password ftp://backupserver/dir
```

To process data from the standard input channel, you can use -T to specify a hyphen as the file name. The following command saves the result of the tar command directly in the name.tgz file on the FTP server:

```
user$ tar czf - dir/ | curl -T - -u usern:pw ftp://bserver/name.tgz
```

curl is also suitable for testing REST APIs as follows:

```
user$ $ curl -X PUT https://httpbin.org/put?para=123 -d @mydata.json \
        -H "Content-Type: application/json"
```

Alternatives

To download files via FTP or HTTP, you can also use wget instead of curl. If, on the other hand, you're more interested in debugging REST APIs, curlie is a good choice (see *https://github.com/rs/curlie*). This curl alternative displays the transferred header by default, formats JSON documents in color, and more.

cut [options] file

cut extracts the columns specified by options from each line of text.

▶ -b *list* or --bytes *list*
Extracts the characters specified in a list. Individual entries may be separated by commas (but not by spaces). Instead of individual characters, entire ranges can also be specified, such as –b 3-6,9,11-15.

▶ -d *character* or --delimiter *character*
Specifies the separator for -f to be used instead of the tab character.

▶ -f *list* or --fields *list*
Extracts like -b, but now for fields (data records) that must be separated by tab characters.

▶ -s or --only-delimited
Eliminates all lines that don't contain any data corresponding to the -f option. This can't be used together with -b.

In practice, the extraction of individual text columns is often much easier via awk than by using cut.

Example

In the following example, cut reads the first, third, and seventh columns from the /etc/
passwd file, that is, the username, the UID, and the default shell. The separator in this file
is the colon.

```
root# cut -d\: -f1,3,7 /etc/passwd
root:0:/bin/bash
bin:1:/sbin/nologin
daemon:2:/sbin/nologin
adm:3:/sbin/nologin
...
```

date [options] [+format]

date returns the current date and time or changes this data. Changes may only be made
by root. date can display the time in various formats.

▶ -d *time span*

Adds or subtracts a time span to or from the current time (e.g., -d +3days or -d
-4hours).

▶ -s *newtime*

Changes the date and/or time of the computer clock.

The following list shows the most important format codes. The format string must be
enclosed in quotation marks.

▶ %Y

The year as a four-digit number

▶ %l

The month as a number (1 to 12)

▶ %m

The month as a two-digit number (01 to 12)

▶ %b

The month as a short character string (Jan to Dec)

▶ %B

The month as a long string (January to December)

▶ %d

The day of the month as a two-digit number (01 to 31)

▶ %e

The day of the month as a number (1 to 31), with single-digit numbers preceded by a
space

▶ %H

The hour as a two-digit number (00 to 23)

▶ %M

The minute as a two-digit number (00 to 59)

Examples

The following three statements provide the current time in ISO format, the date in three weeks, and the time four hours ago:

```
user$ date "+%Y-%m-%d %H:%M:%S"
2025-03-12 17:46:39
user$ date -d +3weeks "+%Y-%m-%d"
2025-04-02
user$ date -d -4hours "+%H:%M:%S"
13:47:25
```

The following script creates a compressed tar archive of the /home/kofler/data directory. The archive is saved under two file names: mydata-day-*dd*.tar.gz and mydata-month-*mm*.tar.gz.

If the script is run daily, you'll have 43 backup versions over time reflecting the state of the backup directory for the past 28 to 31 days and for the past 12 months.

```
#!/bin/bash
fname1=/backup/mydata-day-$(date "+%d").tar.gz
fname2=/backup/mydata-month-$(date "+%m").tar.gz
tar czf $fname1 /home/kofler/data
cp $fname1 $fname2
chmod 600 $fname1 $fname2
```

dconf command

dconf is a low-level command for changing the dconf database, in which Gnome stores countless settings. Instead of dconf, you should generally use the gsettings command, which contains more control mechanisms.

▶ dump/load *dir*

Reads an entire configuration directory recursively and writes it to the standard output (dump) or reads it from the standard input (load).

▶ list *dir*

Outputs the entries of a directory (not recursive).

▶ read *key*

Reads an entry.

▸ reset *key/dir*
Resets an entry or an entire configuration directory to the initial value(s). The -f option must be specified for directories.

▸ write *key value*
Saves or overwrites an entry.

Examples

The following two commands show you how to make a backup of all Gnome settings and restore the saved state later:

```
user dconf dump / > dconf.bak
user dconf load / < dconf.bak
```

The following command causes a total reset of all Gnome settings:

```
user$ dconf reset -f /
```

dd options

dd transfers data between different storage media (SSD, USB flash drive, etc.) and converts the data if required. The command can be used to exchange data between different computer architectures.

Not only can dd copy individual files, but it also can access devices directly. This allows you to copy entire hard disks (partitions), change the boot sector of the hard disk, and so on. No file system needs to be set up on the data medium. If dd is used without options, it reads the data from the standard input (ending with Ctrl+Z) and writes to the standard output. The options of dd are specified without preceding minus signs!

▸ bs=*n*
Determines the block size for the input and output file. (The block size indicates how many bytes are read or written in one pass.) Caution: By default, dd uses a block size of only 512 bytes. That is extremely inefficient! If you read or write to hard disks or SSDs and use dd, you should work with at least bs=1M. The suffix M stands for 1,024 × 1,024 bytes, and K stand for 1,024 bytes.

▸ conv=*mode*
Converts the data during copying. Various settings are permitted for *mode*, including lcase (convert uppercase to lowercase), ucase (convert lowercase to uppercase), swab (swap 2 bytes at a time), and so on. With conv=fsync, dd ends its work with a sync call, that is, waits for the physical completion of the write process. This is particularly important if you suspect that the storage medium may be defective. Only with conv=fsync does dd receive error messages from the kernel in this case.

- count=*n*

  Copies only *n* blocks (and not the entire data).

- ibs=*n*

  Determines the block size of the source file.

- if=*source file*

  Specifies the source file (instead of the standard input).

- obs=*n*

  Determines the block size of the target file.

- of=*target file*

  Specifies the target file (instead of the standard output).

- seek=*n*

  Skips *n* blocks before starting the output.

- skip=*n*

  Skips *n* blocks before starting to read.

- status=progress

  Displays the progress of the copying process. (By default, status=none.)

Note that direct access to devices (e.g., via cat /dev/xxx >) is often faster than dd. You also don't need to worry about the ideal block size.

Examples

The following command transfers the ISO file opensuse.iso directly to the data medium with the device /dev/sdc, for example, to a USB flash drive. You should note that all previously saved data on this data medium will be lost! For this reason, make sure that you enter the correct device name. If the device specified with of is your current hard disk, you'll lose its contents!

```
root# dd if=opensuse.iso of=/dev/sdc bs=1M status=progress
```

The second command creates a 10 GiB raw image file for a virtualization program (e.g., QEMU/KVM). This image file simply consists of zeros that are read from the /dev/zero device.

```
root# dd if=/dev/zero of=image.raw bs=1M count=10000
```

Alternatives

Depending on the intended use, the following two variants are available instead of dd:

- dd_rescue or, depending on the distribution, ddrescue (gddrescue package), attempts to create a copy even if individual data sectors of the source device are defective.

- ▶ dcfldd is optimized for forensic tasks. Among other things, the command can use checksums to verify that the copy is really identical.

You can find a good comparison between dd, dd_rescue, and dcfldd at the following address:

https://superuser.com/questions/355310

declare [options] var[=value]

The bash command declare assigns a new value and/or various properties to shell variables. The command is mainly used in shell scripts. If it's called without parameters, all known variables will be listed with their contents. Note that no spaces may be entered before or after the equal sign!

- ▶ -A
  Declares the variable as an associative array.

- ▶ -r
  Declares the variable as read-only. The variable can only be read, but not changed.

- ▶ -x
  Declares the variable as an environment variable (similar to export). The variable is therefore also available to other commands or in subshells.

delgroup [options] name

delgroup deletes the specified group on Debian and Ubuntu systems, following the rules from /etc/deluser.conf. In Fedora and Red Hat, delgroup is a link to the groupdel command, which has other options.

- ▶ --only-if-empty
  Deletes the group only if no users are assigned to it.

- ▶ --system
  Deletes the group only if it's a system group.

delta [options] file1 file2

delta is a variant of diff. The command from the git-delta package compares the contents of two text files and displays the differences graphically (see also *https://dandavison.github.io/delta*). While diff is optimized for use in scripts and for patch management, delta excels in interactive use.

- ▶ --help
  Displays a help text. (There is no man page.)

▶ --light
Uses colors that are suitable for a terminal with a light background. By default, delta is intended for dark terminals (--dark option).

▶ -s or --side-by-side
Shows the differences next to each other instead of below each other.

deluser [options] name deluser name group

deluser deletes the specified user on Debian and Ubuntu systems and follows the rules from /etc/deluser.conf. As a result, the home directory is usually not deleted. In Fedora and Red Hat, deluser is a link to the userdel command with other options.

▶ --remove-all-files
Deletes the user's home directory and all other files (e.g., various spool and mail files).

▶ --remove-home
Also deletes the user's home directory.

▶ --system
Deletes the user only if it's a system user.

If the command is run as deluser *name group*, it removes the user from the specified group. The command is therefore complementary to adduser *name group*.

depmod [options]

depmod creates the module dependency file modules.dep as well as various *.map files in the /lib/modules/*kernelversion*/ directory. The *-map files specify which kernel module is to be loaded for which hardware component.

▶ -A or --quick
Tests first whether existing *.dep and *.map files are still up-to-date. In this case, depmod doesn't regenerate the files.

▶ -b *directory* or --basedir *directory*
Specifies the directory for which the *.dep and *.map files are to be generated. By default, the command updates the files for the running kernel.

df [options] [directory]

df provides information about where hard disks (partitions) or other drives are mounted in the file tree and how much storage space is available on them. Usually, df returns a list of all active partitions. If you pass a directory to the command, however, it only displays the data of the partition in which this directory is physically located. On systems with many partitions, you can easily determine in which partition a specific directory is saved.

▶ -h

Indicates the storage space in an easily readable form in MiB or GiB.

▶ -i or --inodes

Provides information about the available I-nodes (instead of the free storage space in KiB).

▶ -t *fs*

Considers only file systems of the specified type, such as only ext4 file systems with -t ext4.

▶ -T

Specifies the file system for each partition or data medium.

▶ -x *fs*

Ignores the specified file system.

Example

On the example system, the system partition / still has 15 GiB of free disk space. The other file systems are used for administrative purposes (dynamic display of devices) and for storing Ubuntu-specific snap packages. Due to the -x tmpfs option, numerous temporary file systems, that is, those only located in RAM, aren't taken into account.

```
root# df -h -T -x tmpfs
Filesystem              Type      Size  Used Avail Use% Mounted on
udev                    devtmpfs   16G     0   16G   0% /dev
/dev/nvme0n1p1          vfat     1021M  254M  768M  25% /boot/efi
/dev/nvme0n1p4          ext4      119G   32G   81G  29% /
/dev/mapper/mycryptpart ext4      1.5T  767G  634G  55% /crypt
/dev/loop0              squashfs   62M   62M     0 100% /snap/authy/5
/dev/loop1              squashfs   56M   56M     0 100% /snap/core18/1885
...
```

Alternatives

There are several alternatives to df that present the result more legibly. The duf command from the package of the same name automatically differentiates between "real" and virtual file systems, and it formats the output much more clearly than df (see Figure 1). We also recommend dysk (see the *https://dystroy.org/dysk* project page).

```
┌─────────────────────────────────────────────────────────────────────────┐
│ ⊞                          kofler@host1: ~                    Q  ≡  ×      │
│ kofler@host1:~$ duf                                                        │
│ ┌─────────────────────────────────────────────────────────────────────┐  │
│ │ 2 local devices                                                       │  │
│ │ ┌───────────┬───────┬───────┬───────┬──────────────────┬──────┬─────────────┐│
│ │ │ MOUNTED ON │ SIZE  │ USED  │ AVAIL │      USE%        │ TYPE │ FILESYSTEM  ││
│ │ ├───────────┼───────┼───────┼───────┼──────────────────┼──────┼─────────────┤│
│ │ │ /         │ 61.8G │ 49.4G │  9.7G │ [########...] 79.9% │ ext4 │ /dev/vda1   ││
│ │ │ /localbackup │ 44.1G │ 22.0G │ 20.1G │ [####......]  49.8% │ ext4 │ /dev/vdb    ││
│ │ └───────────┴───────┴───────┴───────┴──────────────────┴──────┴─────────────┘│
│ └─────────────────────────────────────────────────────────────────────┘  │
│ ┌─────────────────────────────────────────────────────────────────────┐  │
│ │ 6 special devices                                                     │  │
│ │ ┌───────────────┬────────┬───────┬────────┬─────────────────┬──────────┬────────────┐│
│ │ │ MOUNTED ON     │  SIZE  │ USED  │ AVAIL  │     USE%        │ TYPE     │ FILESYSTEM ││
│ │ ├───────────────┼────────┼───────┼────────┼─────────────────┼──────────┼────────────┤│
│ │ │ /dev           │  2.8G  │  0B   │  2.8G  │                 │ devtmpfs │ udev       ││
│ │ │ /dev/shm       │  2.8G  │  0B   │  2.8G  │                 │ tmpfs    │ tmpfs      ││
│ │ │ /run           │ 577.9M │ 1.3M  │ 576.7M │ [..........] 0.2% │ tmpfs    │ tmpfs      ││
│ │ │ /run/lock      │  5.0M  │  0B   │  5.0M  │                 │ tmpfs    │ tmpfs      ││
│ │ │ /run/snapd/ns  │ 577.9M │ 1.3M  │ 576.7M │ [..........] 0.2% │ tmpfs    │ tmpfs      ││
│ │ │ /run/user/1000 │ 577.9M │ 4.0K  │ 577.9M │ [..........] 0.0% │ tmpfs    │ tmpfs      ││
│ │ └───────────────┴────────┴───────┴────────┴─────────────────┴──────────┴────────────┘│
│ └─────────────────────────────────────────────────────────────────────┘  │
│ kofler@host1:~$ ▯                                                         │
└─────────────────────────────────────────────────────────────────────────┘
```

Figure 1 "duf" Providing a Good Overview of the Free Storage Space on All Data Media

dhclient interface

The dhclient command from the dhcp-client package obtains the parameters for the network configuration from a DHCP server in most distributions. The command is ideal for an ad hoc network configuration for new server installations. Some distributions use the dhclient3 or dhcpcd commands as alternatives to dhclient.

Example

The following command performs an automatic network configuration for the eth0 interface:

root# **dhclient eth0**

diff [options] file1 file2

diff usually compares two text files. The result is a list of all lines that differ from each other. The command is relatively "intelligent"; that is, if some lines are inserted in one file compared to the other, only this deviation will be reported. Other lines are recognized as identical again, although they now have different line numbers. The command can therefore be used to quickly document the differences between two versions of a program listing.

diff can also compare two directories in the form -r dir1 dir2.

▶ -b

Treats multiple spaces and lines as single spaces or lines.

▶ -q

Only indicates whether two files are different or not. In this variant, diff dispenses with the detailed listing of all differences.

▶ -r

Recursively compares two directories. The result lists which files only exist in the first directory and which only exist in the second directory. For files that exist in both directories but have different content, diff displays the differences. diff -r can be combined well with the -q option.

▶ -w

Ignores spaces and blank lines completely.

Example

The following command compares the contents of your own MySQL configuration file with the configuration suggested by the package. The lines beginning with < show the current status (i.e., text from the first file passed to diff), and the lines beginning with > show the corresponding passage from the second file.

```
root# diff /etc/mysql/my.cnf /etc/mysql/my.cnf.dpkg-dist
100,101c87,88
< server-id              = 100
< log_bin                      = /var/log/mysql/mysql-bin.log
---
> #server-id             = 1
> #log_bin                     = /var/log/mysql/mysql-bin.log
...
```

Alternatives

To put it mildly, "reading" diff results takes some getting used to. It's much more convenient to use the delta command. If you have a graphical user interface, the meld program can be very helpful. Most editors can also highlight the difference between two files (e.g., VS Code).

---

dig [options] [@dnsserver] host/ipaddr [type]

dig runs queries on a domain name server in a similar way to host and whois. Using dig, you can control the process by means of a particularly large number of options that can't even be hinted at here (see man dig).

In its simplest form `dig host` contacts the DNS server set in `/etc/resolv.conf` and returns the A-record. With `@1.2.3.4`, you can specify the IP address of an alternative DNS server. If you're not interested in the A-record but in other data, you must enter its type in the third parameter.

▶ `-4 / -6`
Uses only IPv4 or only IPv6.

▶ `+nocomments`
Doesn't output any comments.

▶ `+nostats`
Doesn't output any statistical data on the query (query time, query size, etc.).

▶ `+short`
Formats the output in a reader-friendly way. (By default, `dig` provides the data in the confusing notation of the DNS server `bind`.)

Example

The following command contacts the public name server of Google, determines the MX entry of `kofler.info` there, and displays it in abbreviated form:

```
user$ dig +short @8.8.4.4 kofler.info MX
10 mail.kofler.info.
```

dircolors [optionen] [file]

`dircolors` provides the program code for setting the colors the `ls` command uses to identify different file types. If a file is passed to `dircolors`, the command shows the settings saved in this file. The file must adhere to the syntax specified by `dircolors -p`.

▶ `-b`
Uses the syntax of the `bash` shell (applies by default).

▶ `-c`
Uses the syntax of the C shell.

▶ `-p`
Uses a very detailed syntax that can be processed by `dircolors` itself. The file contains comments that explain the syntax and in particular the color codes.

Example

In many distributions, the `.dircolors` file is analyzed when the `bash` is started and used to set the colors. To change the `ls` colors, you must first create your own version of this file by using dircolors -p and then make changes to it in an editor. The changed settings take effect as soon as you open a new terminal window or dialog.

```
user$ cd    (change to home directory)
user$ dircolors -p > .dircolors
user$ yourFavoriteEditor .dircolors
```

If changing the .dircolors file isn't sufficient for your distribution, you can use dircolor to create a statement for setting the LS_COLORS variable and add this statement to the end of the .bashrc file:

```
user$ dircolors .dircolors >> .bashrc
```

dirname string

dirname returns the path of a complete file name. Thus, dirname /usr/bin/groff returns /usr/bin.

dirs

The bash command dirs returns the list of all directories saved by pushd. The command is primarily intended for script programming.

disown [options] [jobspec]

The bash command disown removes the specified process from the shell. This means the process continues to run even if the shell is closed. The process is specified either by the process ID (PID) or by the shell job number, whereby the shell job number must be preceded by the % character. Without specifying a job number, the current (background) process is meant.

If you plan to run a program separately from the shell from the outset, you can use nohup at startup to avoid having to call disown later.

▶ -a

Triggers all processes started in the shell.

▶ -r

Triggers all processes started in the shell that are currently running.

Example

In the following example, a longer download process is started first. Ctrl+Z interrupts the download, and bg continues it in the background. disown separates the download process from the shell so that the download continues even after the shell logout.

```
user$ wget -q http://mirror.switch.ch/.../AlmaLinux-n.n.iso
user$ <Ctrl>+<Z>
user$ bg
user$ disown
```

dmesg [options]

dmesg outputs the kernel messages contained in the buffer memory.

► -c

Deletes the buffer memory for the kernel messages after the output. This option requires root permissions.

► -H

Performs the output in color and displays times in an easier to read form (*human readable*).

► -T

Displays the absolute time prior to each message.

► -w

Displays all kernel messages and then continues to run until the command is terminated via Ctrl+C.

► -W or --follow-new

Waits for new kernel messages and displays them until the command is terminated via Ctrl+C.

In some Linux distributions, dmesg requires root permissions. If you want to give all users the right to read kernel messages, you need to enter kernel.dmesg_#restrict=0 in the /etc/sysctl.conf file and then run sysctl -p.

dnf [options] command

dnf is the package management command of Fedora, RHEL, and compatible distributions. dnf has replaced the formerly used yum command.

dnf downloads the packages from the package sources defined in /etc/yum.repos.d/ *.repo. (The name of this directory makes clear the origin of dnf and hasn't changed along with the change from yum to dnf.)

► check-update

Checks whether updates are available for already installed packages.

► clean metadata

Removes the package metadata from the cache. At the next start, dnf updates the data by reloading the metadata of all package sources.

► clean packages
Removes downloaded and already installed package files from the cache.

► config-manager [options]
Manages DNF/YUM package sources:

 – --add-repo *url* adds a package source.

 – --set-enabled *name* or --set-disabled *name* (de)activates a package source.

► copr [enable|disable|remove|list|search] [*package source*]
Manages Copr package sources. Copr is an automated build system used by some developers to offer their own packages to the public. copr isn't a built-in dnf command. It can only be used if the dnf-plugins-core package has been installed.

► downgrade *package*
Replaces a package with a package from an older distribution. You can enter the version number of the distribution using --releasever=n.

► download *package*
Downloads a package to the current directory, but doesn't install it. In combination with the --source option, dnf downloads the source code package. With --resolve, the command also takes dependent packages into account. With --url, it only shows the links to the packages, but doesn't download the packages.

► grouplist, groupinfo, groupinstall, groupupdate, groupremove
Provides information about, installs, updates, or removes package groups. dnf grouplist -v lists the corresponding internal English-language group ID in parentheses together with each package group.

► history
Provides a numbered list of the most recently performed dnf actions. dnf history info *n* provides details on transaction *n*.

► info *name*
Provides information on the specified package.

► install *name1 name2* ...
Searches for the packages name1, name2, and so on from all package sources, and then downloads and installs them. If necessary, additional packages are also loaded and installed or updated to fulfill package dependencies.

► install *rpmfile*
Installs an RPM package that is stored in the local file system. Package dependencies are taken into account. Missing packages are downloaded from the package sources and installed, if necessary.

► install @*package:version*
Installs the package in the specified version. This is a short notation for dnf module install *package:version*.

▶ `list`
Provides a two-part list. The first part contains (in alphabetical order) all packages already installed; the second part contains all packages not yet installed that are available in the package sources. An optional parameter can be used to restrict the result to packages whose name matches a pattern (e.g., `dnf list xorg*`).

▶ `list available/updates/installed/extras/recent`
Restricts the output of `dnf list` to specific packages. For example, `dnf list updates` provides a list of all packages for which updates are available. `dnf list recent` provides an overview of the packages that have recently been added to the package sources.

▶ `makecache [fast]`
Creates a cache with the directory of all packages available on the package sources. With the additional keyword `fast`, `dnf` checks whether the cache is up-to-date and only performs the update if this is really necessary.

Running `dnf makecache` is only necessary in exceptional cases. `dnf` automatically checks the other commands to ensure that the cache isn't too old and updates it itself if necessary.

▶ `module install/update/remove/enable/disable/lock/unlock/list/info ...`
Manages packages that are available in different versions. This extension of the `dnf` command is part of the modules and streams introduced with Fedora 28 and RHEL 8. During installation, you must specify the desired version and profile in addition to the package name, for example, `dnf module install nodejs:13/development`.

▶ `remove name1 name2 ...`
Uninstalls the specified packages.

▶ `repoquery name`
Determines the complete package name for the specified package. It doesn't matter whether or not the package has already been installed. With the additional `-i` option, `dnf repoquery` provides a lot more metadata in a table, such as the version number, license, package source, and project URL. With the `--location` option, the command specifies an address from which the package can be downloaded manually.

▶ `search repo query term`
Returns a list of all packages that contain the search term in the description.

▶ `update`
Displays all installed packages.

▶ `update name1 name2 ...`
Updates only the specified packages.

▶ `upgrade`
Causes the same effect as `update` in combination with the `--obsoletes` option (see the following list).

You can influence the behavior of dnf by using some very rarely needed options:

- --enablerepo=name
  Activates a package source that is deactivated. This option avoids having to change the *.repo file if you only want to install a single package from an inactive package source.

- --exclude=package
  Excludes the specified package from the selected operation (e.g., an update).

- --obsoletes
  Causes dnf to delete packages that are no longer required after an update. This is only useful for a distribution update (version *n* to *n+1*).

- -y
  Answers all questions from dnf with *yes*. This means dnf can be used for installation without further interaction (e.g., in a script).

Examples

The following two commands first update all existing packages and then install the emacs editor:

```
root# dnf update
...
Transaction Summary
Install      1 Package(s)
Update      42 Package(s)
Remove       0 Package(s)
Total download size: 52 M
Is this ok [y/N]: y
...
```

```
root# dnf install emacs
Installing:
 emacs                    aarch64   1:29.2-2.fc39   updates    32 M
Installing dependencies:
 dejavu-sans-mono-fonts   noarch    2.37-20.fc39    updates   469 k
 emacs-common             aarch64   1:29.2-2.fc39   updates    42 M
...
Transaction Summary: Install  20 Packages, total download size: 139 M,
installed size: 728 M
Is this ok [y/N]: y
```

dnf repoquery provides a lot of metadata about a package. The following listing was created on AlmaLinux 9, whereby I had to install the dnf-utils package beforehand:

```
root# dnf repoquery -i joe
Name      : joe
Version   : 4.6
Release   : 12.el9
Summary   : An easy to use, modeless text editor ...
```

The following two commands determine which versions of the nodejs package are available and then install the latest version in the minimal profile:

```
root# dnf module list --all nodejs
AlmaLinux 9 - AppStream
Name       Stream  Profiles                            Summary
nodejs     18      common [d], development, minimal,s2i  Javascript runtime
nodejs     20      common [d], development, minimal,s2i  Javascript runtime
Hint: [d]efault, [e]nabled, [x]disabled, [i]nstalled
root# dnf module install nodejs:20/minimal
```

The last command downloads the source code package of the Gnome text editor:

```
user$ dnf download --source gedit
```

docker [options] command

The docker command manages Docker containers. Before using the command for the first time, you must install docker, ideally from a current Docker package source (see *https://docs.docker.com/get-docker*).

In Fedora and RHEL-compatible distributions, the almost compatible podman command is available instead of docker. Even if there are various differences in the technical implementation, almost all the commands described in the following list work in the same way as with docker:

▶ build *directory/url*
Reads the Dockerfile file located at the specified location and creates a corresponding local image from it. With the -t *name:tag* option, the image can be named and marked with a *tag*.

▶ compose up|down
Evaluates docker-compose.yml (obsolete) or compose.yaml, creates the container setup described there, and executes it or stops it again. The command isn't yet available for podman (as of spring 2025); in the Red Hat world, you can use the podman-compose script instead. A reference for the syntax permitted in docker-compose.yml can be found at the following address: *https://docs.docker.com/compose/compose-file*.

▶ create *imagename/id* [*cmmand*]
Creates a container similar to docker run, but doesn't start it.

▶ exec *containername/id command*
Runs the specified command in an already running container. As with docker run, the -i and -t options enable interactive operation. With -d, the command is executed in the background.

▶ image prune|pull|push|rm [*imagename/id*]
Deletes all unused images, downloads images from or uploads images to the Docker Hub, or deletes a single image. (Docker Hub is a collection of public Docker images, see *https://hub.docker.com*.)

▶ images
Lists local images and specifies their size. With the -q option, only the image IDs are listed, which enables automated further processing.

▶ inspect *containername/id*
Provides detailed information on the specified container.

▶ logs *containername/id*
Displays the logged messages of the container. This only works if the underlying image is configured accordingly.

▶ network create|connect|inspect|rm [*networkname/id*]
Administrates networks for Docker containers.

▶ port *containername/id command*
Lists the port assignments of a container.

▶ ps
Outputs all running containers. With the -a option, stopped containers are also taken into account. If you add -s, the command also displays the container size.

▶ rename *oldname/id newname*
Renames a container.

▶ rm *containername/id*
Deletes the specified container. With -f, the command gets executed even if the container is still running.

▶ rmi *imagename/id*
Deletes the specified image.

▶ run *imagename/id* [command]
Creates and starts a new container from the specified image and then executes either the default command or the specified command in it. The required image is downloaded from the Docker Hub if it's not already available locally. The properties of the new container are controlled by countless options. The most important ones are listed here:

– -d executes the container in the background.

– -e *var=value* defines an environment variable for the container.

– -h *hostname* assigns a hostname to the container.

- -i -t enables interactive operation of the container.
- --name *cname* assigns a name to the container.
- --network *netname* uses the specified Docker network.
- -p *localport:cport* connects a local port with a port of the container.
- --rm deletes the container again as soon as the execution ends.
- -v *localdir:cdir* connects a local directory with a volume of the container.

▶ service create|inspect|logs|ls|ps|remove|update
Administrates the services in a Docker cluster (swarm).

▶ start/stop *containername*
Starts or stops a container that has already been set up.

▶ system df
Lists the space requirements of all images, containers, and volumes.

▶ system info
Displays comprehensive information about the Docker installation.

▶ system prune
Deletes containers that aren't running, unused images, and—with the --volumes option—also volumes.

▶ tag *containername/id* newname:tag
Renames a container.

▶ volume ls|inspect|prune|rm
Manages volumes.

Example

docker run creates a derived container from the official MariaDB image on Docker Hub and gives it the name mydb. If the image hasn't yet been downloaded locally, docker run will take care of this. The command then takes a little longer to execute the first time.

All database files are saved in /home/<username>/varlibmysql. Port 3306 of the container is accessible in the host system via the local port 13306.

```
root# mkdir /home/<username>/varlibmysql
root# docker run -d --name mydb -p 13306:3306 \
        -v /home/<username>/varlibmysql/:/var/lib/mysql mariadb
```

The container runs in the background until it's terminated via docker stop:

```
root# docker stop mydb
```

If you don't want to use the container again, you can now delete it:

```
root# docker rm mydb
```

```
dpkg options [filename/packagename]
```

dpkg takes care of low-level package management in all Debian- and Ubuntu-based distributions. However, instead of using dpkg for package installation, users often use a command based on it, usually apt-get.

▶ --configure *package name*
  Runs the configuration scripts of the specified package. This is usually done during the installation, but, in some cases, it may be necessary to carry out this step explicitly. Some packages have interactive setup programs in addition to the automatic configuration scripts. If you need such a setup program again later, you should run dpkg-reconfigure *package name*.

▶ --get-selections
  Returns a list of all installed packages, similar to --list. However, the result contains less detailed information and is therefore much easier to read. If you redirect the list to a text file and save it, you can install all packages later on a different computer by using --set-selections.

▶ -i or --install *filename.deb*
  Installs the specified package file. If an older version is already installed, it will be uninstalled and replaced by the new version. The configuration scripts provided in the package are also executed during the installation.

  Before the installation starts, it's ensured that all package dependencies are fulfilled. If that's not the case, the error message will usually indicate which packages are missing. Instead of laboriously installing all dependent packages, it's best to run apt install ./filename.deb to perform the installation with apt.

  Usually, dpkg allows only the installation of packages in the appropriate architecture for the distribution. If you use a 64-bit Linux distribution, but a program is only available as a 32-bit package, you can force the installation with --force-architecture. However, whether or not the program actually works is another matter. In any case, you must also install all required 32-bit libraries.

▶ -l or --list
  Provides a list of all installed packages.

▶ -l or --list *'pattern'*
  Provides a list of installed and available packages. The first column of the package list contains a two-letter code.

  The first letter indicates the desired status of the package:

  i = install

  n = do not install

  r/p = remove

  h = hold

The second letter describes the actual status:

i = installed

n = not installed

c = configured

u = unpacked, but not yet configured

f = failed

The information comes from the Debian package database, a collection of files in the /var/lib/dpkg directory. Meta information about all installed and available packages is stored there.

▶ -L or --listfiles *package name*
Returns a list of all files in the specified package. This only works for packages that are already installed. You can determine the contents of uninstalled packages using the command; dpkg-deb --contents *file name*.

▶ -P or --purge *package name*
Removes the specified package including all package files (even if you've modified them).

▶ -r or --remove *package name*
Removes the specified package.

▶ -S or --search *filename*
Determines the package from which the specified file originates.

Example

The first dpkg command installs a new package. dpkg --search determines the package from which the /etc/sensors3.conf file originates. dpkg --listfiles returns a list of all files in this package.

```
root# dpkg --install paketname.deb
root# dpkg --search /etc/sensors3.conf
libsensors-config: /etc/sensors3.conf
root# dpkg --listfiles libsensors-config
...
/.
/etc
/etc/sensors.d
/etc/sensors.d/.placeholder
/etc/sensors3.conf
...
```

dracut [options] initrd file kernel version

dracut is responsible for creating an initrd file on Fedora, RHEL, and (open)SUSE. If dracut is executed without additional parameters, it creates an initrd file named /boot/ initrd-*kernelversion* for the latest kernel in the /boot directory. dracut takes into account the settings in /etc/dracut.conf as well as countless options described in man dracut. Only the three most important options are summarized here:

▶ -d a,b,c
Integrates the kernel modules a, b, and c into the initrd file. Usually, this option isn't required. dracut automatically recognizes which modules are required for the start process.

▶ -f
Overwrites an existing initrd file.

▶ --regenerate-all
Creates new initrd files for all kernel versions installed on the system.

Debian and Ubuntu use update-initramfs instead.

Example

To manually create an initrd file for a self-compiled kernel 5.7.3 in the /boot/ vmlinuz-5.7.3 file, you need to run the following command:

root# **dracut /boot/initrd-5.7.3 5.7.3**

du [options] [directory]

du outputs information about the memory requirements of files or directories. If a file specification is given in the directory parameter (e.g., * or *.tex), du will receive a list indicating the size of all files. If, however, only one directory is specified, du will determine the memory requirements for all child directories. The memory details also include the memory requirements of all child directories. The last numerical value indicates the total memory requirement of all files and subdirectories from the specified directory onward. All figures are in KiB.

▶ -b or --bytes
Displays the sizes in bytes (instead of KiB).

▶ -c or --total
Displays the final total as the final value. This option is only necessary if du applies to files (and not to directories), which makes it relatively easy to determine how much memory all files with a certain ID (e.g., *.pdf) take up.

▶ -h or --human-readable

Displays the sizes in an easy-to-read format. K, M, and G are abbreviations for KiB, MiB, and GiB.

▶ --max-depth=*n* or --max *n*

Outputs the directory size only for the specified number of directory levels.

▶ -s or --summarize

Displays *only* the final total. This option is only useful if the memory requirements of directories are displayed.

▶ -S or --dereference

Displays only the memory requirements directly in the directory. The memory requirements in subdirectories are *not* taken into account.

▶ -x or --one-file-system

Ignores directories where other file systems are mounted. (This option is almost always recommended. On my computer, I've defined a corresponding alias with alias du="du -x".)

Example

du determines which directories take up how much space. With --max 1, subdirectories aren't listed individually but are included in the result.

```
user$ du -h --max 1
15M     ./Downloads
29M     ./Videos
69M     ./.local
...
```

Alternatives

du doesn't provide the option to sort the result. This is exactly what the interactive ncdu command can do. It also provides the option to navigate through subdirectories using the cursor keys. I highly recommend this!

dumpe2fs device

dumpe2fs provides countless internal information about the state of an ext file system, in particular about the superblocks and the organization of the block groups of the data medium.

▶ -b

Lists only defective blocks on the data medium.

▶ -h

Provides a summary of the superblock data.

dvips [options] name.dvi

dvips creates a PostScript file from a *.dvi file. If dvips is used without the -o option, the command redirects the resulting PostScript file to the default printer.

▶ -A

Converts odd pages only.

▶ -B

Converts even pages only.

▶ -D n

Uses a resolution of n dpi (*dots per inch*) when generating LaTeX bitmap fonts. The default resolution is usually 600 dpi (see /etc/texmf/config.ps). Alternatively, n may also be 300, 400, or 1270 (see /usr/bin/mktexpk). This option is only relevant for bitmap fonts. PostScript fonts are always resolution-independent.

▶ -E

Creates an EPS file (*Encapsulated PostScript*) with a bounding box that contains only the part of the page that is actually used. This only makes sense if the DVI file doesn't comprise more than one page and the resulting EPS file is then to be embedded in another document.

▶ -i -S n

Splits the output into files of n pages each. The files are numbered automatically.

▶ -l *last page*

Ends the conversion with the specified page.

▶ -o *target file*

Writes the result to the specified file (instead of forwarding it to the lpr program).

▶ -p *first page*

The conversion starts with the specified page.

▶ -pp n1,n2-n3,n4,n5,n6-n7

Prints the specified pages. Note that there must be no spaces in the page list.

Global default settings for dvips are defined in the config.ps file. Depending on the distribution, this file can be found in the /etc/texmf directory or in /usr/share/texlive/texmf-dist/dvips/config.

e4defrag [options] file/directory/device

e4defrag defragments either a file, the entire contents of a directory, or an entire ext4 file system. Users without root permissions can only defragment their own files.

Defragmentation means that the blocks of a file are reserved on the data medium in as close a sequence as possible. This speeds up read access, especially with conventional hard disks. The effects of defragmentation are less pronounced with SSDs.

▶ -c

Displays a defragmentation counter for the file(s) in question, but doesn't make any changes. If e4defrag is executed with this option for the entire file system, the command returns a list of the most fragmented files after some time.

▶ -v

Displays information before and after the defragmentation of each file.

echo [options] string

The bash command echo outputs the specified character string. If the string contains spaces or special characters, it must be enclosed in double or single quotation marks, depending on whether shell variables are to be output or not.

▶ -e

Interprets various backslash character combinations, for example, \a as a signal tone, \n as the end of a line, and \t as a tab (see also help echo). Thus, echo -e "\a" outputs a warning signal.

▶ -n

Doesn't switch to a new line at the end of the output. The output can be continued by using another echo statement.

Example

The following command displays the current contents of the $PATH variable:

```
user$ echo "The PATH variable contains: $PATH"
```

efibootmgr [options]

In Linux systems that have been installed and started in EFI mode, various EFI settings can be changed using the efibootmgr command. These settings are stored in a nonvolatile memory (NVRAM) on the mainboard. efibootmgr is primarily used to set up new EFI boot entries, remove existing entries, and define the default boot sequence. The command requires the efivars kernel module to be loaded. If this isn't the case, you need to run modprobe efivars.

Note that the efivars module can only be used if Linux was booted in EFI mode (not in BIOS mode!). If necessary, you can use a Linux live system that can be started in EFI mode for repair work.

▶ -b *n* -B

Deletes the specified EFI boot entry. You can determine the numbers of the EFI boot entries by running efibootmgr without parameters.

▶ -c -l \EFI*distrib**grubfile.efi* -L *name*

Creates a new EFI boot entry (-c) and specifies the location of the corresponding boot loader file. The path is relative to the EFI partition /boot/efi, and \ must be used as the directory separator. -L specifies the name of the EFI boot entry (usually the distribution name, which is Linux by default).

The new entry automatically becomes the default entry. Optionally, you can use -p to specify the number of the EFI boot partition (usually 1) and -d to specify the device name of the first hard disk/SSD (usually /dev/sda).

▶ -n *sad*

Determines which EFI boot entry is used at the next restart. The setting isn't permanent, but only applies to the next computer start.

▶ -N

Deletes the setting specified with -n for the next restart.

▶ -o *n1,n2,n3,...*

Sets the new order of the EFI boot entries permanently. If you only want to specify the entry in the first position, it's sufficient to enter its number.

▶ -p *n*

Determines in which partition the EFI boot partition is located (usually in the first partition).

▶ -q

Specifies that no output is required (*quiet*).

▶ -t *n*

Specifies how long EFI should wait before activating the default boot entry after a restart (in seconds). However, it depends on the EFI implementation whether the EFI menu is displayed automatically during this wait time. On one of my test computers, the EFI menu generally gets displayed only when I press a corresponding key ([F8], ASUS mainboard P8H67-M Evo).

▶ -T

Deletes the wait time setting, which has been set using -t; that is, the default boot entry gets activated immediately.

Example

If the command is executed without any further options, it lists the EFI boot entries and some other parameters of the EFI boot loader:

```
root# efibootmgr
BootCurrent: 0000
Timeout: 1 seconds
BootOrder: 0000,0005,0003,0001,0002
Boot0000* ubuntu
Boot0001* Hard Drive
Boot0002* CD/DVD drive
Boot0003* Windows Boot Manager
Boot0005* Fedora
```

On the next reboot, Fedora is supposed to be started (deviating from the saved boot sequence, in which Ubuntu is in first place).

The next command creates a new EFI boot entry. The doubling of the \ characters is necessary because the shell interprets a single \ character as an identifier for special characters.

```
root# efibootmgr -c -l \\EFI\\test\\abc.efi -L abc
```

egrep [options] search pattern file

egrep searches the specified text file for a search pattern and displays the text passages found. The rules of *Extended Regular Expressions* (EREs) apply to the search pattern.

egrep isn't a command in its own right, but merely an alias or a small script that runs the grep command with the -E option. The GNU documentation advises against calling egrep and recommends running grep -E directly. Further information and examples can be found in grep.

enscript [options] source file -p target file

The enscript command from the package of the same name converts a text file into either PostScript, HTML, or RTF format.

▶ --color
Performs syntax highlighting with colors (must be combined with -E).

▶ -E
Performs syntax highlighting for program code (bold, italics).

▶ -f font
Uses the specified font (10-point Courier by default).

▶ -M paper size
Uses the specified paper format (e.g., A3, A4, or Letter).

▶ --n
Formats the text with n columns. Thus, -2 results in two-column formatting.

▶ -r or --landscape
Fills the sheet in landscape format.

▶ -W *format* or --language=*format*
Specifies the desired format. The available options include PostScript (applies by default), html, and rtf.

▶ -X *charset*
Specifies the character set of the text (e.g., ascii, latin *n*). Unfortunately, Unicode isn't supported. To convert Unicode texts to PostScript format, you must use the paps command.

epstopdf [options] file.eps

epstopdf converts the EPS file into a PDF file and saves the result as file.pdf.

▶ --exact
Evaluates the ExactBoundingBox (instead of the regular bounding box).

▶ --hires
Evaluates the HiresBoundingBox (instead of the regular bounding box).

▶ --nocompress
Doesn't compress the PDF data.

▶ --outfile=*name*
Saves the result under the specified file name.

erd [options] files

erd from the erdtree package combines the functions of du, ls in a single command. The output is graphically appealing in colors (see *Figure 2*). The command takes .gitignore into account and doesn't display the files excluded there.

▶ -.
Also displays hidden files.

▶ -d line|word|block
Displays the size of files in lines, words, or blocks.

▶ -H
Indicates the size of files in KiB, MiB, or GiB (instead of bytes).

▶ -i
Ignores .gitignore; that is, it displays *all* files in a Git directory.

▶ -l or --long
Displays access rights, owners, and groups.

- -L or --level *n*

  Tracks subdirectories only to the desired depth. By default, erd runs through all sub-directories.

- -p or --pattern *regex*

  Displays only files that match the regular pattern.

- -s or --sort name|rname|size|rsize|access|raccess|mod|rmod

  Sorts the output according to file name, file size, access, or modification time. The r variants reverse the sort order. Thus -s rsize displays the largest file first.

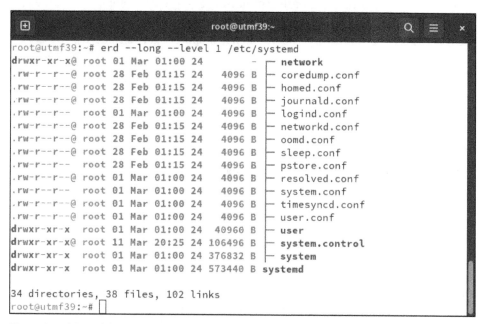

Figure 2 erd Provides an Overview of the /etc/systemd Directory

etherwake [options] mac

The etherwake command from the package of the same name sends a special network packet to a MAC address and attempts to "wake up" the device from sleep mode. The device must support the *Wake On LAN* standard.

- -i *interface*

  Sends the packet via the specified network interface (eth0 by default).

Example

The following command activates the NAS device in my household, which is located in the basement and is normally in sleep mode to save energy:

```
root# etherwake -i wlp0s20f3  00:11:32:12:DD:0D
```

ethtool [options] device [parameter]

ethtool determines or changes (hardware) parameters of Ethernet adapters. If you simply pass an interface name to the command without any further options, you'll receive a list of the most important key data. These include the supported ports, the link modes, the speed, the duplex mode, and so on.

▶ -p *device n*
   Switches the interface LED on for *n* seconds. For servers with multiple adapters, this simplifies the identification of the connector socket. With some adapters, the LED then flashes for the specified time. Other adapters don't support LED control at all—in this case, the error message **Operation not supported** gets displayed.

▶ -s *device para1 value1 para2 value2* ...
   Changes several parameters of the interface. Permitted parameter names include speed, duplex, port, mdix, autoneg, and advertise.

Example

The following command explicitly sets the speed of the network adapter to 100 MBit/second, activates duplex mode, and deactivates the auto-negotiate function:

```
root# ethtool -s eth0 speed 100 duplex full autoneg off
```

eval $var

The bash command eval interprets the contents of the variable as a command line, evaluates this line with all known substitution mechanisms, and finally executes the command. eval is always required if a command stored in a variable is to be executed and this command contains various special characters of the shell.

Example

The command stored in the kom variable can't be executed until eval is used. The first attempt to run the command fails because the bash no longer evaluates the pipe character | after it has replaced $kom with its content.

```
user$ kom="ls | more"
user$ $kom
ls: |: No such file or directory
ls: more: No such file or directory
user$ eval $kom
```

exec command

The bash command exec starts the command specified in the parameter as a replacement for the running shell. The command can be used, for example, to start another shell. This terminates the running shell. With a normal command start without exec, however, the shell continues to run and accepts further input after the end of the command or continues the running script.

exfatlabel device [label]

exfatlabel reads or changes the name of an exFAT file system. The command can be found in many distributions in the exfat-utils package. To set up an exFAT file system, you can use mkfs.exfat.

exiftool [options] file/directory

The Perl script exiftool from the package of the same name can extract and modify the *Exchangeable Image File Format* (EXIF) data from JPEG and TIFF files. EXIF describes how additional metadata, for example, the time a photo was taken, is stored in bitmap files. If the command is called without options, it simply lists all the EXIF information of the specified file.

▶ "-*tag*"
Reads only the specified information, such as the creation date with -CreateDate.

▶ "-*tag=value*"
Overwrites the information with the specified value.

▶ "-*tag1<tag2*"
Overwrites a tag with another tag. A formatting string specified with -d can be taken into account (see the following example).

▶ -lang de
Outputs the EXIF elements in German.

▶ -v
Reveals what's currently happening (*verbose*).

Countless other options are documented at *https://exiftool.org*.

Example

The following command renames all images in the current directory. The new file name is derived from the date of creation (e.g., 2025-12-31-124500) plus a consecutive number for multiple photos created in the same second (format string %.c), plus the original file identifier (format string %%e).

```
user$ exiftool -v '-Filename<CreateDate' -d '%Y-%m-%d-%H%M%S-%.c.%%e' .
user$ ls *.jpg
2025-02-27-161642-0.jpg  2024-10-27-194700-0.jpg ...
```

exit [return value]

The bash command exit terminates a shell script or the running shell. If no return value is specified, the shell returns 0, that is, "OK".

expand [options] file

expand replaces all tab characters in the specified text file with a corresponding number of spaces. If no options are specified, expand assumes a tab spacing of eight characters. The result is sent to the standard output.

▶ -n

  Changes the tab spacing to n characters.

Example

The following command removes the tab characters in test.java and saves the result in test-no-tabs.java. The original file was written with an editor that equated tab characters with four spaces.

```
user$ expand -n 4 test.java > test-no-tabs.java
```

export [options] variable [=value]

The bash command export declares the specified shell variable as an environment variable. This means that the variable is also available in all called commands and subshells. Optionally, variables can also be assigned. If the command is called without parameters, all environment variables are displayed.

▶ -n

  Turns an environment variable back into a normal shell variable. The command therefore has exactly the opposite effect to when used without options.

exportfs [options]

The exportfs -a command reports changes in the NFS configuration file /etc/exports to the running NFS server. The command must therefore be executed for changes made in /etc/exportfs to take effect.

expr expression
expr string : pattern

expr evaluates the specified expression arithmetically or performs a pattern comparison for character strings. There must be spaces between the specified variables, numbers, and operators! A short description of all permissible operators can be found in the man command for expr. Note that many operator characters in the shell must be protected from immediate evaluation by the \ character.

If you use bash as a shell, you can also make calculations in it without using expr: arithmetic expressions can be specified there via $[expression]. The wildcard characters described in grep can be used in the pattern. Usually, the pattern must be placed in apostrophes to prevent the shell from evaluating the special characters.

Examples

The first expr command performs a simple calculation. The second command returns the maximum number of characters that correspond to the pattern expression. The third command extracts the bracketed part of the pattern from a character string.

```
user$ expr 3 + 7 \* 19
136
user$ expr abcdefghi : 'a.*g'
7
user$ expr abc_efg_hij : '.*_\(.*\)_.*'
efg
```

fail2ban-client [options] command

Fail2ban is a security program that monitors logins and blocks the IP address in question for a certain period of time after repeated incorrect logins. If Fail2ban has been set up, its function can be checked and changed via fail2ban-client.

▶ get *parameter*
  Imports a configuration parameter.

▶ reload [*jailname*]
  Reloads the entire configuration or the configuration of a jail. In Fail2ban, a *jail* is a program to be monitored.

- set [*jailname*] *parameter value*
  Changes a general configuration parameter or the parameter of a jail.

- start/stop [*jailname*]
  Starts or stops Fail2ban completely or starts/stops only one jail.

- status [*jailname*]
  Displays the list of all active jails or details of a jail.

If fail2ban-client is executed without commands only with the -d option, the command outputs the entire currently valid configuration, whereby only the filters, jails, and actions that are actually in use are taken into account. This is very helpful in troubleshooting due to the confusing configuration files.

Examples

The following commands show that four jails are active on the test computer. The jail for the SSH server has already temporarily blocked 1,418 IP addresses in the past. Currently, 3 addresses are blocked.

```
root# fail2ban-client status
Status
  Number of jail:   4
  Jail list:        dovecot, postfix, sshd, sshd-ddos
root# fail2ban-client status sshd
Status for the jail: sshd
   Filter
     Currently failed: 4
     Total failed:     151362
     Journal matches:  _SYSTEMD_UNIT=sshd.service + _COMM=sshd
   Actions
     Currently banned: 3
     Total banned:     1418
     Banned IP list:   58.218.nnn.nnn 58.242.nnn.nnn 180.76.nnn.nnn
```

You can unblock an erroneously blocked IP address using fail2ban-client set (here, we do this for the SSH jail):

```
root# fail2ban-client set sshd unbanip 1.2.3.4
```

fc-list [pattern]

fc-list lists all scalable fonts that match the (optional) search pattern. The result can be sorted using | sort.

fdisk [options] [device]

fdisk is an outdated program for partitioning hard disks. It can only process disks with MBR partition tables! For this reason, you should make friends with parted, which is compatible with both the *master boot record* (MBR) and the more modern *GUID Partition Tables* (GPTs).

ffmpeg [inopts] [-i infile] [outopts] outfile

The ffmpeg command from the package of the same name converts video files from one format to another. When specifying options, note that these only apply to the video file specified next. The order of the options is therefore decisive for the correct function of the command. Unless you make different settings, ffmpeg uses the same codecs and settings for the result file as in the source file.

In addition to the options listed here, there are countless others. The ideal setting of the options is a science in itself. If you don't want to deal with this, it's advisable to use a graphical user interface for video recoding, such as HandBrake.

▶ -acodec/-vcodec *type*
Specifies the desired audio or video codec. A list of available codecs is provided by ffmpeg -codecs.

▶ -b *n*
Specifies the desired bit rate in bits per second (200 kBit/s by default).

▶ -f *format*
Uses the specified format.

▶ -formats
Lists the supported formats, codecs, filters, and so on.

▶ -r *n*
Specifies the desired frame rate (25 frames per second by default).

▶ -s *size*
Specifies the desired frame size in the form *wxh* (e.g., 800x600). Alternatively, you can specify a predefined character string (e.g., vga or hd720; see man ffmpeg).

▶ -ss *position*
Starts the conversion at the specified time. The time can be specified either in seconds or in the format hh:mm:ss[.xxx].

▶ -t *duration*
Converts only the specified time span (and not the entire file). The time span is specified in the same way as with the -ss option.

▶ -target *type*
Specifies the desired target file format (e.g., vcd, dvd, etc.).

▶ -vpre *quality*

Selects a file with predefined options (presets) for the specified codec (e.g., default, medium, max, or fast). The preset files available for selection are usually located in the /usr/share/ffmpeg directory.

▶ -y

Overwrites existing files.

Example

The following command creates a movie file in DVD resolution:

```
user$ ffmpeg -i in.avi -target pal-dvd -y out.avi
```

The second example converts an OGV movie file into a YouTube-compatible MPEG file:

```
user$ ffmpeg -i in.ogv -vcodec libx264 -vpre medium -acodec copy -y out.mpg
```

fg [process]

The bash command fg continues a process in the foreground that has been interrupted using Ctrl + Z. If no process is specified, fg applies to the last interrupted process or the last process started in the background. Otherwise, the process must be specified by its name or by the bash-internal job number (not PID).

Instead of fg process, the abbreviation %process is also permitted.

fgconsole [--next-available]

fgconsole outputs the number of the currently active console. With the --next-available option, the command determines the number of the first free (unused) console. To change the active console, you can use the chvt command.

file [options] file

file attempts to determine the file type of the specified file. As a result, file returns a character string with the file name and file type. file doesn't evaluate the file name or its identifier, but the contents of the file!

▶ -z

Attempts to recognize the data type of a compressed file.

Example

As the file name suggests, name.jpg contains a JPEG bitmap file:

```
user$ file name.jpg
name.jpg: JPEG image data, JFIF standard 1.02
```

find [path] [search options]

find helps you search for files. Various search criteria (pattern for the file name, file size, date of creation or last access, etc.) can be taken into account in the search. It's even possible to apply further selection criteria to all files that meet these conditions using another program (e.g., grep). In this way, for example, all *.tex files that have been edited in the preceding three days and that contain the text "Graphics programming" could be found. Due to the large number of possible search criteria, man find provides a description of the command that is several pages long.

The following syntax description explicitly refers to the GNU find implementation commonly used on Linux. It's characterized by some minor syntax differences compared to the find variants of other Unix systems: There, a path *must* be specified, while GNU find uses the current directory as the start directory. In addition, most find implementations require the use of the -print option to display the search results, which isn't necessary with GNU find.

General Options

Unlike most other commands, find automatically searches all subdirectories. If this isn't desired, the number of subdirectories must be limited via -maxdepth.

▶ -depth

First processes the current directory and only then the subdirectories. Depending on where you suspect the file you're looking for is located, this procedure will lead to initial results much more quickly.

▶ -follow

Also edits directories that are captured by symbolic links.

▶ -maxdepth *n*

Restricts the search to *n* directory levels. With -maxdepth 1, no subdirectories are taken into account at all.

Search Criteria

You can specify multiple search criteria at the same time. These criteria are logically linked by using AND. The search is aborted as soon as the first criterion isn't met—the order of the criteria can therefore influence the speed of the command. Criteria can be grouped using \(and \), negated via !, and logically OR-linked using -o (see the examples).

▶ -anewer *reference file*

Works like -newer (see that entry), but takes into account the time of the last read access.

▶ -atime *days* or -cmin *minutes*

Takes into account the time of the last read access (*access*). The specification is in days or minutes (see -mtime).

▶ -cnewer *reference file*

Works like -newer (see that entry), but takes into account the time of the last change to metadata.

▶ -ctime *days* or -cmin *minutes*

Evaluates the time at which the metadata of the file (access rights, etc.) was last changed.

▶ -executable

Finds files that are executable, that is, where the execute bit is set.

▶ -group *crime name* or -nogroup *group name*

Finds files that belong to the specified group (or that don't belong to it).

▶ -mmin *n*

Works like mtime, but the time is given in minutes.

▶ -mtime *days* or -mmin *minutes*

Finds files whose content was last modified exactly *n* days or minutes ago (*modify*). If a + is entered before the number, then all files older than *n* are captured. A preceding - returns files that *are* younger than *n*. -mtime 0 returns files that have been changed in the past 24 hours.

▶ -name *search pattern*

Finds files that match the specified search pattern. If the search pattern contains wildcards, it must be placed in apostrophes. If you don't want find to distinguish between uppercase and lowercase, you should use the -iname option instead of -name.

▶ -newer *reference file*

Finds files that are newer than the reference file, that is, files that were created or modified after the reference file. For example, you can create a timestamp file at the start of a backup using touch. For an incremental further backup at a later point in time, only the files for which –newer *timestamp* applies are taken into account.

▶ -path *search pattern*

Finds files that match the specified search pattern. The option goes beyond -name because the search pattern now affects not only the file name but also the path to it. This option is more flexible than the direct path specification in the first argument of find because here the wildcard characters also include the / character.

▶ -perm *access bits*
Finds files whose access bits correspond exactly to the specified octal value (see chmod). If the octal value is preceded by a -, the file may also have access rights beyond this. If a / is placed in front, it's sufficient if at least one of the searched access bits is set.

▶ -size *file size*
Specifies the desired file size. The default specification is in multiples of 512. -size 3 therefore denotes files between 1,024 and 1,536 bytes. The additional c or k characters can be used to specify the size in bytes or KiB. A prefixed + includes all larger files, and a prefixed - includes all smaller files. -size +10k therefore returns all files that are larger than 10 KiB.

Note the somewhat unique logic of this command when using this option! find -size 1M considers files with a size between 1 byte and 1 MB. -size -1M only finds empty files. With -size +1M, you get files in the result that are larger than 1 MB.

▶ -type *character*
Restricts the search to specific file types. The most important characters are f for regular *files*, d for *directories*, and l for symbolic links.

▶ -user *username* or -nouser *username*
Finds files that belong to the specified user (or that don't belong to them).

Actions When Finding a File

▶ -exec *command [options]* {} \;
Calls the specified command and passes the file name of the file found that has fulfilled all the criteria processed so far. The command can then perform a test to check whether the file meets additional criteria. A typical program that is called by -exec is grep. Alternatively, you can also call commands here that move, copy, or otherwise process the files found.

{} is a placeholder for the file name. ; concludes the command call; that is, additional find options can be specified after it. \ is required within the shell to prevent the interpretation of ; as a special character.

▶ -exec *command [options]* {} +
As in the preceding paragraphs, but the specified command isn't called for every file found. Instead, find tries to pass as many results as possible to the command at once, which is much more efficient. The number of files is only limited by the maximum length of a command. This is approximately 2 million characters in current Linux distributions.

▶ -ls
Provides detailed information on each file found (similar to ls -l).

▶ -print

Displays the file names found on the screen. This option is the default setting unless -exec is used.

▶ -print0

Separates the file names of the result list by 0 bytes. This enables further processing of file names with spaces via xargs.

▶ -printf *format*

Displays the file names found and other information on the screen. The format string can be used to specify the format of the output and the additional information that is also to be output (e.g., the file size or the date of the last change). The syntax for the format string is described in the man pages.

Examples

The first find command returns all files in the current directory (including hidden files, but no files from subdirectories); the second command only returns ordinary files (but no invisible files):

```
user$ find -maxdepth 1 -type f -name '*'
user$ find -maxdepth 1 -type f -name '[!.]*'
```

The following command searches the /etc directory for files that have been changed in the past two weeks. The command must be run with root permissions because some files in /etc may not be read by normal users:

```
root# sudo find /etc -mtime -14
```

The following command deletes all backup files that are identified by the ~ character at the end of the file name. find also takes all subdirectories into account. The list of files to be deleted is redirected to rm by the command substitution with $(command).

```
root# rm $(find . -name '*~')
```

If there is a large number of files, an error will occur when you run the preceding command: the command line with all *~ files will become so long that it exceeds the maximum command line length. In such cases, you must either use the -exec option of the find command or the xargs command. The xargs variant has the advantage that it can also cope with file names that contain spaces.

```
user$ find -name '*~.jpg' -exec rm {} \;
```

In the next example, all files in a directory are to be made readable for everyone (see chmod). It's somewhat more difficult to use find to set the execute bit for all subdirectories (but not for ordinary files) so that everyone on the computer can use cd to change to the directories. The + sign at the end of the command means that chmod isn't called for each directory individually, but that as many directories as possible are passed to chmod at once.

```
root# cd /directory/for/all
root# chmod -R a+r *
root# find . -type d -exec chmod a+x {} +
```

The last example first creates a reference file with the date 1/1/2025 and the directory ../new-images by using touch. All images from the local directory that have been created since 1/1/2025 are now copied there. new-images must not be created in the current directory; otherwise, find will also capture the files copied there. In addition, note the OR link for the file identifiers *.jpg and *.png.

```
user$ touch 2025 -t 202501010000
user$ mkdir ../new-images
user$ find . -newer 2025 \( -name \*.png -o -name \*.jpg \) \
      -exec cp {} ../new-images +;
```

Further find examples can be found in the description of the grep and xargs commands.

Alternatives

find is an important command, but it's difficult to use because nobody can remember all the possible options. Accordingly, there are many attempts to create more convenient alternatives to find. I present two of them in this book:

▶ erd combines the most important functions of du, find, ls, and tree in a single command.

▶ fzf, on the other hand, excels with its interactive selection of files.

Another option is fd (see *https://github.com/sharkdp/fd*), which is very similar to find. Its biggest advantages are that the options are easier to memorize and the search results are displayed in color.

---

findmnt [options] [device/mountpoint]

---

findmnt without parameters returns a tree-shaped list of all file systems integrated in the directory tree. The result also contains various /dev, /proc, /sys, and /run file systems for internal information exchange.

By specifying a device or mount point, the list is reduced to all file systems selected in this way. Other permissible entries are the UUID or the name (label) of a file system or the major and minor number of the device in the notation major:minor.

▶ -A

  Displays all file systems.

▶ -l
Displays the result as a table (not as a tree).

▶ --mtab or --fstab
Takes into account only the entries from /etc/mtab or /etc/fstab. By default, findmnt takes into account the mount table of the kernel in the pseudo file /proc/self/mountinfo.

▶ -o columns or --output columns
Outputs only the information specified in columns column by column. Permitted column codes include SOURCE:

Device

  – TARGET: mount path

  – FSTYPE: File system type

  – OPTIONS: all mount options

  – SIZE: Size of the file system

Further codes are provided by fndmnt --help. The column codes are separated by commas, for example, -o SOURCE,TARGET,FSTYPE. Spaces aren't permitted!

▶ -t typelist or --types typelist
Returns only file systems of the specified type. With -t ext2,ext3,ext4, only ext file systems in versions 2, 3, and 4 are taken into account.

Example

findmnt shows all active file systems on the test computer:

```
root# findmnt
TARGET                   SOURCE       FSTYPE       OPTIONS
/                        /dev/sda1    ext4         rw,relatime,errors=...
 /sys                    sysfs        sysfs        rw,nosuid,nodev,noexec,...
  /sys/kernel/security   securityfs   securityfs   rw,nosuid,nodev,noexec,...
  /sys/fs/cgroup         tmpfs        tmpfs        ro,nosuid,nodev,noexec,...
 ...
```

firewall-cmd options

firewall-cmd reads or changes the firewall configuration in Fedora, RHEL, and SUSE distributions. firewall-cmd communicates with the firewalld program running in the background. This program assigns a zone to each network interface. A zone in the sense of firewalld is a collection of rules for a specific usage purpose. The following list describes the most important zones:

▶ `block`

Blocks all network traffic; the sender receives an ICMP error message.

▶ `drop`

Blocks all network traffic; the sender isn't informed.

▶ `external`

Blocks most ports and activates masquerading (IPv4). The zone is intended for the interface that connects to the internet on a router.

▶ `FedoraWorkstation` and `FedoraServer`

Are used by default for network interfaces of Fedora installations. `FedoraServer` blocks all services except SSH and DHCP client. `FedoraWorkstation` is much more tolerant and also accepts Samba client connections and traffic through all ports between 1025 and 65535.

▶ `home` and `internal`

Blocks most ports, but accepts Samba (only as client), CUPS, and Zeroconf/Avahi/ mdns. The zones are intended for computers on a local network that is considered reasonably secure. If you want to use this zone and share Windows network directories yourself, you must also enable the Samba service.

▶ `public`

Similar to `home`, but also blocks CUPS and Samba client functions. The zone is intended for internet use in insecure networks, for example, in a public wireless network.

▶ `trusted`

Allows any network traffic. This zone is intended for well-secured local networks, but not for Wi-Fi connections.

Usually, `firewall-cmd` only makes the changes dynamically. If changes are to be permanent, you must specify the `--permanent` option. Then, the rules will be saved in the `/etc/ firewalld` directory.

Options

▶ `--add-interface=`*name* or `--remove-interface=`*name*

Assigns the specified interface to the default zone, assigns the zone specified by the `--zone` option, or removes this assignment. Before an interface can be assigned to a new zone, the previous assignment must be explicitly removed!

▶ `--add-prot=`*name* or `--remove-port=`*name*

Allows the use of a TCP port in the zone specified with `--zone` or blocks the port.

▶ `--add-service=`*name* or `--remove-service=`*name*

Allows the use of a network service (e.g., `https` or `ssh`) in the zone specified with `--zone` or blocks the service.

▶ `--change-interface=iface`
Assigns the active zone to the specified network interface. To connect another zone to an interface, you must also pass the `--zone=name` option.

▶ `--get-active-zones`
Lists all zones that are currently in use and indicates which network interfaces are assigned to the zones.

▶ `--get-default-zone`
Determines the default zone.

▶ `--get-services`
Lists all network services known to `firewall-cmd`. You can specify the resulting names (`http`, `ssh`, etc.) with `--add-interface`.

▶ `--get-zones`
Lists all zones known by `firewall-cmd`.

▶ `--get-zone-of-interface=name`
Determines the zone that is assigned to the specified network interface.

▶ `--list-all-zones`
Provides the key data of all defined firewall zones and specifies which network interfaces are assigned to the zones.

▶ `--list-ports`
Lists ports that are allowed (i.e., not blocked) in the currently active firewall zone. Warning: the command only returns ports to which no service name is assigned. Port 22 will never appear in the result because it's assigned to the `ssh` service. You can get nonblocked services via the `--list-all-zones` or `--list-services` options.

▶ `--list-services`
Lists all network services that are permitted (i.e., not blocked) in the currently active firewall zone.

▶ `--permanent`
Saves changes to the firewall configuration permanently. However, the changes aren't applied directly. If you want to make a change immediately *and* permanently, you must run the relevant command *twice*: once with `--permanent` and once without this option.

▶ `--set-default-zone=name`
Defines the default zone. It applies to all network interfaces that aren't explicitly assigned a different zone.

▶ `--state`
Determines whether `firewalld` is running or not.

▶ `--zone=name`
Indicates which zone is to be changed, edited, or read.

Examples

The following commands first determine which zone the network interface enp0s3 is assigned to and then assign the interface to the `trusted` zone. This means that all services can be used via this interface. The changes aren't permanent, so they only apply until the next restart of the computer or the firewall daemon.

```
root# firewall-cmd --get-zone-of-interface=enp0s3
public
root# firewall-cmd --zone=public --remove-interface=enp0s3
success
root# firewall-cmd --zone=trusted --add-interface=enp0s3
success
```

In the second example, the first command is used to determine which zone the network interface enp0s5 is assigned to. The computer is supposed to provide a web server to the outside world via this interface. But instead of generally assigning the interface to the `trusted` zone and thus enabling all other conceivable services, the rules for the zone are now changed (here, FedoraWorkstation) permanently: this zone now allows the HTTP and HTTPS protocols. The `reload` command activates the changes to the zone.

```
root# firewall-cmd --get-zone-of-interface=enp0s5   (determine active zone)
FedoraWorkstation
root# firewall-cmd --permanent --zone=FedoraWorkstation --add-service=http
root# firewall-cmd --permanent --zone=FedoraWorkstation --add-service=https
root# firewall-cmd --reload
```

flatpak [options] command

flatpak helps with the installation, removal, and administration of Flatpak packages. This is a new package format intended to simplify the installation of desktop applications independently of the package sources of a distribution. The largest official Flatpak repository is *https://flathub.org*.

Compared to conventional packages, there are additional security mechanisms, such as sandboxing. However, Flatpak packages are often huge because they include all the necessary libraries and don't take into account any existing traditional packages. The installation of multiple Flatpaks is therefore often associated with large redundancies.

Flatpak packages are usually installed directly in the desktop, for example, in the Gnome program *Software* or in the browser. The program is started in the Start menu or in the Gnome view **Activities**. The flatpak command is therefore rarely required.

Flatpak is available on Fedora and RHEL by default. However, the command can be easily installed and used in other distributions and is intended to become a platform for a distribution-independent distribution of Linux software. As an alternative to Flatpak, Snap packages can be used, which are favored by Ubuntu (see snap).

Although the flatpak command doesn't require root permissions, you must authenticate yourself repeatedly with your own password during execution (sometimes a dozen times for a single operation, such as the installation of a package).

► info *package name*
Displays details of an already installed package. Note that the package name is always composed of several parts, for example, org.gnome.gedit2.

► install *[package source] package name*
Installs the desired package from a previously set up package source.

► list [--system or --user]
Lists all installed Flatpacks (if applicable, only those that have been installed at the system level or for the current user). With the additional -d option, flatpak also displays the package size and a few other details.

► remote-add [--if-not-exists] *package source url*
Sets up a package source. The --if-not-exists option avoids error messages if the package source has already been defined previously.

► remotes
Lists all installed package sources.

► search *Flatpacks term*
Searches the configured package sources for packages that contain the search term in the package name or in the description.

► uninstall *package name*
Removes the specified package without further inquiry. You may need to enter the full package name as displayed by flatpak list, such as flatpak uninstall org.gimp.GIMP/x86_64/stable.

► update [--app *package name*]
Updates all installed flatpacks or only the package specified with --app.

Examples

The following command installs the Visual Studio Code editor at system level so that all users of the computer can use it. Note the insane download sizes for the platform packages. It wasn't long ago that the downloads for an entire Linux distribution were smaller!

```
root# flatpak install  com.visualstudio.code
Required runtime for com.visualstudio.code/aarch64/stable (runtime/org.freed-
esktop.Sdk/aarch64/23.08) found in remote flathub
Do you want to install it? [Y/n]: y
```

```
com.visualstudio.code permissions:
    ipc            network        pulseaudio              ssh-auth
    x11            devices        devel
    file access [1] dbus access [2] system dbus access [3]  tags [4]

    [1] host
    [2] ...

    ID                                  ... Remote    Download
 1. org.freedesktop.Platform.GL.default     flathub   146 MB
 2. org.freedesktop.Platform.GL.default     flathub   146 MB
 3. org.freedesktop.Platform.openh264       flathub   884 kB
 4. org.freedesktop.Sdk.Locale              flathub   366 MB (partial)
 5. org.freedesktop.Sdk                     flathub   534 MB
 6. com.visualstudio.code                   flathub    98 MB
Proceed with these changes to the system installation? [Y/n]: y
```

flatpak search enables you to search for a package in all Flatpak sources. Unfortunately, there is no official Spotify player, but there are various compatible audio players:

```
user$ flatpak search spotify
Name                    Description                     ...  Remotes
ncspot                  Cross-platform ncurses Spotify        flathub
Moosync                 Customizable music player             flathub
Spot                    Listen to music on Spotify            flathub
Oomox theme designer    customize icons, xrdb and GTK         flathub
Tauon Music Box         Play your music with style            flathub
Clementine Music Player Plays music files and internet
fedora,flathub
```

fold [options] file

fold wraps lines of text with a length of 80 characters and displays the result on the screen.

▶ -s or --spaces
   Attempts to wrap at the position of a space (i.e., between two words).

▶ -w *n* or --width *n*
   Sets a maximum text width of *n* characters. Without this option, 80 characters apply by default.

```
for var [in list;] do
    commands
done
```

for creates loops in bash scripts. All list elements are inserted into the specified variable one after the other. The list can also be formed with wildcards for file names or with {...} elements for composing file names. If the list isn't specified, the variable runs through all the parameters that were passed to the shell file (i.e., in $*).

Example

for can be used not only in shell scripts but also in the terminal to apply some commands to a number of files. The following command outputs the file names of all *.jpg files. You can now replace echo with another command that prints, scales, and so on the *.jpg files.

```
user$ for i in *.jpg; do echo $i; done
```

Another for example can be found in the description of cp commands.

```
free [options]
```

free shows how the available memory space (RAM and swap memory) is used.

▶ -h

Selects suitable units for specifying the memory (*human readable*).

▶ -m / -g

Displays the memory requirement in MiB or GiB (instead of KiB).

Example

In the following example, the localization variable LANG is reset to run free. This prevents the lengthy German translations of column names, which confuse the column layout.

In the virtual machine with just under 2 GiB RAM, 1.1 GiB is used. The fact that only 170 MiB are still considered free is because the operating system uses almost 700 MiB for caching files. This memory can be released quickly if required, leaving around 680 MiB available for new processes. The swap memory space is unused except for a few MiB.

```
user LANG= free -h
```

```
          total    used    free    shared  buff/cache  available
Mem:      1.9G     1.1G    171M      10M         686M       677M
Swap:     947M     6.5M    940M
```

fsck [options] device

fsck checks the consistency of the file system and performs repairs if necessary. It can only be run by root. Depending on the type of file system, fsck calls the programs fsck.ext3, fsck.ext4, and more. Be sure to read the man pages of this and the file system–specific command before running fsck!

Before checking a file system, you must run umount. To check the root file system, you first need to execute touch /forcefsck and then restart the computer. Files containing errors or file fragments are saved in the lost+found directory.

▶ -A
Checks all file systems named in /etc/fstab.

▶ -t *type*
Specifies the type of file system (e.g., ext3, xfs).

fsck.ext2 [options] device
fsck.ext3 [options] device
fsck.ext4 [options] device

fsck.ext2/3/4 or e2fsck checks the consistency of an ext file system and performs repairs if necessary. The command can also be started via fsck -t ext2/ext3/ext4.

▶ -b *n*
Reads the alternative superblock *n*.

▶ -n
Answers all queries with n (*no*) and doesn't make any changes.

▶ -p
Performs repairs (changes) in the file system without prompting.

▶ -y
Answers all queries with *yes* and makes changes.

`fstrim` [options] mount directory

`fstrim` runs a trim command for the file system specified by the mount point. By default, all data blocks that are no longer used by the file system are reported to the data medium.

The command is only useful if the data medium is a solid-state disk (SSD), and the file system doesn't automatically run trim commands with every delete operation anyway (see the mount option `discard` for the ext4, Btrfs, and XFS file systems). `fstrim` can be run daily or weekly by a cron job to avoid performance losses due to trim commands being executed too frequently or not at all.

▶ `-l` or `--length` *n*
Reports unused blocks only for the specified area.

▶ `-m` or `--minimum` *n*
Reports only unused contiguous blocks if they are at least *n* bytes in size. This can considerably speed up the execution of the command.

▶ `-o` or `--offset` *n*
Reports only unused blocks from the specified offset (calculated from the start of the file system). The specification can be indicated by a corresponding suffix in binary KiB, MiB, GiB, or TiB (`K`, `M`, `G`, or `T`) or in decimal kilobytes, megabytes, and so on (`KB`, `MB`, `GB`, or `TB`).

▶ `-v` or `--verbose`
Returns the total of the bytes reported using the trim command.

Example

The following command reports all free data blocks of at least 64 KiB to the SSD:

```
root# fstrim -m 64K /
/: 13493051392 bytes were trimmed
```

`ftp` [options] ftpserver

`ftp` establishes a connection to the specified FTP server. After logging in, you can interactively transfer files between the local computer and the FTP server. The most important commands for interactive use are summarized here:

▶ `?`
Displays a list of all FTP commands

▶ `!`
Enables the execution of shell commands.

- ascii
  Switches to text mode.

- binary
  Switches to binary mode.

- bye or quit
  Terminates FTP.

- cd *dir*
  Changes to the specified FTP directory.

- get *file*
  Transfers the file from the FTP archive to the current directory.

- help command
  Displays brief information on the specified command.

- lcd *dir*
  Changes the current directory on the local computer.

- ls
  Displays the list of files on the FTP server.

- lls
  Displays the list of files on the local computer.

- mget *.*pattern*
  Transfers all matching files from the FTP archive to the current directory (see also prompt).

- open
  Establishes the connection to the foreign computer (if it didn't work at the first attempt).

- prompt
  Activates or deactivates the query before the transfer of each file by mget.

- put *file*
  Transfers the file from the current directory to the FTP archive (*upload*).

- reget *file*
  Continues the transfer of a file that has already been partially sent.

> [**function**] name
>   {{commands}}

The optional function keyword defines a subfunction in bash scripts that can be called like a new command. You can use local to define local variables within the function. Functions can also be called recursively. Functions must be declared *before* their first call!

Parameters can be passed to functions. Unlike many programming languages, the parameters must not be placed in parentheses. Within the function, the parameters can be taken from the variables $1, $2, and so on.

Example

The following mini-script outputs *Hello World, abc!*:

```
#!/bin/bash
function myfunc {
  echo "Hello World, $1!"
}

myfunc "abc"
```

fuser file name

fuser determines the program that is accessing the specified file and possibly blocking it for other programs. Usually, fuser returns only the process number (PID) of the program.

▶ -k

Terminates all processes accessing the file.

▶ -m

Lists all processes that access *any* file of the file system that is specified as a device or by a file.

▶ -v

Also specifies which user is executing the process (*verbose mode*).

In the output, the process number is followed by a letter indicating the type of access:

▶ c

The process has set the directory as the *current directory*.

▶ e

The file is being executed (*executable being run*).

▶ f

The file is open for read access (*open file*). This letter is only displayed in *verbose mode*.

▶ F

The file is open for write access. F is also only displayed in *verbose mode*.

▶ m

The file is used as a library or through memory mapping.

▶ r

The directory is used as the *root directory*.

An alternative to fuser is the lsof command.

Example

The attempt to unmount a USB flash drive from the file system fails. This is the fault of process 32664 of the user kofler who is currently reading a text file on the USB flash drive using less.

```
root# umount /media/kofler/usb_backup
umount: /media/kofler/usb_backup: target is busy.
root# fuser -v -m /media/kofler/usb_backup
                              USER     PID ACCESS   COMMAND
/run/media/kofler/usb_backup  kofler   32664 f....  less
```

fwupdmgr command

The command name fwupdmgr stands for *firmware update manager*. The command allows you to initiate firmware, BIOS, or EFI updates in modern Linux distributions. fwupdmgr requires the background service fwupdmgr to be running.

▶ get-devices

Provides a detailed list of all components that can be supplied with updates, as well as a list of which updates were last performed and which are currently available. You then have the option of uploading a status report to the website *https://fwupd.org*.

▶ downgrade *device*

Undoes an update and reactivates an older version.

▶ refresh

Updates the update information (data source: *https://fwupd.org*).

▶ update *device*

Initiates an update for all components or only for the specified component. fwupdmgr only prepares the update. In most cases, the update can't be performed until the next reboot of the computer.

Example

The first command updates the update information. The second command shows a status report for the Lenovo notebook with the type designation 20MD0001GE.

```
root# fwupdmgr refresh
Fetching metadata https://cdn.fwupd.org/downloads/firmware.xml.gz
Fetching signature https://cdn.fwupd.org/downloads/firmware.xml.gz.asc
```

```
Successfully downloaded new metadata: 5 local devices supported

root# fwupdmgr get-devices
20MD0001GE
  Thunderbolt Controller:
        Device ID:          02dcd173032b1c4f136f185c0564b059a2e3d305
        Summary:            Unmatched performance for high-speed I/O
        Current version:    39.00
        ...
    PM981 NVMe Samsung 2048GB: ...
    SAMSUNG MZVLB256HAHQ-000L7: ...
    System Firmware: ...
    UEFI Device Firmware: ...

 Devices that were not updated correctly:

   * ThinkPad P1 Thunderbolt Controller (23.00 -> 39.00)

 Devices that have been updated successfully:

   * UEFI Device Firmware (192.24.1314 -> 192.35.1427)
   * 20MD0001GE System Firmware (0.1.24 -> 0.1.25)
   * UEFI Device Firmware (192.35.1427 -> 192.47.1524)
Uploading firmware reports helps hardware vendors to quickly identify  failing
and successful updates on real devices. Upload report now?
(Requires internet connection):
  0. Do not upload reports at this time, but prompt again for future updates
  1. Do not upload reports, and never ask to upload reports for future updates
  2. Upload reports just this one time, but prompt again for future updates
  3. Upload reports this time and automatically upload reports after
     completing future updates
```

The last command initiates the update for the Thunderbolt controller:

```
root# fwupdmgr update
Upgrade available for Thunderbolt Controller from 39.00 to 50.00
Thunderbolt Controller and all connected devices may not be usable while
updating. Continue with update? [Y|n]: y
```

fx file.json

fx is a command for interactively reading or searching JSON files. The components of the JSON document are color-coded and can be expanded and collapsed.

You can find the extremely useful command as a Docker image or as a binary on the GitHub project page (*https://github.com/antonmedv/fx/releases*). If you want to process JSON files automatically, jq is the better choice.

fzf [options]

fzf from the package of the same name stands for *fuzzy finder* and is a command for the interactive search and selection of files. If the command is started without any other options, it searches the current directory such as find . -type f for files and displays the result list. By entering a search pattern, you can filter the result interactively at lightning speed and then select a file or multiple files using the cursor keys or Tab, respectively. (The multiple selection requires that fzf was started with the -m option.) Enter ends the command and displays the file names.

This basic function sounds rather unspectacular. The strength of fzf comes from its integration into the shell. Provided that fzf is installed correctly, fzf can also be triggered by typing ** Tab, allowing the interactive completion of commands: Using emacs ** Tab, you can open the previously selected file. With kill -9 ** Tab, fzf provides processes for selection, and with ssh ** Tab, it returns host names. Ctrl + T lists all files, Ctrl + R lists recently executed commands, and Alt + C changes to a subdirectory.

fzf has countless other keyboard shortcuts and functions that I can't go into here for reasons of space. Instead, I recommend that you watch one of the many videos on fzf. Even though I'm not usually a video fan, in this case, it's worth it!

fzf therefore doesn't replace find in all its functions. But once you get used to the convenience of fzf, you'll need find much less often.

getcap [options] file name

getcap, from the libcap or libcap-progs package, determines for executable files which operations are permitted for the program (*capabilities*). Although the Linux kernel has supported capabilities for many years, this function is only used sporadically.

Capabilities require a file system with extended attributes. The mount option user_xattr must be used for ext file systems.

Example

Some distributions use capabilities for the ping command. Executing this command actually requires root permissions. Instead of simply setting the setuid bit as it was done in the past, which often leads to security problems, we've switched to assigning the cap_net_raw+ep permission to ping via setcap. This allows the command to use basic network functions, even if it's executed by ordinary users.

The following getcap result was obtained on Fedora. You can see that only very few commands and Gnome programs use capabilities.

```
user$ getcap /usr/bin/*
/usr/bin/arping cap_net_raw=p
/usr/bin/clockdiff cap_net_raw=p
/usr/bin/newgidmap cap_setgid=ep
/usr/bin/newuidmap cap_setuid=ep
```

An example of how to set capabilities yourself is shown in the description of the setcap command.

getenforce

getenforce provides the status of the SELinux system. The possible results are Enforcing, Permissive, or Disabled. To change the status, you can use the setenforce command.

getfacl [options] file name

getfacl determines the extended access rights of the specified files or directories. This only works if the file system supports *Access Control Lists* (ACLs). For ext3/ext4 file systems, the mount option acl must be used. An example can be found in the description of the setfacl command.

▶ -d
Displays the standard ACLs.

▶ -R
Displays the ACLs of all files in all subdirectories.

▶ --skip-base
Returns no results for files to which only the usual Unix access rights apply, but no ACL rules.

▶ --tabular
Displays the ACLs in a table.

getfattr [options] file name

getfattr determines the extended attributes of the specified files or directories. This only works if the file system supports extended attributes (mount option user_xattr for ext file systems). An example can be found in the description of the setfattr command.

▶ -d
Provides a list of all user attributes and their settings.

- ▶ -m *attribute pattern*

  Returns the attributes whose names correspond to the specified pattern.

- ▶ -n *attribute name*

  Returns the value of the specified attribute. The full attribute name must be specified, for example, user.attrname.

- ▶ -R

  Displays the EAs of all files in all subdirectories.

getopts "format"

The getopts command integrated in the bash helps to process passed options and parameters during script programming. You pass a character string to the command that lists several one-letter options and then run getopts in a loop. It returns an option found with each loop pass until all options have been processed (see the following code example).

The following rules apply to the format string:

- ▶ "abc"

  Expects the options -a, -b, and -c in any order, also combined (i.e. -bc instead of -b -c).

- ▶ ":abc"

  Works like "abc", but doesn't return any error messages for invalid parameters.

- ▶ "ab:c:"

  With a colon after b and c, it expects a parameter for each of these options (e.g., -b 1024 -c 3). During evaluation, the parameter passed with an option can be read from the $OPTARGS variable.

Example

During the processing of the options, $OPTIND references the next element of the parameter list. Provided the options are passed first and then the parameters, all processed options can be removed from $* using shift. (shift <n> removes the first n elements from $*. But because $OPTIND already references the next element, only $OPTIND -1 elements may be pushed out of the parameters list.)

```
#!/bin/bash
while getopts ":ab:c:" opt; do
    case $opt in
        a) echo "Option a";;
        b) echo "Option b with parameter $OPTARG";;
        c) echo "Option c with parameter $OPTARG";;
        ?) echo "Invalid option"
```

141

```
            echo "myscript [-a] [-b data] [-c data] [...]"
         exit 2
            ;;
      esac
done
# remove processed options from $*
shift $(( $OPTIND - 1 ))
echo "More parameters: $*"
```

Alternatives

getopts can only handle one-letter options, not long options such as --search. To process such options, you can use the getopt command, which is separate from bash. It's usually included in the util-linux package.

---

git command [options/parameters]

The git command from the package of the same name controls the Git version management system. It was originally designed only for kernel development, but it's now the most widely used version management system outside the open-source world.

Files in Git repositories can assume three states:

▶ *committed*
The local file is stored in the repository and hasn't been changed since.

▶ *modified*
The local file has been changed, but the changes haven't yet been saved in the repository via a commit.

▶ *staged*
A *modified* file is to be included in the subsequent commit.

Commands

▶ add *files/directories*
Adds files or directories to the repository or marks locally modified files already in the repository for the next commit. This changes the status of the files from *modified* to *staged*.

▶ branch [-v]
Lists the local repository branches and marks the current branch with an asterisk in front of it. With the -v option, git also displays the last commit in each branch.

▶ branch *Linux name*
Creates a new repository branch. To switch to this branch, you must also run git checkout *branch name*.

▶ branch -d *branch name*
Deletes the specified branch. The command is only executed if the branch was previously connected to another branch or to HEAD. To delete a branch regardless of the merge status, you can use the -D option.

▶ branch -u *reponame/branch name*
Sets the upstream repository for the current branch permanently and saves the link in .git/config. Alternatively, the upstream repository can also be set during the first push process using git push -u *repo branch*.

As a rule, the local and external branch names are the same, but this isn't mandatory. The link causes the relevant local branches to be automatically connected by git pull and git push and saves you having to repeatedly specify the desired repository for these commands.

▶ checkout [-b] *branch name*
Switches to the specified branch. This sets all files in the working directory to the state they had when they were last committed to this branch. With the -b option, the specified branch is created first and then activated. git checkout -b b1 therefore corresponds to git branch b1; git checkout b1.

▶ clone *repo-url* [*directory*]
Downloads an external Git repository and creates a local copy in the current or specified directory. Depending on whether the communication takes place via HTTPS or via SSH, the URLs have the following format:

https://github.com/<account>/<reponame>.git
git@github.com:<account>/<reponame>.git

Depending on where your repositories are located, you'll need to replace github.com with your Git platform, of course. In the future, the external repository will be addressed under the name origin. Git allows the use of multiple external repositories that are managed remotely with git.

▶ commit [-a] -m '*snapshot description*'
Creates a new local snapshot (a "commit") in the Git repository. All changes since the last commit are saved in it, but only if the status of the files is *staged*. You must therefore first mark all files to be saved as *staged* using git add.

With the additional option -a, you can avoid this trouble as git commit then considers all changes to already versioned files. (However, newly created files are ignored.)

Before the first commit, you must save your name and email address in the Git configuration using git config. This data is stored together with other metadata in the commit.

The last snapshot in the current branch always has the name HEAD. The penultimate snapshot can be addressed with HEAD^, the one before the penultimate snapshot with HEAD^2, and so on. (See Table 1 for more examples of references to other commits.)

▶ config --list
Determines all valid Git settings in the active directory and displays them. With the additional --show-origin option, the command also specifies the origin of the setting, that is, whether the setting was made locally for the repository or globally.

▶ config [--global] *parameter value*
Saves the setting of a Git parameter in the .git/config file of the local Git repository. With --global, the parameter is instead saved in the global Git configuration file .git/config in the home directory and is therefore the default for multiple repositories.

▶ diff
Without any further parameters, git diff shows all changes made since the last commit. By default, git diff compares the project directory with the last commit. If you want to compare the stage area with the last commit instead, you must also specify the --stage option.

▶ git diff displays the changes in a special patch syntax similar to diff. With the x--compact-summary, --numstat, or --shortstat options, you get a compact summary of the number of changed lines instead.

▶ diff *file*
Shows only the changes to the specified file.

▶ diff *revision* [*file*]
Displays the changes between the specified commit and the current status in the project directory.

▶ diff *revision1..revision2* [*file*]
Displays the changes between two commits.

▶ fetch [*reponame*]
Downloads the last changes made from an external Git repository. By default, the repository specified for clone is read, that is, origin. The git fetch command therefore updates the local repository and integrates all new external commits, branches, and tags. To avoid naming conflicts, new external branches are given the name origin/branchname or reponame/branchname.

There's a git pull variant for git fetch that not only downloads the external changes but also merges them with the local branch. git pull corresponds to git fetch; git merge. In doing so, merge connects local branches with the *tracked remote branches* (see also git branch -u).

▶ grep *pattern*
Searches all versioned text files in the project directory for the search expression. *pattern* is a regular expression as in grep.

▶ init
Sets up a Git repository for the current directory.

▶ log [--oneline] [--graph] [*branch/tag* [*^non-branch*]]
Provides a list of the most recent commits. The --oneline option shortens the output per commit to one line. --graph visualizes how branches were connected to each other.

If you specify a branch or a tag as an optional parameter, only those commits that had an influence on this branch will appear. These may well be commits that were made *before* this branch was created because they formed the starting point of the branch. To further reduce the commit list produced by git log *branch name*, you can use ^non-branch to hide commits of the specified branch. git log bugfix ^main thus only returns those commits of the bugfix branch that have *not* yet been connected to main.

The output is guided by less so that you can use the cursor keys to scroll through the results, most of which are several pages long. You can prevent this behavior for all git commands by using the --no-pager option.

The output of git log can be controlled by countless other options. If necessary, you can take a look at *https://git-scm.com/docs/git-log* with approximately 40 pages of text!

▶ merge *branch name*
Connects the specified branch with the current branch. As with git commit, the resulting changes are saved in a new commit for the currently active branch.

If conflicts occur during the merge process, the process will be interrupted. You can fix the conflicts manually and then continue the merge process with git merge --continue. Alternatively, git merge --abort cancels the process.

▶ mv *previously new*
Renames a versioned file or moves it to another directory.

▶ pull [*reponame*]
Transfers all commits from an external repository (like git fetch) and integrates them into the local repository (like git merge). Only the currently active branch is taken into account.

git pull without any additional parameters assumes that the remote repository is configured for the current branch in .git/config.

With the --rebase option, git doesn't perform a normal merge process, but a rebasing.

▶ push [-u] [*reponame* [*branch name*]]
Uploads locally executed commits for *branch name* to the external Git repository *reponame* and merges them there. For example, git push origin main connects all your own commits in the main branch with the branch of the same name in the external origin repository. If *branch name* doesn't exist in the external repository, the branch

gets automatically set up there. The -u option links the local branch and the external branch of the same name (as with git branch -u).

The push command fails if there are external commits that haven't yet been executed locally. For this reason, you should always run git pull before git push.

If you run git push without specifying repository and branch names, git communicates with the default repository (usually origin) and synchronizes the current linked branch (see the branch -u command).

► rebase *branch name*
Variant of git merge that changes your own commits (i.e., converts them into new commits) as if they had been created from the beginning on the basis of the commits of the external branch. git merge has the advantage that the resulting commit sequence looks "nicer," that is, isn't interrupted by constant merge commits. However, local commits are created that never existed originally. This makes it difficult to trace errors.

► remote [-v]
Lists all external repositories. With -v, git also displays the URL of each repository. The root repository—origin—is defined by git clone. Additional repositories can be managed with commands such as git remote add or git remote remove.

► reset HEAD - *file*
Changes the status of a file mistakenly marked as *staged* back to *modified*.

► reset [--hard] *revision*
Undoes changes and reverts to an old version of the repository. *revision* specifies the desired version: HEAD leads back to the last commit, HEAD^ to the penultimate one, HEAD^2 to the one before the penultimate one, and so on. Alternatively, you can also specify the initial digits of the hexadecimal SHA1 checksum of the commit (see Table 1). reset origin/main undoes all local changes and synchronizes the local branch with the main branch of the default repository. The --hard option causes local changes to be overwritten irrevocably.

► reset *file*
Removes the file from the stage area and thus undoes git add *file*.

► restore *file*
Restores the state of the file from the last commit. Changes made since then are overwritten without further inquiry. This command has only been available since Git version 2.23 (August 2019). If you specify a directory instead of a file, the command applies to all files changed in the directory.

► rm *file*
Deletes the file from the local directory and removes it from the *staging* area. If the file has been modified since the last commit, you must force the deletion using --force.

► show *revision*:*file*
Shows the file in the state it was in when the specified commit was made.

► status [-s]
Provides information about the current Git branch and the status of modified files. The -s option makes the status output more compact. All file names are preceded by two letters, with the first letter indicating the *staging* state and the second one the *modifying* state.

► tag
Without any further parameters, git tag lists all tags (markers, version markers).

► tag -a '*marker-text*' [-m '*comment*'] [*revision*]
Adds a tag to the last commit, that is, a marker, which is a kind of bookmark. This function is ideal for marking a release version, for example. Specifying a revision makes it possible to subsequently mark an older commit.

Git distinguishes between two types of tags: the -a or -m options create an annotated tag that also stores metadata. Without this option, a lightweight tag without metadata is created.

Instead of just one man page for git, a whole lot are available as each git command is documented on its own page. For example, man git-clone provides a comprehensive description of git clone.

Revision Syntax

In git commands, you can refer to past commits or revisions in a variety of ways (see Table 1). The latest commit of a branch is considered the *head*. The notation @{xxx} refers to the local reference log and can therefore only be used for actions last performed locally.

Example	Meaning
HEAD	Head of the current branch
@	Head of the current branch (short notation)
develop	Head of different branch
develop:readme.txt	Readme file at the head of another branch
refs/remots/origin/feature	Head of a remote branch
v1.3	Commit with tag v1.3
HEAD@{2 days ago}	Commit in the current branch two days ago

Table 1 Examples of the Git Revision Syntax

Example	Meaning
main@{7}	Commit in the main branch prior to two actions
HEAD~	Predecessor (parent) of the last commit
HEAD~3	Pre-predecessor of the last commit
v1.3~3	Pre-pre-predecessor of the last commit with tag v1.3
234ae33^	First parent of merge commit 234ae33
234ae33^^	Second parent of merge commit 234ae33

Table 1 Examples of the Git Revision Syntax (Cont.)

Examples

In the first example, you first save your name and email address in your personal Git configuration file, if you haven't already done so. Then, you download an existing repository from GitHub:

```
user$ git config --global user.name  "Howard Hawks"
user$ git config --global user.email "hawks@my-company.com"
user$ git clone https://github.com/git-book/hello-world.git
user$ cd hello-world
```

The second example shows how to quickly perform a bugfix in the main branch of an existing project. You can use git checkout to switch to the main branch. git pull downloads any changes saved in the remote repository. Now, you need to correct an error, save your changes using git commit, and transfer the changes back to the origin repository via git push:

```
user$ cd directory/with/existing/git/repo
user$ git checkout main user$ git pull
user$ my-favorite-editor files ...
user$ git commit -a -m 'Bugfix login authentication'
user$ git push
```

glances [options]

Depending on the application, the glances command from the package of the same name is a modern alternative to top or a simple monitoring program that runs in client/server mode. The latter requires a corresponding configuration but also enables access to the monitoring data via a web browser.

When called directly, the command also provides an overview of network and I/O activities as well as the status of the file systems compared to top. The color display is optimized for a dark terminal. If you've set a light background color, you run `glances` with the `--theme-white` option (see Figure 3).

Various shortcut keys sort the process list according to different criteria: C by CPU usage, I by I/O activity, M by memory usage, P by name, or T by CPU runtime. By pressing Enter, you can filter the process list according to a search term. Other shortcuts show and hide information or activate different display modes. For example, U switches between the current network throughput and cumulative data volumes. The help page, which you can call by pressing H, shows even more shortcuts.

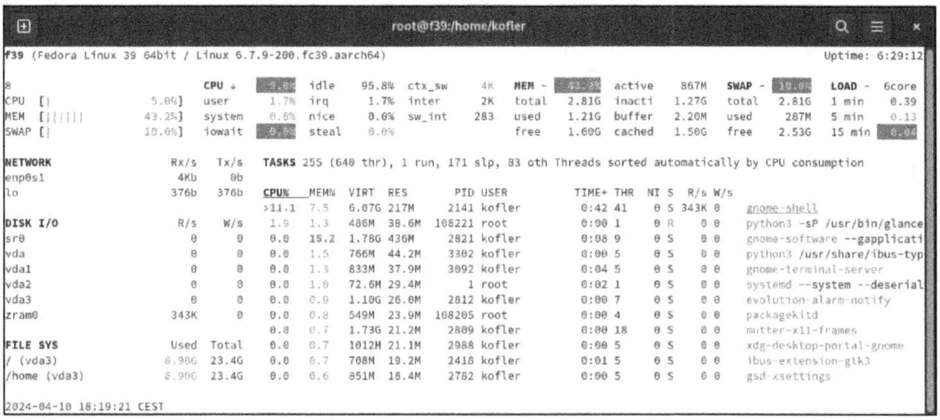

Figure 3 Simple Monitoring with "glances"

gnome-session-quit [options]

The `gnome-session-quit` program displays a dialog in which you can log off, switch off, or restart the computer.

▶ `--force`
Performs the logout even if there are still unsaved files. This option can only be used in combination with `--logout`.

▶ `--logout`
Asks the user if they want to log out.

▶ `--no-prompt`
Performs the logout without prompting. This option can only be used in combination with `--logout`.

▶ `--power-off`
Asks the user whether they want to switch off the computer.

▶ `--reboot`
Asks the user whether they want to reboot the computer.

gpasswd [options] group

gpasswd is actually used to set group passwords. However, passwords that must be known by several members of a group are considered insecure and are therefore no longer in use.

Nevertheless, gpasswd fulfills an important function. In the following variant, it removes a user from a group:

root# **gpasswd -d user group**

This makes gpasswd -d the counterpart to usermod -a –G group user, which adds a user to a group.

gpg or gpg2 [options/commands]

gpg encrypts, decrypts, or signs files and manages keys. gpg is controlled by countless options and is only to a limited extent suitable for manual use. The command is used by email clients for encrypting and signing messages and by package management tools for verifying the integrity of packages and package sources.

With the -c and -d options, gpg can be used for symmetric encryption and decryption of files. This is simple but not ideal in terms of safety. All other options assume that you want to use asymmetric encryption: encryption is then performed using the public part of your communication partner's key, and decryption is done using the private part of your own key. This assumes that you also use gpg for key management.

gpg is intended for server and embedded use and relies on a few external libraries. gpg2, on the other hand, is designed for desktop use. This version also supports S/MIME. Apart from that, both commands support the same options.

▶ --batch
Activates the batch mode. Interactive commands aren't permitted in this, and there's no interaction. Batch mode is intended for script programming.

▶ -c *file* or --symmetric *file*
Prompts you to enter a password, then encrypts the specified file, and finally saves the result as file.gpg.

▶ --cipher-algo zip/zlib/bzip2/none
Defines the encryption algorithm. By default, gpg uses the CAST5 algorithm. The gpg --version command determines the algorithms available for selection.

▶ --compress-algo zip/zlib/bzip2/none
Specifies whether or how the data should be compressed. The ZIP procedure is used as standard. For compatibility reasons, settings other than zip or none aren't recommended. none is recommended if the source data is already compressed. Recompression then takes unnecessary time without reducing the space required.

▶ -d *file* or --decrypt *file*
Prompts you to enter a password, then decrypts the specified file, and finally writes the result to the standard output.

▶ --decrypt *file* > *plain text*
Decrypts the specified file and redirects the output to the *plain text* file. The file must be encrypted with the public part of your own key. To decrypt it, you must enter the password for your key.

▶ --encrypt *file*
Encrypts the specified file and saves it as *file.gpg*. You enter the email address of the key interactively. The key may have to be saved in the key database beforehand using --import. If you also specify --armor before the --encrypt option, the encrypted file will be saved in ASCII format. This requires more space, but simplifies forwarding by email.

▶ --export *name@host*
Saves a binary representation of the public part of the key for *name@host* in the specified file. You must pass this file on to people who will send you encrypted files or emails. If you also specify --armor before the --export option, the public key is saved in an email-compatible text format.

▶ --gen-key
Creates a new key and saves it in .gnupg/pubring.gpg. You'll need to enter various data: the desired type and length of the key, as well as your name, email address, password, and so on.

▶ --import public-key-file
Imports the public part of another person's key into your own key collection. This is necessary to be able to send encrypted files or emails to that person.

▶ -k or --list-keys
Lists all keys stored in .gnupg/pubring.gpg.

▶ -no-tty
Suppresses the output of warnings.

▶ --passphrase-file *file*
Reads the encryption password from the specified file.

▶ --passphrase-repeat *n*
Specifies how often the password entry must be repeated (once by default). If the password is read from a file in scripts (see the --passphrase-repeat option), you need to deactivate the repeated entry with the parameter 0.

▶ -q or --quiet
Dispenses (to a large extent) with status outputs.

▶ --verify *signature file try*
Checks whether the signature file and the source file match.

▶ --version
Displays the version of gpg and provides a list of all supported encryption and hash algorithms.

Examples

The following command performs a symmetric encryption of the secret.odt document:

```
user gpg -c secret.odt
Enter the passphrase: ***********
Enter the passphrase again: ***********
```

gpg creates the new secret.odt.gpg file. The original secret.odt file doesn't get changed. If you want to prevent access to this file in the local file system, you must explicitly delete the file and its backups! To restore the encrypted file, proceed as follows:

```
user gpg -d secret.odt.gpg > secret.odt
Enter the passphrase: ***********
```

In the second example, gpg is used in two scripts to encrypt and decrypt files with the AES256 algorithm. A key file is required for the scripts, which is generated using openssl. (While 32 bytes may seem small, that's 256 bits. For symmetrical procedures, 128 bits are already considered sufficiently secure.)

```
root# openssl rand  32 > key.txt
root# chmod 400 key.txt
```

The mycrypt encryption script reads as follows:

```
#!/bin/sh
# Usage: mycrypt < in > out.crypt
gpg -c --batch --cipher-algo AES256 --compress-algo none \
  --passphrase-file key.txt --passphrase-repeat 0
```

To decrypt the encrypted file again, you must use myuncrypt:

```
#!/bin/sh
# Usage: myuncrypt < in.crypt > out
gpg -d --batch --no-tty -q --cipher-algo AES256 --compress-algo none \
  --passphrase-file key.txt --passphrase-repeat 0
```

The application looks like this:

```
root# ./mycrypt   < backup.tar.gz       > backup.tar.gz.crypt
root# ./myuncrypt < backup.tar.gz.crypt > backup-copy.tar.gz
root# file backup.tar.gz.crypt
backup.tar.gz.crypt: GPG symmetrically encrypted data (AES256 cipher)
root# diff backup.tar.gz backup-copy.tar.gz
(no output, the files are identical)
```

The last example shows the verification of signatures. On many websites, text files with checksums are specified in addition to the actual downloads. But how can you be sure that you're not being tricked into false downloads and that your checksums aren't being manipulated? On the Kali Linux download page, the checksum file is therefore signed with a key from a Kali developer. To verify the signature, you first need to import this key and then call gpg --verify:

```
user$ wget -q -O - https://www.kali.org/archive-key.asc | gpg --import
  gpg: Key 7D8D0BF6: Public key
    "Kali Linux Repository <devel@kali.org>" imported ...
  Primary key fingerprint: 44C6 513A 8E4F B3D3 0875 F758 ED44 4FF0 7D8D 0BF6
user$ gpg --verify SHA256SUMS.gpg SHA256SUMS
  Signature made Thu 09 Nov 2017 05:21:22 PM CET
  using RSA key ID 7D8D0BF6
  Good signature from "Kali Linux Repository <devel@kali.org>"
  WARNING: This key is not certified with a trusted signature!
  There is no indication that the signature belongs to the owner.
  Primary key fingerprint: 44C6 513A 8E4F B3D3 0875 F758 ED44 4FF0 7D8D 0BF6
```

At first glance, the preceding issue doesn't inspire confidence. In fact, however, everything is in good order: The gpg command confirms that the signature and checksum file match. The third line of the gpg output starts with Good signature. It would be fatal if the message Wrong signature were displayed at this point!

The warning that appears refers to the fact that the Kali developers' key is self-signed, but this signature couldn't be verified. You can ignore this warning because you've downloaded the key from *www.kali.org* and made sure that its fingerprint (i.e., the hexadecimal number sequence 44C6 . . . for the preceding commands) matches the expected values. Furthermore, to verify the signature of the key itself, you and the Kali developers would have to exchange your personal GPG keys either in a face-to-face meeting or via the *web of trust*:

```
gpioget [options] chipnr gpio1 gpio2 ...
gpioset [options] chipnr gpio1=status1 gpio2=status2 ...
```

With gpioset, you can set the *general purpose input/output* state of the Raspberry Pi. gpioget reads the state. The chip number must be transferred to the commands as the first parameter. It's 0 for Raspberry Pi models 1 to 4. A new hardware chip was installed with model 5. The number 4 must be transferred so that this is addressed correctly (internally, /dev/gpiochip4).

GPIO numbers are passed as additional parameters, whereby the internal designation of the GPIO is relevant rather than the pin number on the GPIO contact strip:

```
user$ chip=4            (for Raspberry Pi 5)
user$ gpioset $chip 7=1  (GPIO 7 = set pin 26 to high)
user$ gpioset $chip 7=0  (GPIO 7 = set pin 26 to low)
```

You can control multiple outputs at once (here, it's GPIO 7, 8, and 25):

```
user$ gpioset $chip 7=0 8=1 25=0
```

Various options allow you to control additional functions:

▶ --bias=as-is|disable|pull-down|pull-up
Activates or deactivates the internal pull-up or pull-down resistor of a GPIO.

▶ --mode=exit|wait|time|signal
Specifies how long the command should run. The default setting is exit, so the command is terminated immediately. With wait, the program waits until the user presses [Enter]. Using the time setting, you can set the desired wait time with --sec= n or --usec=n. signal means that the program continues to run until it's terminated using [Ctrl]+[C].

▶ --background
Continues the command as a background service.

Alternatives

gpioget and gpioset are part of the gpiod package, which is installed by default on the Raspberry Pi OS. The package also contains the gpiodetect and gpioinfo commands, which enable you to determine information about the hardware of the minicomputer. Instead of gpioset, the pinctrl command can also be used to control GPIOs.

```
grep [options] search pattern file
grep -R [options] search pattern
```

grep searches the specified text file for a search pattern and displays the text passages found or simply indicates where or in how many lines the pattern was found. grep can be combined with find to search all files matching certain conditions for texts.

▶ -n
Displays not only the line with the text found but also the n immediately preceding and following lines (e.g., grep -3).

▶ -c

Indicates only the number of lines in which the search pattern was found, but not the lines themselves.

▶ --color=auto

Indicates the matches in color during output. This only works if you display the grep result directly in a console, but not if you process the result with another command (e.g., with sort or less). To embed the color information in the output in this case as well, you must use --color=always.

▶ -E

Activates *Extended Regular Expressions* (EREs). This extended syntax allows the characters ? (not at all or once), + (at least once), and | (logical OR) in the search pattern. For example, grep -E 'pattern1|pattern2' file returns all lines of the specified file containing one of the two patterns.

▶ -f *file*

Reads the options listed here for the specified file (for complex or frequently used search patterns).

▶ -i

Doesn't distinguish between uppercase and lowercase.

▶ -l

Displays only the file names in which the search pattern was found.

▶ -n

Also indicates the line number of each line in the output.

▶ -o

Doesn't output the entire line, but only the part of the line that matches the search pattern (*only matching*).

▶ -P

Interprets the regular pattern according to the *Perl Compatible Regular Expressions* (PCRE2).

▶ -q

Doesn't provide any screen output and only returns the return value 0 (search text found) or 1 (not found). This option is useful if grep is called by other programs (see the example in find).

▶ -R

Applies grep recursively to all files in the directory tree.

▶ -v

Applies the inverse effect: grep returns all lines that *do not* match the search pattern. grep -v '^#' file | cat -s returns all lines of the file that don't begin with # (where cat -s reduces several empty lines to a single one).

▶ -w

Finds only entire words. If this option is specified, the search pattern "the" in the word "these" is no longer recognized.

The search pattern consists of two components: the specification of what is being searched for and the specification of how often the search expression may occur. Table 2 summarizes the meanings of the most important characters. The abbreviation ERE marks characters for the Extended Regular Expressions (-E option or egrep command).

Character	Meaning
abc	The character string abc.
[abc]	One of the characters a, b, or c.
[^abc]	None of the characters a, b, or c (i.e., any other character).
[a-z]	One of the characters between a and z.
.	Any character.
?	The preceding character may not appear at all or may appear once (ERE).
*	The character may appear any number of times (or not at all).
+	The character may appear any number of times, but must appear at least once (ERE).
{n}	The character must appear exactly n times.
{,n}	The character may appear a maximum of n times.
{n,}	The character must appear at least n times.
{n,m}	The character must appear at least n times and not more than m times.
^	Start of line.
$	End of line.
|	Logical OR (ERE).

Table 2 Structure of "grep" Search Patterns (ERE = Extended Regular Expression)

If special characters such as ?, *, +, [,], (,), #, or ! will be used in the search pattern, they must be preceded by \. Patterns are already predefined for some character groups, such as [:digit:] for digits or [:space:] for *white space* (i.e., blank and tab characters; see man page).

Some examples of search patterns follow:

▶ 'abc' searches for exactly this character string.

- `'[a-z][0-9]+'` searches for strings starting with a lowercase letter and followed by one or more digits.

- `'\(.*\)'` searches for any character strings that are enclosed in two parentheses.

grep only applies the search pattern line by line. Text passages that are interrupted by a line break can't be recognized.

Examples

The following command searches the Apache logging file /var/log/apache2/access.log for all lines containing IP address 1.2.3.4. The command must be executed with root permissions because regular users don't have access rights to the Apache logging files.

```
root# grep 1.2.3.4 /var/log/apache2/access.log
```

The following command determines how often the mysql string has been used in a PHP file in both uppercase and lowercase:

```
user$ grep -c -i mysql code.php
```

In the next example, all Apache configuration files are searched recursively starting from a start directory. grep displays the file names in which the ErrorDocument keyword appears:

```
root# cd /etc/apache2
root# grep -R -l ErrorDocument
```

The following grep command filters out all lines from the kernel messages that contain the search pattern eth in uppercase or lowercase:

```
root# dmesg | grep -i eth
e1000: eth0 NIC Link is Down
e1000: eth0 NIC Link is Up 1000 Mbps Full Duplex, Flow Control: RX ...
```

The output of the df command contains various entries for temporary file systems, which are often uninteresting. grep -v eliminates these lines. (df -x tmpfs will also lead to the desired result.)

```
root# df | grep -v tmpfs
File system              1K blocks    Used   Available Used% Mounted to
/dev/mapper/fedora-root  16080304 4820060   10420360   32% /
/dev/sda1                  487652  109669     348287   24% /boot
```

Some configuration files in the /etc directory look as if the authors have confused the file with the manual. Hidden among hundreds, sometimes thousands, of lines of comments are a few settings that are actually effective. The following commands eliminate all lines that start with # or ; or that are empty:

```
root# cd /etc/samba
root# cp smb.conf smb.conf.orig
root# grep -Ev '^#|^;|^$' smb.conf.orig > smb.conf
```

To also eliminate comments where # or ; isn't at the beginning of the line, as well as empty lines with white space (spaces, tab characters), you must formulate the grep command as follows:

```
root# grep -Ev '^[[:space:]]*#|^[[:space:]]*;|^[[:space:]]*$' \
      smb.conf.orig > smb.conf
```

The following example searches an Apache configuration file for lines containing the search pattern SSL*File:

```
user$ grep 'SSL.*File' /etc/httpd/conf.d/ssl.conf
SSLCertificateFile        /etc/letsencrypt/live/pi-book.info/cert.pem
SSLCertificateKeyFile     /etc/letsencrypt/live/pi-book.info/privkey.pem
SSLCertificateChainFile   /etc/letsencrypt/live/pi-book.info/chain.pem
```

You can also combine find and grep to perform particularly effective searches. In the following command, find searches for *.php files, and grep checks whether they contain the mysql string. If they do, the respective file name will be displayed on the screen. Note that you must specify the -print option after -exec.

```
root# find -name '*.php' -exec grep -q -i mysql {} \; -print
```

Variants of grep

▶ ack and ag are grep variants optimized for programmers. They ignore backup files, binary files, and directories of version management systems.

▶ grepmail helps to search email archives in mbox format for character strings. The advantage over grep is that the entire email is extracted from the file rather than a single line.

▶ zgrep searches compressed files.

```
grim [options] filename.png
```

grim stands for *grab image* and creates a screenshot if your desktop is running on Wayland. By default, the screenshot covers the entire screen and is saved in PNG format. Various options allow you to define the screen section, change the output format (JPEG or PPM), and integrate the mouse pointer into the screenshot.

Examples

If you capture screenshots from the terminal, it's often useful to wait a few seconds before doing so. The easiest way to do this is via `sleep`:

```
user$ sleep 5; grim my-new-screenshot.png
```

`grim` can be combined very well with other commands. For the following command, we use `slurp` from the package of the same name to mark a rectangular area on the screen. (Unfortunately, this doesn't work with every Wayland compositor.)

The coordinates of this area are passed to `grim` with the `-g` (*geometry*) option. Because `grim` ends with `-`, it doesn't write the screenshot to a file, but to the clipboard. `wl-copy` from the `wl-clipboard` package takes the data and inserts it into the Wayland clipboard. From there, you can copy the screenshot into a graphics program (e.g., GIMP).

```
user$ grim -g "$(slurp -d)" - | wl-copy
```

If you want to create screenshots in the X graphics system from the terminal, `scrot` (for *SCReen shOT*) provides functions that are similar to `grim`.

groupadd name

`groupadd` sets up a new group.

▶ `-g` *n*
Uses *n* as the new GID (*group identifier*).

groupdel name

`groupdel` deletes the specified group.

groupmod [options] name

`groupmod` changes the GID and/or the name of the group.

▶ `-g` *n*
Determines the new *group identifier* (GID).

▶ `-n` *name*
Determines the new group name.

Example

The following command gives the group `mysqladmins` the new name `dbadmins`:

```
root# groupmod -n dbadmins mysqladmins
```

---

groups [username]

groups displays a list of all groups to which the current or specified user belongs. These are the main group specified in /etc/passwd and the optional groups specified in /etc/group.

Example

The current user belongs to the groups kofler, docuteam, and wheel. Members of the wheel group may perform administrative tasks in RHEL 6.

```
user$ groups
kofler docuteam wheel
```

---

grub-install [device]
grub2-install [device]

The grub-install command (Ubuntu, Debian) or grub2-install (Fedora, SUSE, RHEL) installs the boot loader in the boot sector of the specified hard disk device or, for EFI computers, in the EFI partition. grub-install requires that the GRUB configuration file /boot/grub/grub.cfg has been set up beforehand, usually by calling update-grub (Debian, Ubuntu) or via grub2-mkconfig (Fedora, SUSE, RHEL).

Example

On Ubuntu, the update-grub command creates the GRUB configuration file /boot/grub/grub.cfg, automatically creating entries for all operating systems found on the computer. grub-install installs GRUB 2 on BIOS computers in the boot sector of the first hard disk:

```
root# update-grub
root# grub-install /dev/sda
```

On an EFI computer, the device specification is omitted. grub-install requires that the EFI partition in the /boot/efi directory is included in the directory tree.

```
root# update-grub
root# grub-install /dev/sda
```

On Fedora, SUSE, and RHEL, you have to run these commands instead, with or without the device specification for grub2-install, depending on whether it's a BIOS or an EFI installation:

```
root# grub2-mkconfig -o /boot/grub2/grub.cfg
root# grub2-install [/dev/sda]
```

grub-mkconfig [options]
grub2-mkconfig [options]

grub-mkconfig (Ubuntu, Debian) or grub2-mkconfig (Fedora, SUSE, RHEL) evaluates the GRUB configuration files in /etc/grub.d and analyzes all hard disk partitions. The command creates a new GRUB configuration file from this information and outputs it.

▶ -o file
Saves the resulting configuration in the specified file.

gsettings command

gsettings reads values from the dconf database or saves settings there. The dconf database is usually located in the .config/dconf/user file and is available in binary format. The database is used to store various settings of newer Gnome programs.

▶ get *schema key*
Determines the value of the key parameter for the specified schema. *schema* denotes the software provider, the program name, and (optionally) a settings directory. For example, all Nautilus settings for the icon view are assigned to the org.gnome.nautilus.icon-view schema.

▶ list-keys *schema*
Returns a list of all keys for the specified schema.

▶ list-schemas
Provides an unsorted list of all schemas.

▶ set *schema key value*
Assigns a new value to the key parameter.

Example

The following command sets the text scaling factor in such a way that it enlarges all texts by 66%. This is useful for a HiDPI or Retina monitor, for example, on which the text would otherwise be illegibly small.

```
user$ gsettings set org.gnome.desktop.interface text-scaling-factor 1.66
```

gunzip file

gunzip decompresses the specified file, regardless of whether it was compressed using gzip or using compress. The .gz or .Z identifier is automatically removed from the file name. gunzip is a link to gzip, whereby the -d option is automatically activated.

`gzip` [options] file

gzip compresses or decompresses the specified file. Compressed files are automatically appended with the identifier .gz. gzip is only suitable for processing individual files. If you want to save multiple files or entire directories in one file, you can use the tar command.

▶ -1 to -9, --fast, --best
Controls the speed and quality of the compression. -1 corresponds to --fast and results in fast compression, but slightly larger files. -9 corresponds to --best and results in higher computing times but smaller files. The default setting is -6.

▶ -c, --stdout, or --to-stdout
Leaves the file to be (de)compressed unchanged and redirects the result to the standard output (usually to the screen). From there, it can be redirected to any file by using > (see example below).

▶ -d, --decompress, or --uncompress
Decompresses the specified file instead of compressing it (corresponds to gunzip).

▶ -r or --recursive
Also (de)compresses files in all subdirectories.

Examples

The following command compresses all *.tif files in the current directory. The result is nothing but *.tif.gz files.

```
user$ gzip *.tif
```

In the second example, gzip compresses the specified file, but leaves it unchanged and writes the result to backup.gz:

```
user$ gzip -c file > backup.gz
```

`halt` [options]

halt terminates all running processes and then shuts down the computer. halt corresponds to shutdown -h now.

▶ -p
Switches the computer off after shutdown (*power-off*). Many systems switch off automatically. The option is only required if this doesn't work, for example, in many virtual machines.

hash [option]

The bash command hash displays the contents of the hash table. This is a table in which the shell remembers the path names of all commands that have been executed previously. This speeds up the repeated execution of an already known command because it's no longer necessary to search all PATH directories for the program.

▶ -r

Deletes the hash table of the bash. This is necessary if the directory of a program in the hash table changes. The bash won't find the command otherwise. In tcsh, the rehash command must be used instead of hash -r.

hciconfig [hcidevice] [command]

hciconfig from the bluez package helps with the configuration of local Bluetooth adapters via the host controller interface (HCI). If no HCI device name is specified (usually hci0 or hci1), hciconfig communicates with all local Bluetooth devices. If the command is executed without parameters, it lists information about all local Bluetooth adapters.

▶ reset

Initiates a restart of the Bluetooth adapter.

▶ up/down

Activates or deactivates the Bluetooth adapter. If the error message *Operation not possible due to RF-kill* is displayed, you can try to switch on the adapter beforehand using rfkill unblock.

hcitool [options] [command]

The hcitool command from the bluez package helps with scanning and setting up Bluetooth devices.

▶ -h

Lists all supported commands.

▶ -i hciX

Applies the next command to the specified Bluetooth device. Without this option, the command is sent to the first available device.

Commands

▶ cc bt-mac

Establishes a connection to the Bluetooth device with the specified MAC address.

▶ dc *bt-mac*

Terminates the specified connection again.

▶ dev

Provides a list of local Bluetooth devices. Normally, this is the built-in Bluetooth adapter, which is usually assigned the device name hci0.

▶ scan

Lists all external Bluetooth devices within radio range, including their MAC addresses.

Example

There's only one local Bluetooth adapter with the device name hci0. All other commands are therefore automatically sent to this adapter; the -i option can be omitted. A smartphone and a mouse are also within wireless range.

```
user$ hcitool dev
Devices:
        hci0    00:1F:CF:41:00:A2
user$ hcitool scan
  60:FB:42:FC:BB:8C    Logitech Mouse
  10:86:F3:52:86:18    iPhone 13
  ...
```

head [options] file

head displays the first 10 lines of a text file on the screen.

▶ -n *lines*

Outputs the specified number of lines.

▶ -n -*lines*

Returns the entire file except for the last lines. Thus, head -n -3 returns the entire text except for the last three lines.

Alternatives

tail fulfills the reverse function. For example, tail -n 5 file returns the last five lines of a file. If you want to skip the first five lines, you can use tail -n +6 (returns all lines from the sixth line onwards).

help name

help displays a short description of the specified bash command. help only works for commands that are integrated into the bash, such as alias, cd, or type.

history [options] [n]

The bash command history displays the most recently executed commands with a consecutive number. The n parameter limits the output to the commands that were run last. With !n, you can repeat the command with the specified number: !-1, !-2, and so on runs the last command, the penultimate one, and so on, respectively.

▶ -c

Deletes the sequence of stored commands in RAM. (.bash_history is retained.)

▶ -r [filename]

Loads the commands stored in the specified file or in .bash_history into the history memory.

▶ -w [filename]

Saves the command sequence (by default in .bash_history).

host [options] name/ip-address

host returns the IP address for the specified network name or the network name for the specified IP address. In many distributions, host is a component of a bind package, such as bind-utils or bind9-host.

▶ -a

Provides additional information that may help to find errors in the name server configuration.

▶ -t type

Returns DNS entries of the desired type.

Example

The IP addresses 5.9.22.29 and 2a01:4f8:161:107::4 are assigned to the host name; kofler.info. The domain name of the mail server is mail.kofler.info. The SPF information of the mail server can be found in the DNS text entry.

```
user$ host kofler.info
kofler.info has address 5.9.22.29
kofler.info has IPv6 address 2a01:4f8:161:107::4
kofler.info mail is handled by 10 mail.kofler.info.
user$ host 5.9.22.29
```

28.22.9.5.in-addr.arpa domain name pointer kofler.info.
user$ **host 2a01:4f8:161:107::4**
4.0.0.0.0.0.0.0.0.0.0.0.0.0.0.0.0.7.0.1.0.1.6.1.0.8.f.4.0.1.0.a.2.ip6.arpa
  domain name pointer kofler.info.
user$ **host -t txt kofler.info**
kofler.info descriptive text "v=spf1 ip4:5.9.22.29 ip6:2a01:4f8:161:107::4
-all"

---

hostname [name]

hostname returns the current network name of the system or changes it until the next restart. For this reason, the change isn't saved permanently! If you want to do this, you must run hostnamectl or change the corresponding configuration file.

▶ -d

Returns the domain name instead of the host name.

▶ -i

Returns the IP address assigned to the host name. As it's possible that several IP addresses are assigned to the host, you should use the -I option instead of this option if possible. It often happens that hostname -i returns the address 127.0.0.1, which isn't very helpful.

▶ -I

Returns all IP addresses assigned to the host name in a list and separated by spaces. The localhost address and local IPv6 addresses aren't taken into account. However, the -I option isn't available for all hostname versions. You'll have to do without -I on Arch Linux and Alpine Linux in particular.

---

hostnamectl [options] [command]

The host name can be reset for distributions with the systemd init system using the hostnamectl command. Unlike the hostname command, the new setting is saved permanently.

▶ set-hostname *name*

Sets the new host name. You must log out and log in again for the change to take effect. The new host name is saved in /etc/hostname at the same time.

▶ status

Displays the current host name and various details about the running system, including the distribution name and the kernel version.

Example

The following command sets the host name workstation and the domain name mylan:

```
root# hostnamectl set-hostname workstation.mylan
```

htop [options]

htop from the package of the same name is a variant of top. The main advantage over top is that you can scroll horizontally and vertically through the process list using the cursor keys. The selected process can be easily ended by pressing K. You can also use the function buttons to choose between different display formats.

▶ -d *n*

Specifies after how many tenths of a second the process list should be updated.

▶ -p *n1,n2,n3,...*

Shows only the processes with the specified PIDs.

▶ -s *name*

Sorts the process list according to the specified criterion. A list of all permitted keywords is provided by htop -s help.

▶ -u *name*

Shows only processes of the selected user.

htpasswd [options] password file username [password]

htpasswd or, in some distributions, htpasswd2, creates a password file for the Apache web server or changes entries in an existing password file.

The file can be used for basic authentication (AuthType Basic). Its location must be specified in an Apache configuration file (e.g., httpd.conf or .htaccess) with the AuthUserFile keyword. For security reasons, you should make sure that the web server can read the file but isn't allowed to deliver it via HTTP!

▶ -b

Expects the password as a parameter. This simplifies the creation of password entries via script, but it's insecure.

▶ -c

Creates a new password file.

▶ -D

Deletes a user from the password file.

▶ -l

Temporarily blocks the account.

167

▶ -u
Reactivates a blocked account.

Example

The following commands create the new password file `passwords.pwd` and insert an entry for user `name1`. Additional username/password pairs are added without the `-c` option:

```
user$ htpasswd -c passwords.pwd name1
New password: ********
user$ htpasswd passwords.pwd name2
New password: ********
```

hwclock [options]

Without additional parameters, `hwclock` reads the time from the computer's hardware clock and displays it.

▶ -s or --hctosys
Reads the hardware clock and uses it to set the computer's time.

▶ -w or --systohc
Saves the current time of the computer in the hardware clock.

The command also supports a number of special functions that are described in `man hwclock`.

hydra [options] [hostname/ipaddress] service

The hacking or penetration testing command `hydra` from the package of the same name reads passwords from a file or generates them itself and attempts to use them to log in to a network service.

`hydra` supports a wide range of services, including FTP, HTTP(S), IMAP, MySQL, Microsoft SQL, POP3, PostgreSQL, SMTP, Telnet, and VNC. The command can also attempt logins in web forms (GET, PUT, POST). You can determine the list of permitted service names using `hydra -h`.

▶ -6
Uses IPv6, if possible.

▶ -C *filename*
Uses the login name and password combinations specified in the file. The logins and passwords must be contained line by line in the format `login:password`.

▶ -e nsr

Also tries an empty password (n as in *null*), the login name as password (s as in *same*), and the reverse login name (r as in *reverse*).

▶ -f

Terminates the command as soon as a valid login/password combination is found.

▶ -l *loginname*

Uses the specified login name.

▶ -L *user file*

Reads the login names line by line from the specified text file.

▶ -m *options*

Transfers additional options that are specific to the network service. You can determine permitted options using hydra -U *service*, such as hydra -U http-get for logins that are supposed to be done with an HTTP GET request.

▶ -M *host file*

Reads the host names or IP addresses to be attacked from the file and attacks all hosts in parallel.

▶ -o *result file*

Saves the successful login password combinations in the specified file instead of displaying them in the standard output.

▶ -p *password*

Uses the specified password.

▶ -P *pwfile*

Tries out the passwords from the specified text file one after the other.

▶ -R

Resumes the hydra call last interrupted via ⌈Ctrl⌉+⌈C⌉, provided the hydra.restore file exists. No further options need to be specified; they are included in hydra. restore.

▶ -s *portnr*

Uses the specified port instead of the default port of the respective service.

▶ -t *n*

Runs *n* tasks (threads) in parallel. The default setting is 16. This can be too high because some services block the login if there are too many parallel requests (from the same IP address at that).

▶ -x *min:max:chars*

Generates passwords that are between *min* and *max* characters long and contain the specified characters. Here, a is shorthand for lowercase letters, A for uppercase letters, and 1 for numerals. All other characters, including äöüß, must be entered individually.

Example: With -x '4:6:aA1-_$%', hydra uses passwords that are four to six characters long and contain the characters -, _, $, and % in addition to letters and numbers. With -x '4:4:1', hydra tries all four-digit numbers. This results in 10,000 possibilities.

The -x option is only useful in exceptional cases, namely if you have (almost) an infinite amount of time and the target computer tolerates an unlimited number of login attempts.

Example

In the following example, hydra tries to find an account on a Linux server with a trivial password or no password at all for an SSH login. To do this, cut first generates a list of all accounts. Ideally, you should run this command on a computer running the same distribution as the target computer.

```
user$ cut -d: -f1 /etc/passwd > logins.txt
```

After that, hydra is supposed to try an SSH login for all accounts stored in logins.txt, using the account name, the reversed account name, and an empty character string as the password:

```
user$ hydra -L logins.txt -t 4 -e nsr 10.0.0.36 targethost
```

Alternatives

If you know passwords in the form of hash codes, you can try to find out the plain-text passwords with the offline password cracker (see john). hashcat is even faster. However, this command requires the sometimes-complicated installation of suitable GPU drivers.

---

iconv -f charset1 -t charset2 in.txt > out.txt

iconv performs a character set conversion from character set 1 to character set 2. iconv --list provides a comprehensive list of all supported character sets. The following command creates a corresponding UTF-8 file from a Latin-1 encoded text file:

```
user$ iconv -f latin1 -t utf-8 latin1dat > utf8dat
```

---

id

id specifies the name and ID of the user, their primary group, and the other assigned groups. On Fedora and RHEL, the command also provides the SELinux context.

Example

The user `kofler` has the UID 1000, belongs to the primary group `kofler` with the GID 1000, and is a member of the `wheel` group with GID 10:

```
root# id
uid=1000(kofler) gid=1000(kofler) groups=1000(kofler),10(wheel)
  context=unconfined_u:unconfined_r:unconfined_t:s0-s0:c0.c1023
```

```
if condition; then
  commands
[elif condition; then
  commands]
[else
  commands]
fi
```

`if` creates branches in bash scripts. The block after `then` is only executed if the condition is met. Otherwise, any number of `elif` conditions are evaluated. If necessary, the `else` block, which is also optional, is executed.

Several commands can be specified as a condition. The last command must be followed by a semicolon. The return value of the last command is used as the criterion. Comparisons and other tests can be performed using the `test` command. Instead of `test`, a short form in square brackets is also permitted. However, a space must be entered after [and before].

```
ifconfig interface [options] [ip-address]
```

`ifconfig` helps to determine or change the current network configuration. However, the command has been outdated for years and isn't compatible with all network functions of the Linux kernel. Use `ip`!

```
iftop [options]
```

`iftop` monitors the network traffic of one or all network interfaces and shows on a page (updated every three seconds) to which hosts or IP addresses most of the data is flowing. `iftop` runs like `top` until it's terminated by pressing Q.

▶ `-B`
Calculates in bytes/s instead of bits/s.

▶ `-F ipadr/mask`
Considers only traffic from or to the specified address.

- ▶ -G *ip6adr/mask*

  Considers only traffic from or to the specified IPv6 address.

- ▶ -i *name*

  Takes only the specified network interface into account.

- ▶ -n

  Displays IP addresses instead of host names.

ifup interface
ifdown interface

ifup activates the specified interface, ifdown deactivates it again. The commands are called by the init system for network initialization and access the distribution-specific configuration files. For this reason, the implementation of the commands varies depending on the distribution; the available options and their meaning also depend on the distribution (see man ifup/ifdown). In some current distributions (Ubuntu, Raspberry Pi OS), the commands are no longer available or only work with major restrictions.

Example

The following two commands first shut down the eth0 network interface and then bring it up again - for example, to activate a changed configuration:

```
root# ifdown eth0
root# ifup eth0
```

info [command name]

info launches the online help system of the same name. To navigate in the help text, you can use the keyboard shortcuts summarized in Table 8. Alternatively, you can also read info texts via the pinfo command from the package of the same name, with the emacs editor or in the Gnome and KDE help systems. All variants are more convenient than the original.

- ▶ -f *file*

  Loads the specified file instead of a file from /usr/share/info. If the info text is spread over several files, the first file must be specified (e.g., elisp-1.gz).

init [n]

init activates the runlevel specified by *n*. This only works with distributions that use the traditional Init-V system. For distributions with the systemd init system, you need to change the runlevel or, more precisely, the *target* with the `systemctl isolate` command.

`inotifywait` [options] [files/directories]

`inotifywait` from the `inotify-tools` package monitors changes to files and their metadata. In the simplest form, you pass one or more file names to the command. In this case, the command waits until an `inotify` event occurs for one of these files, such as a change to the file or a read access. This terminates the command.

The command is often used in scripts to automatically respond to changes to files. Alternatively, you can also perform monitoring indefinitely and log the events that occur.

▶ `-d` or `-m`
Works as a background process (`-d`, *daemon*) or in the foreground (`-m`, *monitor*). `inotifywait` doesn't end with the first occurring event, but runs until it's explicitly terminated, for example, via `kill` or Ctrl+C.

▶ `-e` *event*
Responds only to the specified event. By default, the command processes all `inotify` events. The `-e` option can be specified multiple times to select multiple events. The most important events include `access`, `close`, `create`, `delete`, `modify`, `move`, and `open`. A detailed description of all events can be found on the `man` page.

▶ `--fromfile` `file.txt`
Reads the list of files or directories to be monitored line by line from `file.txt`.

▶ `-q`
Avoids unnecessary outputs (*quiet*).

▶ `-r`
Also monitors all subdirectories of the specified start directory recursively. An `inotify` monitoring is set up for each individual file. For directories with many files, this takes a while and requires a relatively high number of resources. The maximum number of monitors is normally set at 8,192. You can change this value in the `/proc/sys/fs/inotify/max_user_watches` file if required.

▶ `-t` *n*
Always ends after *n* seconds, even if no event occurs.

Example

The following shell script monitors the `*.md` files in the current directory. Each time a change is made to one of these files, it checks whether there is an `*.md` file that is more up-to-date than the corresponding `*.pdf` file. In this case, the PDF document in question is recreated with pandoc. The script runs endlessly until it's terminated via Ctrl+C.

```
#!/bin/bash
while :
do
  for mdfile in *.md; do
    pdffile=${mdfile%.md}.pdf
    if [ $mdfile -nt $pdffile ]; then
      echo $mdfile
      pandoc -t beamer -H header.tex $mdfile -o $pdffile
    fi
  done
  inotifywait -e modify -q *.md
done
```

insmod [options] module file [parameter=value ...]

insmod loads the specified kernel module. The complete file name must be transferred. In addition, parameters (options) can be passed to the module. If you want to specify hexadecimal values, you must prefix them with 0x, such as option=0xff. You can determine the module parameters available for selection using modinfo.

▶ -f

Attempts to load the module even if it hasn't been compiled for the current kernel version. Whether this actually works depends on whether there are any incompatibilities between the kernel version and the module version. This option is particularly useful if hardware manufacturers only provide a module as a binary version (without any source code). However, the option is of course no guarantee that the module is actually compatible with your kernel version.

install [options] source target

install copies a file or an entire directory to a new location and can simultaneously set the owner, the group assignment, and the access rights. install thus combines the functions of cp, chmod, and chown. As the name suggests, the command is primarily intended for installation scripts.

▶ -d

Recursively copies an entire directory, not just one file.

▶ -g *group*

Defines the group assignment of the new files.

▶ -m *oct_permissions*

Defines the access rights with an octal code (see chmod).

▶ -o *owner*

Determines the owner of the target file.

▶ -p or --preserve-timestamps

Adopts the time data of the source files unchanged. (By default, the current time is saved as the modification time for the target files).

Example

The following command copies the login_sample file from the current directory to /usr/share/myproject/logindata. The target directory, in this case /usr/share/myproject, must already exist. The target file belongs to the apache user and can only be read and changed by that user. (On Debian and Ubuntu systems, you must use the www-data user instead of apache.)

```
root# install -m 600 -o apache login_sample /usr/share/myproject/logindata
```

inxi [options]

The Perl script inxi from the package of the same name collects information about the hardware, drivers, and software installed on the computer. Without further options, it only outputs the key data of the computer. In the terminal, keywords of the output are highlighted in color, which makes the text more readable than in the following listings.

▶ -A

Provides information about the audio system.

▶ -B

Provides information on the battery.

▶ -D

Provides information on the data media (*disks*).

▶ --export json|xml

Formats the results as a JSON or XML document.

▶ -E

Provides information on the Bluetooth setup.

▶ -G

Provides information on the graphics system.

▶ -F

Returns all data the command is able to determine (*full*).

Example

```
user$ inxi
CPU: 4x 1-core AMD EPYC-Rome (-SMP-) speed: 3593 MHz
Kernel: 5.15.0-100-generic x86_64
Up: 2d 5h 7m Mem: 2186/5779 MiB (37.8%) Storage: 108 GiB (66.3% used)
Procs: 179 Shell: Bash
user$ inxi -G
Device-1: Red Hat Virtio 1.0 GPU driver: virtio-pci v: 1
Display: server: X.Org v: 23.2.4 with: Xwayland v: 23.2.4 driver:
  dri: swrast gpu: virtio-pci resolution: 1: 800x600~60Hz 2: 1280x800~75Hz
API: OpenGL v: 4.5 vendor: mesa v: 23.3.6 renderer: llvmpipe (LLVM
  17.0.6 128 bits)
API: EGL Message: EGL data requires eglinfo. Check --recommends.
```

ionice [options] [command]

ionice runs the specified command with a changed I/O priority. ionice thus has a function that's similar to nice, but influences I/O operations and not the CPU utilization.

▶ -c *n*

Specifies the desired scheduling class. Permitted settings are:

– 0: No preferences.

– 1: *Real time*, that is, maximum I/O speed.

– 2: *Best effort*, applies by default.

– 3: *Idle*, that is, only perform I/O operations when the system isn't under load.

▶ -n *ionic*

Indicates the priority level within the selected scheduling class. The permissible values range from 0 (maximum priority) to 7 (minimum priority). Priority levels are only provided for scheduling classes 1 and 2 and allow differentiation within the class.

▶ -p *pid*

Changes the I/O priority of the process specified by the ID.

Example

The following command starts a backup script with minimum I/O priority:

```
root# ionice -c 3 backupscript
```

iotop

iotop from the package of the same name displays the I/O activity of all running processes. This helps when searching for processes that place a particularly high load on the hard disk or other data media.

▶ -o

Shows only processes that are actually I/O-active (and not all running processes by default).

▶ -u or --user=*user*

Shows only the processes of the specified user.

ip [options] object command

ip is an extremely versatile command that determines or changes information about network devices, tunnels, routing rules, and so on. ip should be used instead of ifconfig and route because these two commands are considered obsolete.

▶ -f *fam* or -family *fam*

Determines the desired network protocol (inet, inet6, or link). Instead of -f inet, the short form -4 is permitted; instead of -f inet6, the option -6; and instead of -f link, the option -0.

▶ -o or -oneline

Summarizes related outputs in one line. This reduces the readability, but simplifies further processing via grep or wc.

▶ -r or -resolve

Resolves IP addresses and displays the host names instead. This requires a name server.

One of the following keywords must be specified as the object: addr, addrlabel, link (i.e., a network interface), maddr (a multicast address), mroute, monitor, neighbor (an ARP or NDISC cache entry), route, rule, or tunnel. These keywords can be abbreviated. The add, delete, and list = show commands are available for most objects. The other commands are object-specific. In the following reference, I'll limit myself to the most important commands for the addr, link, and route objects:

▶ ip addr [show dev *xxx*]

Displays the IP addresses of all interfaces. The output normally comprises multiple lines. The line beginning with link/ether specifies the MAC address of the interface. The line beginning with inet contains the IPv4 address, including a mask in the short notation (/*n*) as well as the broadcast address. The lines beginning with inet6 specify the IPv6 addresses and can have multiple lines.

With -4 or -6, the output can be restricted to IPv4 or IPv6. ip addr show dev *xxx* only provides information on the specified interface.

► ip addr add *n/m* dev *xxx*

Adds the IP address *n* with the mask *m* to the interface *xxx*. A permissible IPv4 address, including mask, would be, for example, 10.0.45.34/24.

► ip addr del *n/m* dev *xxx*

Cancels the address assignment to interface *xxx*. Exactly the same parameters must be specified as for ip addr add.

► ip addr flush dev *xxx*

Deletes *all* address assignments of interface *xxx*.

► ip link [show dev *xxx*]

Provides a list of all network interfaces, in contrast to ip addr show, but without specifying IP addresses.

► ip link set *xxx* up/down

Activates or deactivates the network interface.

► ip neigh

Provides a list of all other IP addresses known in the local network, that is, an enumeration of the "neighbors."

► ip route [list]

Outputs the IPv4 routing table. If you want IPv6 data, you must specify the -6 option. The gateway address can be found in the line beginning with default.

► ip route add default via *n*

Specifies the IP address *n* as the default gateway.

► ip route add *n1/m* via *n2* dev *xxx*

Defines the routing address *n2* for the address range *n1/m*. The IP packets are routed via the *xxx* interface.

► ip route del ...

Removes the specified routing entry. The parameters must exactly match those of the ip route add command.

The documentation of the ip command extends over several man pages. man ip only provides an overview. man ip-address provides details on ip addr, man ip-route on ip route, and so on. You can also use ip object command help to determine a syntax description of a specific command, such as the syntax for resolving an address assignment via ip addr del.

Example

The following command displays the current routing table. In many modern distributions, interface names such as enp0s3 are used instead of eth0.

```
user$ ip route show
10.0.0.0/24 dev eth0  proto kernel  scope link  src 10.0.0.41  metric 1
default via 10.0.0.138 dev eth0  proto static
```

To assign the address 10.0.0.41 to the eth0 interface and set up the gateway 10.0.0.138, you must run the following commands:

```
root# ip route add 10.0.0.41/24 dev eth0
root# ip route add default via 10.0.0.138 dev eth0
```

The following commands show an IPv6 configuration:

```
root# ip -6 addr add 2a01:4f8:161:107::2/64 dev eth0
root# ip -6 route add default via fe80::1 dev eth0
```

A compact list of all network interfaces is provided by ip -o link:

```
root# ip -o link
1: lo:     <LOOPBACK,UP,LOWER_UP>               mtu 16436 ...
2: eth0:   <BROADCAST,MULTICAST,UP,LOWER_UP>    mtu 1500 ...
3: br0:    <BROADCAST,MULTICAST,UP,LOWER_UP>    mtu 1500 ...
4: virbr0: <NO-CARRIER,BROADCAST,MULTICAST,UP> mtu 1500 ...
5: vnet0:  <BROADCAST,MULTICAST,UP,LOWER_UP>    mtu 1500 ...
```

ipcalc ipadress [netmask]

ipcalc from the package of the same name determines all other parameters from a given IPv4 address and the netmask, that is, the network address, the broadcast address, and so on.

```
user$ ipcalc 10.11.12.13/16
Address:   10.11.12.13      00001010.00001011. 00001100.00001101
Netmask:   255.255.0.0 = 16 11111111.11111111. 00000000.00000000
Wildcard:  0.0.255.255      00000000.00000000. 11111111.11111111

Network:   10.11.0.0/16     00001010.00001011. 00000000.00000000
HostMin:   10.11.0.1        00001010.00001011. 00000000.00000001
HostMax:   10.11.255.254    00001010.00001011. 11111111.11111110
Broadcast: 10.11.255.255    00001010.00001011. 11111111.11111111
Hosts/Net: 65534              Class A, Private Internet
```

iptables [options]
ip6tables [options]

iptables configures the filter for network packets (*Netfilter* for short) of the Linux kernel. The iptables options follow a simple schema: An option in capital letters indicates the action to be performed (e.g., -P to set the default behavior). Additional options in lowercase control the details of this action. This syntax summary is organized by action.

Note that the Linux kernel manages entirely separate filter tables for IPv4 and IPv6. Accordingly, there are also two configuration commands: iptables for IPv4 and ip6tables for IPv6. The following description applies equally to both command variants.

Before you call iptables, you should make sure your distribution doesn't have its own firewall system. Fedora, RHEL, and SUSE, for example, rely on *Firewalld* (see firewall-cmd). Even defined firewall rules can then easily result in conflicts.

Modern distributions use the new *nftables* firewall system instead of *Netfilter*. Fortunately, this change doesn't make the iptables command obsolete—it can still be used thanks to a compatibility layer. (Otherwise, countless firewall scripts would no longer work). However, there is of course also the option of controlling nftables directly by using the nft command.

iptables -P chain policy [-t table]

iptables -P*(policy)* defines the default behavior for the specified chain of rules. Possible behaviors are given in the following table.

ACCEPT: Forward packet (default setting)

DROP: Delete packet

RETURN: Return packet (rare)

QUEUE: Forward packet to a program outside the kernel (rare)

By default, the command applies to *filter* chains or custom-defined chains. If a *NAT* or *mangle* chain of rules is to be changed, the table name must be specified with the -t option, such as iptables -P POSTROUTING ACCEPT -t nat.

It's not possible to define a standard behavior for custom defined chains. However, you can define the default behavior with the last rule if required, for example, via iptables -A mychain -j DROP.

iptables -A chain [-t table] options

iptables -A *(add)* adds a new rule to the specified chain. In general, one rule applies to all possible cases (i.e., for all IP protocols, for all ports, for all sender and destination addresses, for all interfaces, etc.).
The validity can be restricted using options. Most options can also be used in the negative with an exclamation mark. With -p udp, for example, a rule only applies to UDP packets. With -p ! udp, however, it applies to all packets except for UDP packets.

Not all possible combinations of options are permitted. For example, the -d and -s options can only be used for tcp packets, that is, in combination with -p tcp.

► -d *ipaddress*
Specifies the destination address (*destination*). Address ranges can be specified in the format; 192.168.0.0/24 or 192.168.0.0/255.255.255.0. In both cases, all IP numbers 192.168.0.* are meant.

► [--dport *port[:port]*]
Specifies the port or port range of the destination address (e.g., 0:1023).

► -i *interface*
Specifies the interface from which the IP packet originates (only for *input, forward,* and *prerouting* chains). In the interface name, the special character + is permitted as a placeholder for all interface numbers, that is, ppp+ for ppp0, ppp1, and so on.

► -j ACCEPT/DROP/mychain/..
Specifies what should happen to the packet (*jump*). One of the predefined procedures (ACCEPT, DROP, etc.) is usually specified here. For special applications, iptables provides REDIRECT or MASQUERADE.

Instead of one of the predefined keywords, you can also specify a custom defined chain of rules. In this case, all rules of this chain will be applied. If no rule in the custom defined chain applies, the next rule in the original chain will be applied instead. In procedural programming, this would correspond to a subroutine call.

► -m *module*
Indicates that an additional module will be used. Subsequently, special options defined by this additional module may be used. A particularly important additional module is state, which allows packages to be selected according to their connection status. For example, a rule with -m state --state NEW only applies to IP packets that initiate new connections. The following table shows the status keywords that can be specified with --state.

NEW: The packet initiates a new connection.

ESTABLISHED: The packet belongs to an existing connection.

RELATED: The packet initiates a new connection but belongs to an existing connection.

INVALID: The packet doesn't belong to any existing connection and doesn't initiate a new connection.

▶ `-o interface`
Specifies the interface to which the IP packet is traveling (only for *output*, *forward*, and *post-routing* chains).

▶ `-p protocol`
Determines the protocol (e.g., `tcp`, `udp`, or `icmp`).

▶ `-s ipaddress`
Specifies the sender address (*source*).

▶ `--sport port[:port]]`
Specifies the port or port range for the sender.

▶ `--syn`
Specifies that the rule should only apply to TCP packets for which the SYN bit is set. Such packets are used to initiate a connection (e.g., for all TCP wrapper functions, for HTTP, etc.).

`iptables` provides the option of logging the effectiveness of individual rules using `syslogd`. To do this, you can enter `LOG` as the action for the rule. For a rule to be both effective and logged, it must be specified twice: once with `-j LOG` and a second time with `-j ACCEPT` or `-J DROP`! Note that logging rules can very quickly result in huge log files.

`iptables -N mychain`

`iptables -N` (*new*) creates a new chain of rules named *mychain*.

`iptables -L [chain] [-t table] [-v]`

`iptables -L` (*list*) returns a list of all rules for the three chains of rules of the filter table as well as for all custom-defined chains without any further options.

You can use the other options to precisely specify the desired chain (e.g., `iptables -L mychain` or `iptables -L POSTROUTING -t nat`). The additional `-v` option provides more detailed information. `-n` causes IP and port numbers to be displayed in the output (instead of network or port names).

`iptables -D chain [-t table] options`

`iptables -D` (*delete*) deletes the rule from the chain. Exactly the same options must be specified as for `iptables -A`.

`iptables -F` `chain [-t table]`

iptables -F (*flush*) deletes all rules from the specified chain.

`iptables -X` `[mychain]`

iptables -X deletes the specified custom chain. If no chain is specified, all custom-defined chains will be deleted.

Example

The following iptable commands define a mini firewall for IPv4. Incoming packets are only accepted if they are either assigned to an existing connection or *don't* originate from the eth0 interface via which the computer is connected to the internet.

The example assumes that the iptables tables are initially in the default state, that is, accept all packets, and that there is only one interface to the internet.

```
root# iptables -N wall
root# iptables -A wall -m state --state ESTABLISHED,RELATED -j ACCEPT
root# iptables -A wall -m state --state NEW ! -i eth0 -j ACCEPT
root# iptables -A wall -j DROP
root# iptables -A INPUT -j wall
root# iptables -A FORWARD -j wall
```

`ip[6]tables-save` `[options]`
`ip[6]tables-restore` `[options]`
`ip[6]tables-xml` `[options]`

The iptables-xxx or ip6tables-xxx commands help to save the rules of a packet filter firewall in a file or to read them from it. Each command briefly described below is also available in an IPv6 variant.

▶ iptables-save outputs all rules of all firewall filters. An easy-to-read syntax is used. The output can be redirected to a file by using >. The -t name option causes only the rules of the specified filter to be saved.

▶ iptables-restore reads rules from the standard input. The rules normally replace existing filter chains. If you don't want this to happen, you must specify the -n option. With -T name, only the rules of the specified filter are taken into account.

▶ iptables-xml works in a similar way to iptables-save, but creates an XML document.

iw `object command`

You can use the `iw` command to control Wi-Fi adapters that support the nl80211 interface. This is the case with most current Wi-Fi adapters whose drivers are based on the mac80211 framework.

There are multiple ways to specify the object you want to control. The abbreviation `dev` or `phy` can be omitted if the interface or device name is unique.

▸ dev *name*
  Specifies the interface name (e.g., `dev wlan0`).

▸ phy *name* or `phy #n`
  Indicates the name or index number of the device. For notebooks with a Wi-Fi adapter, the device name is always `phy0`.

▸ `reg`
  Controls the *regulatory agent*, that is, a set of rules for national radio standards.

The commands available for selection depend on the object type. I'll only present a few selected commands here:

▸ dev *name* connect *ssid*
  Establishes a connection to the specified Wi-Fi network. This is only possible with networks that don't use encryption. If the wireless network is secured by WEP, you must specify the key using the optional `keys` parameter (e.g., `keys 0:0011223344`). The key is specified either in the form of 5 or 13 ASCII characters or 10 or 26 hexadecimal digits.

  If the network is secured by WPA, you must specify the key in `/etc/wpa_supplicant/wpa_supplicant.conf` before running `iw dev connect` and make sure that `wpa_supplicant` is running as a background service.

▸ dev *name* del
  Removes (deletes) the interface. If the interface is to be used again later, it must be set up again via `interface add`.

▸ dev *name* disconnect
  Terminates the connection.

▸ dev *name* info
  Provides general information about the interface.

▸ dev *name* link
  Provides information on the active network connection or *not connected*.

▸ dev *name* scan
  Provides detailed information on all wireless networks within range.

▸ phy phy0 `interface add` wlan0 `type` managed
  Sets up the `wlan0` interface for the `phy0` device. Other interface types are `monitor`, `wds`,

mesh, or mp, as well as ibss or adhoc. The new interface must then be activated using ifconfig wlan0 up.

For troubleshooting, it's often useful to run the iw event command in a second window or console. This command returns all status and error messages until the end by pressing [Ctrl]+[C].

Example

The following commands manually establish a network connection to a wireless network that isn't protected by a password:

```
root# iw phy phy0 interface add wlan0 type managed
root# ifconfig wlan0 up
root# iw dev wlan0 connect hotel-wlan
root# dhclient wlan0
```

In many current distributions, the Wi-Fi interfaces are no longer called wlan0, wlan1, and so on, but *wlpnsm*.

j directory

The j command from the autojump package helps to change to another directory in a particularly efficient way. j is, so to speak, a learning variant of the cd command. For example, if you've executed j /etc/X11/xorg.conf.d once, j xorg.conf.d is sufficient the second time, that is, the last part of the directory path (provided it's unique).

You can further shorten the input with the [Tab] completion, for example, as j xorg [Tab]. If multiple directories match, you can simply press [Tab] several times. If required, the jumpstats command provides statistics on all recently visited directories.

A modern alternative to j is z.

john [options] [hashfile]

The program *John the Ripper* (command and package name john) is an offline password cracker. To use the program, you must have the passwords as hash codes. john then tests whether self-generated passwords or passwords from a predefined list correspond to the hash codes.

john can handle most common hash algorithms. In the simplest case, you simply pass the name of a text file containing hash codes line by line to the command. The lines of the text file can also have the following format: name:hashcode:xxx.

john first tries the account name as password (if available), then passwords from a built-in word list, and finally passwords that it generates itself (`--single`, `--wordlist`, and `--incremental` mode). Cracked passwords are stored in `.john/john.pot`.

▶ `--format=hashname`
Specifies which hash method john should use. The command supports the following formats, among others: `afs`, `bcrypt`, `bsdicrypt`, `crypt`, `descrypt`, `dummy`, `lm md5crypt`, and `tripcode`. This option is only required if john doesn't recognize the hash method itself.

A brief description of the hash formats can be found on the following website. Note that this website also describes formats that are only included in the unofficial "jumbo" version (see the "Alternatives" section).
https://pentestmonkey.net/cheat-sheet/john-the-ripper-hash-formats

▶ `--incremental[:lower|lowernum|alpha|digits|alnum]`
Generates passwords itself, first trying out short and then increasingly longer passwords. Note that, in this mode, john runs endlessly, unless the passwords you're looking for are trivial.

By default, the characters a–z, A–Z, and 0–9 are taken into account in this mode (corresponds to `alnum`, 62 characters). Optionally, you can restrict the character set: `lower` corresponds to a–z, `lowernum` includes a–z and 0–9, `alpha` corresponds to a–z and A–Z, and `digits` includes digits only. Some other modes that can be defined are provided in `/etc/john/john.conf`. (On Ubuntu, many modes don't work because the corresponding character set files are missing in `/usr/share/john`.)

Finally, you can use john `--make-charset` to generate your own character set file, which you can then use with `--incremental=charsetfile`. The following website provides some useful tips on this: *https://security.stackexchange.com/questions/66106*.

▶ `--restore`
Continues the execution of john that was interrupted by Ctrl+C.

▶ `--show`
Displays passwords that have already been cracked. For this purpose, the `.john/john.pot` file gets evaluated.

▶ `--wordlist filename`
Tries out the passwords contained in the file line by line.

Example

The starting point for this example is a Linux computer with multiple accounts. The unshadow command supplied with john combines `/etc/passwd` and `/etc/shadow` into a new hashes file. (The access to the shadow file requires root permissions.) The hash codes in the resulting file are abbreviated here for reasons of space:

```
root# unshadow /etc/passwd /etc/shadow > hashes
root# chown user hashes
user$ cat hashes
...
peter:$6$U.zGFBlF$LdNTE...:1001:1001::/home/peter:/bin/bash
maria:$6$gSJg6.d8$mN.en...:1002:1002::/home/maria:/bin/bash
hugh:$6$UinuQqJY$iD59.N...:1003:1003::/home/hugh:/bin/bash
```

john finds two particularly insecure passwords within seconds. peter has used his own name as his password, and hugh used the popular password "123456". However, Maria's password can't be cracked straight away, which is why the process gets stopped after a while by pressing Ctrl+C:

```
user$ john hashes
Loaded 3 password hashes with 3 different salts
Press 'q' or Ctrl-C to abort, almost any other key for status
peter           (peter)
123456          (hugh)
  <Strg>+<C>
```

You can download lists of popular passwords from GitHub using git. (Note that the space required for the lists is around 750 MiB!) One of these dictionaries actually contains Maria's password! It reads secret:

```
user$ git clone https://github.com/danielmiessler/SecLists.git
user$ john --wordlist=SecLists/Passwords/Common-Credentials/ \
                 10-million-password-list-top-10000.txt  hashes
Loaded 3 password hashes with 3 different salts
Remaining 1 password hash with 1 different salt
secret          (maria)
```

Alternatives

On GitHub there is the greatly expanded, community-maintained "jumbo" version of john: *https://github.com/openwall/john*.

If you want to significantly speed up the cracking of hash codes with GPU support, you should consider using hashcat. The biggest hurdle in this case is the installation of suitable GPU drivers. To check network services for insecure passwords, you can use the online password cracker: hydra.

journalctl [options] [search expression]

The *journal* is the syslog service belonging to systemd. It now runs on most common Linux distributions, sometimes in parallel with a conventional logging system (usually rsyslogd).

The journal files are saved in a binary format and can only be read using the journalctl command described here.

One or more search expressions can be formulated as a path to a program or in the field=value syntax. journalctl then only shows logging entries to which all search expressions apply. Search expressions with logical OR can be formulated with +, that is, field1=value1 + field2=value2.

The MESSAGE, PRIORITY, ERRNO, and _PID, _UID, _GID, or _SELINUX_CONTEXT keywords are permitted for field. A reference of additional search parameters and a detailed description of their meanings can be found with man systemd.journal-fields.

When called without further parameters, journalctl usually returns an almost endless list of all logged messages, but you can use options to filter the output:

▶ -b
   Shows only the messages since the last reboot of the computer.

▶ --disk-usage
   Shows how much space the journal files take up in /var/log.

▶ -e
   Jumps immediately to the end of the messages to be displayed.

▶ -f
   Starts journalctl in continuous operation, constantly displaying the currently arriving messages. [Ctrl]+[C] terminates the command.

▶ -k
   Shows only kernel messages.

▶ -n *systemd. Journal*
   Shows only the last *n* lines.

▶ --no-pager
   Writes the output directly to the standard output instead of using a pager (normally, the less command) for display.

▶ -p *n*
   Shows only messages in a certain priority level. The number range is from 0 to 7 for emerg, alert, crit, err, warning, notice, info, and debug. -p 2 means that only messages of levels 0, 1, and 2 are taken into account.

▶ -r
   Shows the latest messages first.

▶ --since 2025-12-31 19:30:00

Shows only messages that were logged after the specified time.

▶ -t *name*

Shows only messages for the specified syslog keyword (tag, as in logger -t).

▶ -u *name*

Shows only messages for the specified systemd service (unit, e.g., avahi-daemon).

▶ --user

Displays messages from the user journal instead of the system journal. The user journal is used in modern Linux distributions to log messages from the graphics system, Gnome, or other desktop systems, as well as the programs running in them.

▶ --until 2025-12-31 19:30:00

Shows only messages that have been logged up to the specified time.

▶ --vacuum-size=500M

Deletes old logging files to such an extent that the total memory requirement is a maximum of 500 MiB.

Examples

The following command shows all messages from the Open SSH server in reverse order, that is, the latest messages first:

user$ **journalctl -u sshd -r**

If you want to follow the logging messages live, you should call journalctl as follows:

user$ **journalctl -u sshd -f**

The third example shows all messages in which the IP address 10.0.0.2 appears:

user$ **journalctl | grep 10.0.0.2**

jq [options] filter expression [file.json]

jq stands for *JSON query* and helps to extract data from JSON files. The command can be found in the package of the same name, which must be installed before the first use. jq reads the JSON document from the standard input or from the files passed as parameters. The simplest application is to format and indent JSON documents in a readable way. To do this, you simply need to enter a dot as the filter expression: jq . file.json.

The following list provides some more examples of filter expressions:

▶ .key1

Returns the value for the specified key.

▶ .[]

Returns all array elements of the current level.

- ▸ `.[].key1.key2`
  Returns all corresponding elements of the array.

- ▸ `.[n]`
  Returns the *n*th element.

- ▸ `filter | {json expression}`
  Creates a new JSON document from the elements found.

- ▸ `filter | select {.key1.key2=="xxx"}`
  Filters elements that have a specific property.

Examples

```
user$ echo '{"key1":"123","key2":"456"}' | jq .
{
    "key1": "123",
    "key2": "456"
}
user$ echo '{"key1":"123","key2":"456"}' | jq .key1
"123"
```

You can find more complex application examples in the `jq` tutorial:

https://jqlang.org/tutorial/

kbdrate [options]

`kbdrate` sets the delay time and the speed of key repetitions on the keyboard. Without options, the command displays the currently valid settings.

- ▸ `-d n` or `--delay=n`
  Indicates after how many milliseconds the key repetitions start. Permissible values are 250, 500, 750, and 1000.

- ▸ `-r n` or `--rate=n`
  Controls how many characters per second the keyboard forwards for a longer keystroke. *n* can be a float, but not every value is permitted (see `man kbdrate`). The command automatically selects the next permissible value.

kexec [options]

The `kexec` command from the `kexec-tools` package enables you to activate a new kernel without rebooting the computer. All processes and the init system must be shut down completely. Compared to a real reboot, a kernel reboot with `kexec` is still a few seconds faster: In particular, BIOS/EFI initialization and the GRUB process are no longer required. However, there may be problems when reinitializing various hardware components.

kexec is used in two steps: first you need to prepare the kernel reboot using `kexec -l`, then you actually perform the kernel exchange using `kexec -e` or `systemctl isolate kexec`.

▶ `--append=kernel parameters`
   Passes the parameters to the kernel in addition to `--command-line` or `--reuse-cmdline`.

▶ `--command-line=kernel parameters`
   Passes the parameters to the kernel.

▶ `-e`
   Restarts the previously installed kernel (*exec*).

▶ `--initrd=initrd file`
   Specifies the location of the initrd file.

▶ `-l kernel file`
   Specifies the location of the new kernel to be loaded (*load*).

▶ `--reuse-cmdline`
   Adopts the parameters used during the last kernel start.

Example

The following command prepares the kernel restart:

```
root# kexec -l /boot/vmlinuz-6.6.7-300 \
          --initrd=/boot/initrd.img-6.6.7-300 --reuse-cmdline
```

In the second step, you must activate the new kernel—either immediately via `kexec -e` or after you've shut down all running services via systemd. On a server, the second option is definitely preferable! This is the only way to ensure that a database server, for example, can complete all open transactions and close all files properly.

```
root# kexec -e                    (immediate restart)
root# systemctl isolate kexec     (first shut down services, then restart)
```

kill [-s signal] processno

The bash command `kill` sends signals to a running process. If `kill` is used without the `-s` option, the SIGTERM signal (15) is sent by default to terminate (*kill*, hence the name of the command) the process. In particularly obstinate cases, `-9`, `-s SIGKILL`, or `-KILL` will help. However, the process then has no chance of performing any cleanup work.

However, `kill` can also be used to send more harmless signals. Quite often, `-1`, or `-s SIGHUPv`, or `-HUP` is used to request a daemon to reload its configuration files. In this way, you can activate a new configuration for some programs without having to completely stop and restart the daemon.

The easiest way to make the required process number (PID) `kill` a convenient variant is by using the `ps` command: the program that is to be terminated can simply be "shot down" with the mouse.

`killall` [-signal] process name

`killall` works almost like the `kill` command. The difference is that it's not the process number (PID) that is specified, but the name of the process. If there are multiple processes with this name, they all receive the specified signal (again SIGTERM by default). The desired signal is specified either as the number -*n* or with a name such as `-HUP`. You can obtain a list of all signal names via `killall -1`.

Example

The following example terminates all running Firefox instances of the current user. If the `killall` command is executed by `root`, it terminates all running Firefox processes of *all* users.

```
user$ killall firefox
```

`kpartx` [diskdevice]

The low-level `kpartx` command from the package of the same name determines all partitions of the specified data medium and creates the corresponding device files. Usually, the device files are created automatically by the `udev` system as soon as a new data medium is detected, for example, when a USB hard disk gets connected. `kpartx` is primarily intended for editing virtual data media or virtual machine image files.

▶ `-a`
Creates new device files for the specified data medium (*add*).

▶ `-d`
Removes the device files for the data medium (*delete*).

▶ `-1`
Reads the partitions of the data medium, but doesn't create any device files.

▶ `-u`
Updates the device files for a changed data medium (*update*).

▶ -v

Provides information on the actions performed.

Example

The following command connects all partitions contained in the raw image file with loop devices:

```
root# kpartx -av image.raw
add map loop0p1 (252:12): 0 1024000 linear /dev/loop0 2048
add map loop0p2 (252:13): 0 19945472 linear /dev/loop0 1026048
```

The entire virtual hard disk can then be accessed via the /dev/mapper/loop0 device.

kvm [options] [imagefile]

The kvm script, which is still commonly used on Ubuntu, runs a virtual machine. The qemu emulator and the Linux virtualization system *kernel-based virtual machines* (KVMs) are used. In most other distributions, virtual machines are started directly via the qemu command, which is why I describe the numerous identical options there.

l2ping [options] bluetoothmac

l2ping from the bluez package sends L2CAP echo requests to the Bluetooth device specified by its MAC address. In this way, you can test whether a connection between the local Bluetooth adapter and the external device is possible. l2ping must be run with root permissions. You can determine the MAC address of external Bluetooth devices using hcitool scan.

lame [options] in out.mp3

The LAME acronym stands for *LAME Ain't an MP3 Encoder*, but the lame command is actually used to convert WAV files into compressed audio files that are compatible with the MP3 format. Alternatives to lame are toolame or twolame, both of which create files in MPEG-1 Layer 2 format (i.e., MP2). This type of file can also be played by most MP3 players without any problems.

▶ -r

The source file is in raw format (not as a WAV file).

▶ -s 8/11.025/12/16/22.05/24/32/44.1/48

Specifies the sampling frequency within the raw file. This option isn't required for WAV files.

Example

The following command creates the MP3-compatible audio file `title.mp3`:

```
user$ lame title.wav title.mp3
```

`last` [options]

`last` provides a list of users who were last logged on to this computer. The command evaluates the `/var/log/wtmp` file.

- `-i`

  Displays the IP address instead of the host name of the remote computer for SSH logins.

- `-n` or `-n wimpy`

  Controls how many entries will be displayed, for example, 100 with `-100`.

- `-t YYYYMMDDhhmmss`

  Indicates who was logged in at the specified time.

Example

The last user to work on the test computer was `kofler`:

```
user$ last
kofler   pts/0  62.47.230.2   Tue Sep 24 13:51   still logged in
kofler   pts/0  212.183.46.83 Mon Sep 23 19:50 - 20:22  (00:32)
kofler   pts/1  91.115.236.11 Fri Sep 13 15:39 - 17:51  (02:11)
kofler   pts/0  91.115.236.11 Fri Sep 13 15:31 - 17:51  (02:20)
```

`lastb` [options]

`lastb` shows which login attempts have recently failed. The command is controlled with the same options as `last`. It can only be run by `root` and evaluates the `/var/log/btmp` file.

`ldconfig`

`ldconfig` updates the links to all libraries and creates the cache file `/etc/ld.so.cache`, which helps with the efficient search for libraries. `ldconfig` evaluates the configuration file `/etc/ld.so.conf`. The command must be executed after the manual installation of libraries.

`ldd` program

ldd provides a list of all libraries required to run the specified program. You can also use the command to determine whether all the necessary libraries are available on the computer.

Example

The Gnome terminal accesses around 80 libraries on Fedora! The following lines show the first four of these in alphabetical order:

```
root# ldd /usr/bin/gnome-terminal | sort
  libatk-1.0.so.0 => /lib64/libatk-1.0.so.0 (0x0000ffffa4010000)
  libatk-bridge-2.0.so.0 => /lib64/libatk-bridge-2.0.so.0 (0x0000ffffa3d20000)
  libatspi.so.0 => /lib64/libatspi.so.0 (0x0000ffffa2aa0000)
  libblkid.so.1 => /lib64/libblkid.so.1 (0x0000ffffa3390000)
  ...
```

`less` [options] file

less displays the specified text file page by page. The command is often used as a filter, such as ls -l | less, to display a long file list on a page-by-page basis.

▶ -m

  Shows the current text position as a percentage in the status bar.

▶ -M

  Displays the file name and the text position in lines in the status bar.

▶ -p *search text*

  Shows the first line in which the text to be searched for was found.

▶ -r

  Displays formatting and color control codes (*raw*). The option is required in rare cases to correctly display a file that contains such control codes.

▶ -s

  Reduces multiple empty lines into one line.

While less is running, pressing H displays a help text. The cursor keys allow you to navigate through the text. The space bar scrolls one page forward, B moves one page back. < and > jump to the beginning or end of the text, respectively. / allows you to enter a search text. Q terminates less.

`lftp` [options] [site]

`lftp` is an interactive FTP client. `lftp` can also be used in scripts and controlled by commands.

▶ `-c "command"`
Runs the FTP commands specified in a character string and separated by semicolons (e.g., `lftp -c "open -u user,passw server; mirror -R dir"`).

▶ `-f file`
Reads the commands to be executed line by line from a file.

`lftp` login settings can be saved line by line for various servers in the `.netrc` file. Entries in this file look like the following pattern:

```
machine backup.hostname.com   login u123456    password x234bOCT2xCb
```

Example

The following command uploads a file to an FTP server:

```
root# lftp -c "open -u username,password backupserver; put file"
```

To transfer an entire directory to the backup server instead of a file, you can use the `mirror -R` command. `mirror` usually copies directories from the FTP server to the local computer. The `-R` command reverses the transfer direction. Here's another example:

```
root# lftp -c "open -u usern,passw bserver; mirror -R directory"
```

`libcamera-still` [options]
`libcamera-vid` [options]

`libcamera-still` and `libcamera-vid` take photos and videos on the Raspberry Pi. As a matter of fact, the `rpicam-still` and `rpicam-vid` commands are actually used for this. The syntax of the commands is described there. The command names `libcamera-still` and `libcamera-vid` are obsolete and are only available for compatibility reasons.

`ll` [options] `files`

`ll` is a predefined alias in many distributions that run the `ls` command with the `-l` option and possibly other options.

```
ln [options] source [target]
ln [options] files target directory
```

ln sets up fixed or symbolic links to files and directories. The cp command has the same functionality as ln if the options -l or -s are specified there.

▶ -b or --backup

Renames existing files with the same name to back up files (name plus ~ character) instead of overwriting them.

▶ -d or --directory

Creates a fixed link for a directory. This operation is only permitted for root. All other users can create symbolic links to directories.

▶ -s or --symbolic

Generates symbolic links. (Without this option, ln returns fixed links.)

Example

The following command creates the symbolic link xyz to the existing file abc:

```
user$ ln -s abc xyz
```

```
loadkeys [options] file name
```

The low-level loadkeys command loads a keyboard table for text mode. The required settings files are usually located in the /lib/kbd/keymaps directory. If you enter a file name without a path, the command itself attempts to find a suitable file with the .map.gz identifier.

For this reason, the keyboard settings changed in this way only apply if you work in a text console. The settings for the graphics mode, however, must be changed as part of the X configuration; for distributions with systemd, you can also use the localectl command.

▶ -d

Loads the default keyboard table, which usually is the file /lib/kbd/keymaps/defkeymap.map.

Example

The following command activates the German keyboard layout in the text consoles:

```
root# loadkeys de-latin1
Load /lib/kbd/keymaps/i386/qwertz/de.map.gz
```

local var[=value]

The bash command local defines a local variable in functions within shell scripts. The command can only be used in custom-defined functions (see function). Note that you must not enter any spaces before or after the equal sign.

localectl [options] [command]

The localectl command is used to control the language and keyboard settings in distributions with the systemd init system. It changes the following files: /etc/locale.conf, /etc/vconsole.conf, and /etc/X11/xorg.conf/00-keyboard.conf.

Options

▶ --no-convert
Changes only the keyboard for the console (set-keymap command) or for the graphics mode (set-x11-keymap command).

By default, localectl attempts to change the keyboard settings for both text and graphics mode at the same time. However, the relevant files have a different syntax, which is why the conversion of the parameters sometimes fails.

▶ --no-pager
Runs output directly on the standard output instead of using less.

Commands

▶ list-keymaps
Determines a list of all possible keyboard layouts for text mode.

▶ list-locales
Determines a list of all possible language settings.

▶ list-x11-keymap-models
list-x11-keymap-layouts
list-x11-keymap-variants [layout]
list-x11-keymap-options
Determines the permissible parameters for setting the keyboard layout in graphics system X.

▶ set-keymap *name*
Activates the specified keyboard layout.

▶ set-locale *name*
Activates the specified language setting.

▶ set-x11-keymap *layout* [*model variant options*]
Activates the specified keyboard layout for the graphics mode.

▶ status
Shows the current language and keyboard settings.

Example

The following two commands set the German language, the UTF-8 character set, and a German keyboard layout, taking into account the special features of an Apple keyboard:

```
root# localectl set-locale de_DE.UTF-8
root# localectl set-x11-keymap de de mac grp:alt_shift_toggle
```

locate pattern

locate enables a particularly fast search for files. It searches a file database, usually updated once a day, in which the specified pattern occurs in the full file name (including path). However, files that were created or changed after the last database update can't be found.

In most cases, locate and the underlying updatedb system must be installed separately (mlocate package) and often also configured.

logger [options] message

logger logs a message with the syslog service. Depending on the distribution, this is usually rsyslogd or, in more modern distributions, the journal (see journalctl).

▶ -p *n*
Logs the message in the specified priority level between 0 = emerg and 7 = debug.

▶ -p *facility.level*
Logs the message for the specified service and at the specified priority level. By default, logger uses the user.notice setting.

▶ -t *tag*
Saves the specified keyword together with the message. This simplifies a later search for corresponding messages.

Example

The following logger command could be positioned at the end of a backup script. It logs the successful end of the backup.

```
user$ logger -p local0.notice -t mybackuptool "Backup completed"
```

`loginctl` [options] [command] [name]

Like systemctl and journalctl, loginctl belongs to the systemd family. You can use the command to control the login manager (see man systemd-logind.service) or query data. Without options or parameters, the command simply displays all active logins (sessions).

▶ -p *property*
Specifies which property will be queried. Without this option, subcommands such as show-session or show-user provide a great deal of detail.

▶ active *sessionname*
Activates the specified session. The command may also change the active console, similar to the following shortcut keys: [Ctrl]+[Alt]+[F1], [Alt]+[F2], and so on.

▶ enable-linger *loginname*
Allows processes to continue after a user logout. This is useful for continuing long-lasting processes (downloads, calculations) to their end after a logout. loginctl disable-linger deactivates the function again.

▶ list-sessions
Shows all active sessions (default behavior).

▶ list-users
Shows all logged-in users.

▶ session-status *sessionname*
Shows which processes were started within a session and which logging messages were logged.

▶ show-session *sessionname*
Shows details of a session.

▶ show-user *username*
Shows details about a user.

▶ terminate-session *sessionname*
Ends a session. All assigned processes receive a kill signal.

Example

There are four sessions on the sample computer. On console 1 (tty1), the Gnome Display Manager is running and displays a login box. A user is logged in locally on console 7 (tty7). There are also two other logins that have been made via SSH (the output of loginctl doesn't reveal these).

```
root# loginctl
    SESSION    UID    USER      SEAT     TTY
         5    1000    kofler
        c1     119    gdm       seat0    tty1
         2    1000    kofler    seat0    tty7
         4    1000    kofler
```

The second loginctl command shows that the graphics system in console 7 uses Wayland. If Xorg is used instead, the output is Type=x11.

root# **loginctl show-session 2 -p Type**   Type=wayland

logname

logname displays the login name (username).

logout

logout or even shorter simply Ctrl+D terminates the session in a console or terminal window.

lpadmin [options]

lpadmin sets up a new printer for the CUPS printing system, changes its access rights for network operation, or deletes it again.

▶ -d *name*
  Defines the specified printer as the default printer.

▶ -E *name*
  Activates the specified printer.

▶ -p *name*
  Sets up a new printer. You can specify the configuration parameters with various other options (see man lpadmin).

▶ -x *name*
  Deletes the configuration for the specified printer.

lpinfo [options]

lpinfo lists the devices and drivers available for CUPS.

▶ -l
  Provides particularly detailed information. The option must be combined with -m or -v.

▶ -m

Provides a list of all available printer drivers.

▶ -v

Provides a list of all known printing devices.

lpoptions [options]

lpoptions displays or changes the options of CUPS printers.

▶ -l

Provides a list of the available options and their current setting.

▶ -o *option name=value*

Changes the setting of the specified option.

▶ -p *name*

Specifies the desired printer. (Without the option, lpoptions refers to the default printer.)

lpq [options]

lpq provides a list of all cached files or print jobs. The size of the file and a job number are also specified. You can specify this job number as a parameter of lprm to remove a file from the printer spooler.

▶ -a

Displays the print jobs of all queues.

▶ -P*name*

Displays the print jobs in the specified *name* queue.

lpr file

lpr prints the specified file.

▶ -l

Bypasses the filter system and sends the printer data to the printer unchanged. This option is useful if a print file is already available in the printer-specific format.

▶ -o options

Passes various additional parameters, such as -o media=A4 or –o page-ranges= 23-27,29,31. Numerous examples can be found in the CUPS documentation: *www. cups.org/doc/options.html*.

▶ -P*name*

Uses the *name* queue instead of the default printer. Note that the option isn't followed by a space!

lprm [options] [id]

lprm cancels the current print job or the print job specified by the ID.

▶ -Pname

Specifies the queue.

lpstat [options]

lpstat displays information about CUPS classes, printers, and their print jobs. lpstat works for both local and network printers.

▶ -a

Indicates for all printers whether they are ready to accept print jobs.

▶ -c

Displays all classes.

▶ -d

Displays the default printer.

▶ -s

Displays a status overview (default printer, list of all classes and printers, etc.).

▶ -t

Displays all available information.

▶ -v

Displays all printers.

ls [-options] [path]

ls displays a list of all files and directories. If ls is used without additional parameters or options, the command returns a multicolumn table sorted by file name in which all files, links, and directories in the current directory are displayed.

▶ -a or -all

Also displays files whose name begins with . Option -A has a very similar effect. The only difference is that the . and .. files (references to the current and parent directory, respectively) won't be displayed.

▶ --color [=never|auto|always]

Uses different colors for different file types (links, directories, etc.) or deactivates the color display. You can configure the colors yourself using dircolors.

▶ -d or --directory

Shows only the name of the directory, not its contents. This option is useful, for example, if a directory name is specified as the path, and the access rights of this directory are to be checked (and not its content).

► -h

Displays the file sizes in KiB, MiB, and GiB (*human readable*).

► -i or --inode

Shows the I-node of the file in addition to the other information. (The I-node is an internal identification number of the file, which is required for Linux-internal file management.) The option can be used to recognize hard links because files linked in this way have the same I-node.

► -I *file* or --ignore *file*

Excludes the specified files from being displayed. For example, -I*ps prevents files with the extension ps from being displayed. If a file pattern rather than a single file is specified after -I, no space may be entered between -I and the pattern!

► -l or --format=long or --format=verbose

Displays further information in addition to the file name, such as the file size in bytes, access rights, and more. A separate line is used to display each file (instead of the space-saving multicolumn list).

► -L or --dereference

Doesn't display the path of the link but the content of the original directory for a symbolic link to a directory.

► -o or --no-color

Doesn't use different colors or fonts.

► -p or -F

Appends a special character to the file name to indicate the type of file. In some Linux distributions, this option is activated by default in /etc/profile by means of an alias abbreviation. The most important special characters are / for directories, @ for symbolic links, * for executable files, and | for FIFOs.

► -r or --reverse

Reverses the sort order. The option is often used in combination with -t or -S.

► -R or --recursive

Also captures files in subdirectories.

► -S or --sort=size

Sorts the files according to their size (the largest file first).

► -t or --sort=time

Sorts the files according to the date and time of the last change (the newest file first).

► -u or --sort=access

Sorts the files based on the date and time of the last read access. The option must be specified together with -t (otherwise, ls doesn't sort at all).

▶ -v

Sorts numbers in file names in a reasonable way, such as tty2 before tty10.

▶ -X or --sort=extension

Sorts the files according to their identifier (i.e., according to the letter combination that follows the last . in the file name).

▶ -Z or --context

Shows the SELinux context information. More information is provided by --lcontext; less information is provided by --scontext.

Unfortunately, ls doesn't display the total memory requirements of all listed data. This task is performed by the du command. ls isn't able to display the access bits of a file in octal format. To do this, you can use the stat command.

Example

The following command displays all files in the current directory and sorts them by date (the newest file last):

```
user$ ls -ltr
-rw-r--r-- 1 kofler kofler 1681276 2025-03-02 10:01 cimg3079.jpg
-rw-r--r-- 1 kofler kofler 1582496 2025-03-02 10:01 cimg3014.jpg
-rw-r--r-- 1 kofler kofler 1615070 2025-03-02 10:01 cimg2965.jpg
...
```

Briefly, some comments on the interpretation of the ls result: The 10 characters at the beginning of the line indicate the file type and the access bits. The following file types are possible: the hyphen - for a normal file, d for a *directory*, b or c for a device file (*block* or *char*), or l for a symbolic link.

The next three characters (rwx) indicate whether the owner is allowed to read, write, and execute the file. Analogous information follows for the members of the group as well as for all other system users.

The number following the 10 type and access characters indicates how many hard links reference the file. The other columns indicate the owner and the group of the file (in this case, kofler), the size of the file, the date and time of the last change, and finally the file name.

Alternatives

ls is one of the most frequently executed Linux commands. Accordingly, there are many attempts to replace ls with more modern commands that are easier to use, display the results in a more visually appealing way, or integrate additional functions. In this book, I'll only go into detail about erd, which combines the functions of du, find, ls, and tree. The eza (formerly exa) and lsd commands are interesting as well:

https://github.com/eza-community/eza

https://github.com/lsd-rs/lsd

`lsattr` [options] files

lsattr displays the status of the additional attributes of files or directories in Linux file systems (ext2 to ext4, Btrfs, XFS, etc.). These additional attributes control, for example, whether files should be compressed, whether changes to files should be made using the copy-on-write procedure, and more. It depends on the implementation of the respective file system which attributes are actually taken into account—for most file systems those are only very few (see also the description of the chattr command for changing these attributes).

▶ -a
  Also displays hidden files.

▶ -d
  Shows the attributes of directories, not the attributes of the files in the directories.

▶ -R
  Recursively displays the attributes of all files in the directory tree.

Example

On the test system (an openSUSE Leap installation with a Btrfs file system), the C attribute is set for the /var/lib/mysql and /var/lib/libvirt/images directories. It deactivates the copy-on-write function for files in this directory.

```
root# lsattr -d /var /var/lib /var/lib/libvirt/images /var/lib/mariadb
---------------- /var
---------------- /var/lib/
---------------C /var/lib/libvirt/images
---------------C /var/lib/mariadb/
```

`lsblk` [options] [iodevice]

lsblk provides a hierarchical list of all block devices or all devices located on a data medium. The command is ideal for quickly obtaining a good overview of all partitions and logical volumes.

▶ -a
  Also includes empty devices in the result.

▶ -b

Specifies sizes in bytes.

▶ -f

Also shows the file system type, the label, and the UUID for each device.

▶ -p

Specifies complete device names, such as dev/mapper/fedora-swap for logical volumes.

Example

The following lsblk result was created on a server with two hard disks and RAID configuration:

```
root# lsblk
NAME     MAJ:MIN RM   SIZE RO TYPE  MOUNTPOINT
sda        8:0   1   3.7T  0 disk
  sda1     8:1   1    16G  0 part
    md0    9:0   0    16G  0 raid1 [SWAP]
  sda2     8:2   1   512M  0 part
    md1    9:1   0 511.4M  0 raid1 /boot
  sda3     8:3   1     2T  0 part
    md2    9:2   0     2T  0 raid1 /
sdb        8:16  1   3.7T  0 disk
  sdb1     8:17  1    16G  0 part
    md0    9:0   0    16G  0 raid1 [SWAP]
  sdb2     8:18  1   512M  0 part
    md1    9:1   0 511.4M  0 raid1 /boot
  sdb3     8:19  1     2T  0 part
    md2    9:2   0     2T  0 raid1 /
```

lsb_release [options]

lsb_release determines distribution-specific information, such as the name and version number of the running Linux system. LSB stands for *Linux Standard Base*. If no option is passed, -v applies.

▶ -a

Displays all available information: the distribution name, the full description, the version number, and the code name (see the example later in this entry).

▶ -v

Displays the LSB version number, but only if the LSB modules are installed in addition to the command (lsb-core package). This is often not the case—the command then returns the error message *no LSB modules are available*. In practice, however, the LSB version number is rarely of interest. If you only want to know the version of the installed Linux system, you should use the -a option.

Example

In the following example, lsb_release was run on an installation of Ubuntu 24.04:

```
root# lsb_release -a
Distributor ID: Ubuntu
Description:    Ubuntu Noble Numbat (development branch)
Release:        24.04
Codename:       noble
```

lscpu [options]

lscpu displays detailed information about the CPU, its cores and cache, and—if applicable—the virtualization system.

▶ -e
Displays only a summary of the key data in an easy-to-read format.

▶ -p
Like -e, but provides the data in a format that can be easily processed in scripts.

lshw [options]

The lshw command from the package of the same name creates a hierarchical list of all hardware components on the computer.

▶ -businfo
Lists the components line by line and specifies an ID for the SCSI, USB, or PCI bus used for each component.

▶ -html
Formats the output as an HTML document.

▶ -short
Provides only a compact summary.

▶ -xml
Formats the output as an XML document.

Example

The following hardware overview was created on a Raspberry Pi 4B:

```
root# lshw -businfo
Bus info        Device  Class     Description
==================================================
                        system    Raspberry Pi 4 Model B Rev 1.1
                        bus       Motherboard
cpu@0                   processor cpu
cpu@1                   processor cpu
cpu@2                   processor cpu
cpu@3                   processor cpu
                        memory    855MiB System memory
pci@0000:00:00.0        bridge    Broadcom Limited
pci@0000:01:00.0        bus       VL805 USB 3.0 Host Controller
usb@1           usb1    bus       xHCI Host Controller
usb@1:1                 bus       USB2.0 Hub
usb@1:1.4               bus       Keyboard Hub
usb@1:1.4.2             input     Apple Keyboard
usb@1:1.4.3             input     USB-PS/2 Optical Mouse
usb@2           usb2    bus       xHCI Host Controller
                eth0    network   Ethernet interface
                wlan0   network   Wireless interface
```

lsmod

lsmod provides a list of all modules that are currently loaded into the kernel.

lsof [options] [file/interface]

Without any additional parameters, lsof provides a usually very long list of all currently open files or interfaces and the processes assigned to them. lsof file provides information on the process that keeps this file open.

▶ -i *address*

Provides information on processes that use the specified network address. The address is composed in the following format: [46][protocol][@hostname|hostaddr] [:service|port]. For example, -i 4tcp provides information on all processes that use TCP in IPv4.

▶ -n

Doesn't resolve network names.

▶ -N

Takes NFS files into account as well.

▶ -u *user*

Provides only information on files/interfaces that are used by the specified user. Multiple users can also be specified (name or UID, separated by commas).

▶ -X

Ignores all open TCP and UDP files.

Examples

The following two commands show all processes that use UDP or port 22:

```
root# lsof -i udp
ntpd        3696      ntp    16u  IPv4   9026       UDP *:ntp
portmap   4745  daemon    3u  IPv4  12931       UDP *:sunrpc
rpc.statd 4764    statd    5u  IPv4  12962       UDP *:700

...
root# lsof -i :22
COMMAND   PID USER    FD   TYPE DEVICE SIZE NODE NAME
sshd      5559 root    3u  IPv6  14097       TCP *:ssh (LISTEN)
sshd      7729 root    3r  IPv6  33146       TCP mars.sol:ssh->merkur.sol:45368
                                                              (ESTABLISHED)
```

lspci [options]

lspci provides information about the PCI bus and all devices connected to it.

▶ -k

Shows which kernel module is responsible for each PCI device.

▶ -tv

Provides a tree-shaped device list that clearly shows how the devices are connected to each other.

▶ -v or -vv or -vvv

Provides even more details.

lsscsi [options]

lsscsi provides information about all connected SCSI and SATA devices.

▶ -c

Displays the data in the same format as /proc/scsi/scsi.

▶ -H

Provides a list of SCSI hosts (instead of SCSI devices).

▶ -l

Provides detailed information on each device.

Example

The following result was created on a computer with two PCIe SSDs and an external USB hard disk:

```
root# lsscsi
[0:0:0:0]    disk    WD Elements 25A2   1026          /dev/sda
[N:0:4:1]    disk    PM981 NVMe Samsung 2048GB__1    /dev/nvme0n1
[N:1:4:1]    disk    SAMSUNG MZVLB256HAHQ-000L7__1   /dev/nvme1n1
```

lsusb [options]

lsusb provides information about all connected USB devices.

▶ -t

Indents the output like a tree and thus makes it clear which device is connected to which bus.

▶ -v

Provides detailed information on each device.

Example

The following command shows that an Apple keyboard is connected to the computer:

```
user$ lsusb
Bus 004 Device 001: ID 1d6b:0003 Linux Foundation 3.0 root hub
Bus 003 Device 001: ID 1d6b:0002 Linux Foundation 2.0 root hub
Bus 002 Device 004: ID 1058:25a2 Western Digital Technologies, Inc. Elements
25A2
Bus 002 Device 001: ID 1d6b:0003 Linux Foundation 3.0 root hub
Bus 001 Device 007: ID 06cb:009a Synaptics, Inc.
Bus 001 Device 019: ID 05ac:021e Apple, Inc. Aluminum Mini Keyboard (ISO) ...
```

lvcreate [options] name

The LVM command lvcreate creates a new logical volume (LV) within the volume group (VG) that has been specified by name.

▶ -i n

Distributes the LV evenly over the specified number of physical volumes (PVs). This assumes that the VG consists of at least as many PVs. If the PVs are located on different hard disks, you'll achieve a similar effect as with RAID-0 (striping).

▶ -L *size*

Indicates the desired size. By default, the size is specified in MiB, but with the corresponding suffix also in bytes, KiB, GiB, or TiB (e.g., -L 2G).

▶ -n *lvname*

Indicates the LV name. If this option is missing, the command uses the name *lvoln*.

▶ -s

Creates a snapshot, that is, a static copy of an existing LV. In this case, name doesn't specify the VG, but the underlying LV. The size specification -L indicates the memory required for copied sectors if the underlying LV changes. Once this memory is exhausted, the life of the snapshot ends.

Example

The following command creates a new logical volume in the myvg1 VG named myvol1 with a size of 50 GiB:

```
root# lvcreate -L 50G -n myvol1 myvg1
  Logical volume "myvol1" created
```

The command simultaneously creates the /dev/myvg1/myvol2 file. This is a link to the /dev/mapper/myvg1-myvol2 file. The LV can then be used under one of these two device names like an ordinary hard disk partition. The following command sets up a file system there:

```
root# mkfs.ext4 /dev/myvg1/myvol1
```

---

lvdisplay [options] lvname

---

lvdisplay shows detailed information on an LV.

---

lvextend [options] lvname [pvname]

---

lvextend increases the memory for the specified LV device. You can specify the new total amount of memory via the -L option (see lvcreate). Alternatively, you can use the notation -L +size to specify the desired change.

The required memory is reserved within the same VG in which the LV is located. If the VG is composed of multiple physical volumes (PVs), you can use the optional specification of pvname to select which PV provides the storage.

An example of the use of lvextend can be found in the description of the resize2fs command, which enables you to change the size of ext file systems.

lvm [command]

lvm is the central administration command for the Logical Volume Manager (LVM). lvm is usually located in the lvm2 package and must be installed separately for many distributions if an LVM system hasn't already been set up during installation.

If you run lvm without any other commands, you'll be taken to a shell in which you can execute LVM commands interactively. help provides a list of all commands available for selection. Alternatively, you can pass a command to lvm, which is then executed immediately.

All LVM commands can be run either via lvm or directly. The lvm lvcreate xxx and lvcreate xxx commands are therefore equivalent.

For reasons of clarity, I've decided to describe the most important LVM commands separately in this book. Commands for editing logical volumes start with the initial letters lv, those for administrating physical volumes start with pv, and commands for managing volume groups start with vg.

lvreduce [options] lvname

lvreduce reduces the memory space of the specified LV device. If the LV contains a file system, this must *first* be reduced in size, or data will be lost! You can either enter the new total size in the -L size format or the desired change in the -L -size format.

lvremove [options] lvname

lvremove deletes the specified LV.

▶ -f

Suppresses system queries.

lvrename oldlvname newlvname

lvrename gives the LV a new name.

lvscan [options]

lvscan lists all LVs.

```
lz4 [options] [filename]
lzop [options] [filename]
```

The lz4 and lzop commands from the packages of the same name compress files. Although they achieve lower compression rates than gzip, bzip2, or xz, they are *much* faster (although lz4 is still a little faster than lzop). The syntax of lz4 and lzop is based on the gzip command.

lz4 or lzop are perfect if you want to save files in a space-saving way (e.g., for a backup) but don't want to invest too much CPU power. You'll mostly use the commands in pipe mode, as shown in the two tar examples later in this entry.

- ▶ -c or --stdout or --to-stdout
  Redirects the result to the standard output.

- ▶ -d, --decompress, or --uncompress
  Decompresses the specified file instead of compressing it.

- ▶ -*n*
  Controls the speed and quality of the compression. -1 is the fastest, and -9 delivers the smallest files with lzop. With lz4, -12 provides the best compression.

Examples

The first command compresses the file and returns file.lzo; the second command decompresses the file again:

```
user$ lzop file
user$ lzop -d file.lzw
```

The following two examples show the use of lzop in combination with tar:

```
user$ tar cf -c verz | lzop -c > backup.tar.lzo     (create archive)
user$ lzop -d < backup.tar.lzo | tar xf -           (unpack archive)
```

```
magick [options] imageold imagenew
```

magick from the ImageMagick package converts image files from one format to another. In this book, the most important commands and options are described in convert, which is the command name that's valid up to version 6. As of version 7, it's recommended to call the syntactically equivalent magick command instead of convert although convert still works with most distributions.

```
mail [options] mail@hostname [< message text]
```

You can use the mail command from the mailx or mailutils package (Debian, Ubuntu) to read local emails stored in /var/mail/name. However, the operation takes some getting used to, to say the least. If you're looking for a mail client for text mode, it's better to install mutt.

The strength of mail lies in the fact that you can easily transfer emails to a local mail server, either interactively or in scripts. You can simply run mail recipient@site interactively. You can now enter additional recipients (CC addresses), the subject and finally the message text, which you conclude by pressing [Ctrl]+[D]. This transfers the email to the local mail server, using loginname@hostname as the sender.

For a script-controlled mail dispatch, you'll need to familiarize yourself with the mail options, of which I'll only present the three most important ones here:

▶ -a "header"
Specifies an additional header line. This can be used to add a BCC address or to set the character set of the message (see the example next).

mail doesn't recognize the correct mail host name in many configurations and then composes the sender address incorrectly. This even applies to Debian and Ubuntu systems, where /etc/mailname would be very easy to evaluate. This issue can be solved by explicitly specifying the sender address with -a "From:name@myhost.com".

▶ -A file
Adds a file to the email.

▶ -s "subject"
Specifies the text for the subject line.

Example

In its simplest form, mail is completely uncomplicated to use:

```
user$ echo "message" | mail -s "subject" name@a-company.com
```

The following command sends the UTF-8 text from message.txt to the specified recipients:

```
user$ mail name1@company-abc.com,name1@company-abc.com \
      -s "Subject 1 2 3 äöü" \
      -a "BCC: blindcopy@company-abc.com" \
      -a "Content-Type: text/plain; charset=UTF-8" < message.txt
```

makepasswd [options]

makepasswd from the package of the same name generates a new, random password. The command is well suited if you need a password for a new account but don't want to stretch your own imagination.

On CentOS, Fedora, and RHEL, makepasswd isn't available. Instead, you can install the expect package that contains the mkpasswd command. On (open)SUSE, mkpasswd also isn't available. There, you can switch to pwgen.

▶ --chars *n*

Specifies the desired password length (10 characters by default).

▶ --count *n*

Returns *n* passwords (the default setting is only one).

▶ --strings *string*

Uses the characters contained in the string to compose the password. By default, the command only uses uppercase and lowercase letters (ASCII) and numbers, but no special characters.

Example

The following command returns a password with 12 characters:

```
user$ makepasswd --chars 12
TaQKD1VGYmax
```

man [group] [options] name

man displays online information on the specified command or file. The search can be restricted by specifying a group. Important groups are 1 (user commands), 5 (configuration files), and 8 (commands for system administration). In most distributions, the man texts are displayed by less; that is, the keyboard shortcuts apply for scrolling or searching in the help text, as summarized in the less command in Table 11.

▶ -a

Displays all man pages of the same name in sequence. Without this option, only the first of several files with the same name from different subject areas is usually displayed. Some Linux distributions use -a as the default setting.

▶ -f *keyword*

Displays the meaning of a keyword (a single-line text). With this option, man corresponds to the whatis *theme* command.

▶ -k *keyword*

Displays a list of all existing man texts in which the keyword occurs. However, no full text search is performed. Instead, only the keywords of each man text are analyzed. With this option, man corresponds to the apropos *thema* command.

md5sum files

md5sum calculates checksums for all specified files. *Message-Digest Algorithm 5* (MD5) checksums are often used to ensure that a file is unchanged after it has been transferred. Note that the MD5 algorithm is no longer considered secure. The shasum commands are cryptographically better.

mdadm [options]

The mdadm command from the package of the same name helps with the management of software RAID connections. The command is controlled by countless options. An option usually specifies the mode (e.g., --create or -C to set up a new RAID array). The other options control the details of the desired operation.

▶ *mddevice* --add *device*

Adds a partition to the network.

▶ --assemble *mddevice device1 device2 ... devicem*

Reassembles the *mddevice* network from the hard disk partitions *device1* to *devicem*. This only works if *device1* to *devicem* were previously initialized using --create as components of a RAID array.

▶ --assemble --scan *mddevice*

Composes the RAID partition *mddevice* from the hard disk partitions specified in mdadm.conf.

▶ --create *mddevice* --level=*n* --raid-devices=*m device1 device2 ... devicem*

Creates the new *mddevice* array for RAID level *n*, which consists of the *m* hard disk partitions *device1* to *devicem*.

▶ --detail *mddevice*

Provides detailed information about the RAID partition.

▶ --examine *device*

Provides detailed information about the specified hard disk partition, which must be part of a RAID array.

▶ --examine --scan

Provides a summary of all available RAID partitions in the syntax of mdadm.conf.

▶ *mddevice* --fail *device*

Marks the hard disk partition *device* as faulty and deactivates it.

- --grow *mddevice* --size=max

  Enlarges the network so that the underlying hard disk partitions are used in an optimal way. The command only makes sense in this form after enlarging the hard disk partitions.

- --grow *mddevice* --raid-devices=*n*

  Increases the number of hard disk partitions integrated into the array (only for RAID levels 1, 5, and 6).

- --monitor *device*

  Monitors the active RAID arrays. The command is usually started when the computer is booted and sends an email to the administrator (-m option) in the event of problems.

- --query *device*

  Tests whether the partition is a RAID partition or part of a RAID array.

- *mddevice* --remove *device*

  Removes the hard disk partition *device* from the RAID array. *device* must not be active, so it may have to be deactivated beforehand by using --fail.

- --stop *mddevice*

  Deactivates the specified network. The array can now be reassembled using --assemble.

- --zero-superblock *device*

  Deletes the RAID metadata of a hard disk partition.

Examples

The following two commands create a RAID-0 array from the two partitions /dev/sda3 and /dev/sdb3 and set up an ext4 file system there. The partitions must first be marked as RAID partitions. If you use fdisk for partitioning, you need to set the partition ID to the hexadecimal value fd using the T command. In parted, you must run set *partition number* raid on.

```
root# mdadm --create /dev/md0 --level=0 --raid-devices=2 /dev/sda3 /dev/sdb3
mdadm: array /dev/md0 started.
root# mkfs.ext4 /dev/md0
```

The procedure for setting up a RAID-1 array is exactly the same as for RAID-0. Only the command for setting up the RAID system looks a little different and now contains --level=1 instead of --level=0:

```
root# mdadm --create /dev/md0 --level=1 --raid-devices=2 /dev/sda3 /dev/sdb3
mdadm: array /dev/md0 started.
root# mkfs.ext4 /dev/md0
```

To test the functionality of a RAID-1 array, you should mark a partition as defective:

```
root# mdadm /dev/md0 --fail /dev/sdb3
```

You can continue to use the network; however, all changes are now only saved on the remaining hard disk partition. To add /dev/sdb3 back to /dev/md0, you must first explicitly remove the partition marked as defective:

```
root# mdadm --remove /dev/md0 /dev/sdb3
root# mdadm --add    /dev/md0 /dev/sdb3
```

Now the automatic resynchronization of the two partitions starts, which takes some time depending on the size of the network (approximately 20 minutes per 100 GiB for conventional SATA hard drives). After all, you can continue to work during this time. However, the file system will respond more slowly than usual. cat /proc/mdstat provides information on how far the synchronization process has progressed.

To save all changes made in the RAID configuration file /etc/mdadm/mdadm.conf, you need to run the following command. You must then use an editor to remove any previously entered RAID arrays from the configuration file so that the file is free of duplicates.

```
root# mdadm --examine --scan >> /etc/mdadm/mdadm.conf
```

mkdir directory

mkdir creates a new directory. The most important options are listed here:

▶ -m *mode* or --mode=*mode*
   Sets the access rights of the new directory as specified by *mode* (see chmod).

▶ -p or --parents
   Also creates intermediate directories. If you run mkdir -p a/b/c and the directories a and a/b don't yet exist, these directories will be created.

▶ -Z
   Activates the default context of SELinux for the new directory.

mkfifo file

mkfifo sets up a *First In, First Out* (FIFO) file. Basically, FIFO files work like pipes and enable data to be exchanged between two programs.

Example

In the following example, mkfifo sets up a FIFO file. ls redirects the table of contents to this file. more reads it from there. ls must be started with & as a background process because the ls process isn't finished until more has read all the data from fifo.

```
user$ mkfifo fifo
user$ ls -l > fifo &
user$ more < fifo
```

Displaying the table of contents via more could of course be done much more easily without a FIFO file, namely by using ls -l | more.

mkfs [options] device [blocks]

mkfs sets up a file system on a previously partitioned hard disk. mkfs can only be run by root. Depending on the specified file system, the program branches to the mkfs.*file system type* command.

▶ -t *file system type*
Specifies the type of file system. For example, ext4 and xfs are possible. The -t option must be specified as the first option! All other options are transferred to the command that actually sets up the file system. They depend on the type of file system.

The most important file system–specific mkfs.xxx commands are described in the next six entries.

mkfs.btrfs [options] device1 [device2 device3 ...]

mkfs.btrfs sets up a Btrfs file system. Depending on the distribution, the command is part of the btrfs-tools, btrfs-progs, or btrfsprogs package, which may have to be installed separately.

▶ -A or --alloc-start *n*
Leaves *n* bytes unused at the beginning of the device (by default, *n*=0).

▶ -d or --data *type*
Specifies the desired RAID type for the data. Permitted values are raid0, raid1, raid10, raid1c3, raid1c4, raid5, raid6, or single (applies by default). If you set up a Btrfs RAID array, you must specify a corresponding number of devices (two for RAID-1, four for RAID-10).

▶ -L or --label *name*
Gives the file system a name.

▶ -m or --data *type*
Specifies the desired RAID type for the metadata. In addition to the RAID variants specified for the -d option, there is also dup. This variant applies by default for file systems that only include one device. All metadata is therefore stored twice for security reasons. SSDs are an exception, where single is the default for single-device file systems. dup isn't useful here because the internal optimization of many SSDs

recognizes the redundancy of the data and simply stores it. This means that the security gain intended by the duplication is lost again.

If you distribute a Btrfs file system across multiple devices without specifying any further options, the RAID type raid1 applies by default. This means that the data is simply saved, but the metadata is saved multiple times!

When creating a RAID array, you usually specify the same RAID type using -m as with the -d option. If you want your file system to be as fast as possible, you can use -d raid0 -m raid0 to eliminate all redundancies.

Example

The following command can be used to set up a RAID 1 file system that uses the two equally sized partitions /dev/sdb1 and /dev/sdc1:

```
root# mkfs.btrfs -d raid1 -m raid1 /dev/sdb1 /dev/sdc1
```

If a device fails, you can continue to use the file system with the additional mount option, degraded:

```
root# mount -o degraded /dev/sdb1 /media/btrfs
```

To restore the RAID array, you should add a new device to the file system. In the following example, this is again /dev/sdc1, but this time, the partition comes from a new hard disk or SSD. To redistribute the file system across both devices and thus restore RAID-1 redundancy, you must also run btrfs filesystem balance. For large file systems, the execution of this command naturally takes a long time. After all, the file system can be used during this time, albeit at a greatly reduced speed.

```
root# btrfs device add /dev/sdc1 /media/btrfs
root# btrfs filesystem balance /media/btrfs
```

Only now can the defective device be removed from the file system. You can use the missing keyword to specify the device:

```
root# btrfs device delete missing /media/btrfs
```

mkfs.exfat [options] device

mkfs.exfat from the exfat-utils package sets up an exFAT file system. This file system is specially optimized for large SD cards. It's used by default with SDXC memory cards and is supported by SDXC-compatible digital cameras and smartphones.

▶ -i id

Determines a 32-bit ID for the file system. The number is given as a hexadecimal code. If the command is executed without the option, a random ID is set.

► -n *name*

Defines the name (label) for the file system. The name remains empty by default. If necessary, the name can be changed subsequently using exfatlabel.

```
mkfs.ext2 [options] device [blocks]
mkfs.ext3 [options] device [blocks]
mkfs.ext4 [options] device [blocks]
mke2fs [options] device [blocks]
```

mke2fs, mkfs.ext2, mkfs.ext3, and mkfs.ext4 set up an ext2, ext3, or ext4 file system. All four commands reference the same program via links, but they use different default options that are stored in /etc/mk2efs.conf.

► -b *n*

Determines the block size (usually 4,096 bytes). The values 1024, 2048, 4096, and up to a maximum of 65536 are permitted for *n* (i.e., 2^{10}, 2^{11}, etc. up to 2^{16}).

► -c

Tests whether defective blocks exist before setting up the data medium with badblocks. This type of blocks isn't used for the file system. The test requires that *each* data block be changed and read again, and therefore considerably extends the time required to set up the file system.

► -i *n*

Specifies the number of bytes after which an I-node is set up. I-nodes are internal management units of a file system. All administrative information of a file is stored in an I-node with the exception of the name, that is, access bits, owner, date of last access, and so on. The number of I-nodes is defined during formatting. For example, with *n=4096*, there are 1,048,576 ÷ 4,096 = 256 I-nodes per MiB; that is, a maximum of 256 files or directories can be stored per MiB, regardless of how small the files are.

If you want to save a large number of very small files, you should make *n* smaller. The minimum value is *1024* (this corresponds to 2,048 files per MiB). It doesn't make sense to select a value smaller than the block size, as only one file can be saved per block anyway. However, if you want to save only a few very large files, you can also select a larger *n*. This reduces the administrative overhead.

► -J

Allows you to pass additional parameters, for example, to set the size of the journaling file or to set up the journaling file in another device. This can possibly improve the speed of the file system a little, but at the same time makes administration and restoring the file system after a crash more complicated. The available options are described in detail in the manual page of mke2fs.

► -m *n*

Specifies what percentage of the disk should be reserved for root data (5% by default). This reserve memory enables root to continue working even if the file system is already completely full for all other users. This is therefore a safety reserve.

▶ -t ext2/ext3/ext4
Specifies the desired ext version (only required for mke2fs).

Example

The following two commands first set up a new ext4 file system in the /dev/sda5 partition and then deactivate the automatic file system check (see also the description of the tune2fs command):

```
root# mkfs.ext4 /dev/sda5
root# tune2fs -i 0 -c 0 /dev/sda5
```

mkfs.ntfs [options] device

mkfs.ntfs sets up an NTFS file system that is compatible with all current Windows versions. The command is part of the ntfsprogs package in many distributions and must be installed separately.

▶ -f or --fast
Doesn't fill each data block with 0 bytes. This option speeds up formatting many times over and is highly recommended.

mkfs.vfat [options] device

mkfs.vfat sets up a Windows file system that is compatible with all common Windows versions. This format is particularly suitable for USB flash drives and SD cards that are also used on Windows or in cameras.

▶ -F 12/16/32
Specifies the type of FAT tables (12-, 16-, or 32-bit). Usually, mkfs.vfat automatically selects the 32-bit mode for large data media. If not, this option will help.

▶ -I
Allows formatting the entire disk without creating a partition table (mkfs.vfat -I /dev/sdg). This method is sometimes used for USB flash drives.

▶ -n name
Assigns a name (label) to the file system with a maximum length of 11 characters.

Example

The following command sets up a 32-bit VFAT file system named MyPhotos in the /dev/sdd1 partition:

```
root# mkfs.vfat -F 32 -n MyPhotos /dev/sdd1
```

mkfs.xfs [options] device

mkfs.xfs sets up an XFS file system. This process can be controlled by countless options (man mkfs.xfs), the most important of which are listed here:

► -b size=*n*
Specifies the desired block size (4,096 bytes by default).

► -l logdev=*device*
Saves the journaling data in a separate partition.

► -l size=*n*
Specifies the size of the area for the journaling data. Without this option, mkfs.xfs chooses a suitable size itself.

► -L *label*
Gives the file system the desired name (maximum 12 characters).

mknod devicefile [b|c] major minor

mknod sets up a new device file. Device files are located in the /dev directory and enable access to various hardware components. Device files are characterized by three pieces of information: major and minor use two numbers to indicate the driver that can be used to access the devices (major and minor device number). The b or c characters indicate whether the device is buffered or unbuffered.

In all current Linux distributions, device files are set up automatically by the udev system. A manual execution of mknod is therefore not necessary. A list of the most important Linux devices and the corresponding device numbers can be found here:

www.kernel.org/doc/Documentation/admin-guide/devices.txt

mkpasswd [options]

mkpasswd from the expect package (CentOS, Fedora, RHEL) generates random passwords.

► -c *n*
Specifies the minimum number of lowercase letters that must be contained in the password (default: 2).

► -C *n*
Specifies the minimum number of capital letters that must be included (default: 2).

► -d *n*
Specifies the minimum number of digits that must be included (default: 2).

▶ -l *n*

Specifies the desired number of characters for the password (default: 9).

▶ -s *n*

Specifies the minimum number of special characters that must be included (default: 1).

Example

The following command generates a six-character password without special characters:

```
root# mkpasswd -l 6 -s 0
SX39vz
```

mkswap device/file

mkswap sets up a device (e.g., a hard disk partition) or a file as a swap area. mkswap can only be run by root. The swap area can then be activated using swapon. The swap area must be entered in the /etc/fstab file so that it's used automatically each time the computer is started.

Example

The following commands first fill a 512 MiB file with zeros, then make the file a swap file, and finally activate the file:

```
root# dd if=/dev/zero of=swapfile bs=1M count=512
root# chmod 600 swapfile
root# mkswap swapfile
root# swapon swapfile
```

modinfo module name

modinfo provides information about the specified module. This includes the full file name, author, license, dependent modules, short description, and list of all parameters the module knows.

modprobe [options] module name [parameter=value ...]

modprobe loads the specified module into the kernel. The command is an extended variant of insmod. It takes into account the module dependencies defined in /usr/lib/modules/kernel-version/modules.dep and, if necessary, also loads modules that are required by the desired module. In addition, it includes the module parameters specified in /etc/modprobe.d/*.

► -c

Provides an almost endless list of all currently valid module options and module settings. The parameters are derived both from the source code of the modutils package and from the settings in /etc/modprobe.d/*.

► -f

Forces the module to be loaded even if it was compiled for a different kernel version. Whether this actually works depends on whether there are any incompatibilities between the kernel version and the module version.

► -r

Removes the module from the kernel (instead of loading it).

---

mogrify [options] image file

mogrify changes parameters of an image file, for example, the resolution or the number of colors. It uses the same options as are used for convert. In contrast to convert, which creates a new file, mogrify overwrites the original file. The command is part of the ImageMagick package.

Example

The following command reduces the resolution of an image to 800 × 600 pixels:

```
user$ mogrify -resize 800x600 photo.jpg
```

---

more file

more displays the contents of a text file page by page. The display is interrupted after each page. more is now waiting for a keyboard input. The most important input options are ⎡Enter⎤ (one line down), ⎡ ⎤ (one page down), ⎡B⎤ (one page up), and ⎡Q⎤ (quit). Instead of more, the more powerful less command is usually used.

---

mount
mount [options] device directory

mount without parameters returns a list of all currently mounted file systems, including the file system type and the mount options. In the second syntax variant, mount integrates a data medium (partition of a hard disk, USB flash drive, etc.) into the Linux file system. You must specify the device name of the data medium (e.g., /dev/nvme0n1p3) and the target directory in the directory tree. Instead of the device, you can also specify the UUID of the file system (i.e., mount UIID=1234 directory).

The mounting of file systems when Linux is started is controlled by the /etc/fstab file. For all devices listed in /etc/fstab, mount can be used in a short form in which only the device file or the mount directory is specified. mount automatically reads the missing data and options from fstab.

Data media can be removed from the file system using umount. As a rule, both mount and umount can only be run by root. Drives that are marked in fstab with the option user or users are exceptions.

The following description of general and file system–specific mount options extends over several pages and is followed by a few examples.

General mount Options

▶ --move *olddir newdir*
Changes the directory in which a file system is mounted.

▶ -o options
Allows you to specify additional options, for example, in the format -o acl,user_ xattr. The options specified in this way must not be separated by spaces! See the next subsection for a reference of the file system options.

▶ -r
Prevents write operations on the data medium (*read-only*).

▶ -t *file system*
Specifies the file system. Possible options include ext2/3/4 and btrfs or xfs for Linux partitions, vfat and ntfs for Windows partitions, and iso9660 and udf for data CDs and DVDs.

General File System Options

Using -o, you can pass countless options that influence the behavior of the mounted file system. The keywords for general options are summarized in the following list. File system–specific options follow a little further down.

All options for mount -o are also permitted in the fourth column of /etc/fstab. This file is evaluated during the boot process; on Ubuntu systems, this is done by the mountall command. For this reason, the following compilation also applies to /etc/fstab and mountall and has therefore been supplemented by some fstab-specific and mountall-specific keywords that aren't directly relevant for the mount command. These include defaults, noauto, nobootwait, optional, owner, user, and users.

▶ atime / noatime / relatime / strictatime
Specifies the circumstances under which the *inode access time* of a file is updated:

with every read access (`atime` or `strictatime`, POSIX-compliant), at most once a day (`relatime`), or never (`noatime`). The inode access time always gets changed for write accesses.

The option has a relatively large influence on the speed of read operations and on the service life of flash data storage devices. The default setting for current kernel versions is `relatime`.

▶ `defaults`
Specifies that the file system is to be included in the directory tree with the default options. `defaults` is specified in the fourth column of `/etc/fstab` if no other options are to be used. (The column must not be left blank.)

▶ `dev` / `nodev`
Causes or prevents device files from being interpreted as such. On Linux, files can be marked as block or character devices. This option is useful for external data media for security reasons.

▶ `discard`
Activates the trim function for the ext4, Btrfs, and XFS file systems. The kernel now informs the SSD about which data blocks have been deleted, giving the SSD the opportunity to optimize the internal management of the memory cells. The trim function also works in combination with LVM and software RAID (`mdraid`).

Whether the `discard` option should be used is controversial: Although it helps with the long-term optimization of the SSD, it can lead to a significant slowdown in delete operations (e.g., if many files are to be deleted in a short time, see *https://forums. freebsd.org/threads/ssd-trim-maintenance.56951/#post-328912*). A good alternative to the `discard` option is a cron job that runs the `fstrim` command once a day or once a week and in this way reports all deleted sectors to the SSD.

▶ `exec` / `noexec`
Determines whether programs on the file system may be run on Linux or not. For most file systems, the `exec` setting applies by default. However, security-conscious administrators will specify the `noexec` option for CD/DVD drives and external data media. (If you use the `user` or `users` option, `noexec` applies. This can be changed again using an `exec` option.)

▶ `noauto`
Is only useful in `/etc/fstab` and ensures that the data medium named in this line *isn't* automatically mounted at system startup. Nevertheless, it makes sense to enter the data medium in `fstab` because users can then conveniently perform `mount name` without explicitly specifying all other `mount` options. `noauto` is used, for example, for rarely required data partitions.

▶ `nobootwait`
Doesn't stop the boot process if the device isn't available. The keyword is evaluated by `mountall` on Ubuntu and can't be used in other distributions. It's particularly

suitable for external data media that aren't always connected, for example, for a USB hard disk.

▶ nofail

Doesn't report an error if the device doesn't exist. This option is useful if you want mount to try to access a USB flash drive or a network directory in a script.

▶ optional

Ignores the fstab entry if the file system type isn't available during the boot process. The keyword is evaluated by mountall on Ubuntu and can't be used in other distributions.

▶ owner

Allows each user to mount or unmount the relevant file system themselves. The option is only useful in /etc/fstab. The difference from user is that mount or umount may only be run if the user has access rights to the relevant device file (e.g., /dev/fd0).

▶ remount

Changes the options of a file system already included in the directory tree. This option can only be specified for a direct call of mount and isn't permitted in /etc/fstab.

▶ ro / rw

Specifies that files may only be read or modified. Usually, rw (i.e., *read/write*) applies to most file system types.

▶ suid / nosuid

Allows or prevents the evaluation of the SID and GID access bit. These access bits allow ordinary users to execute programs with root permissions, which often is a safety risk. The nosuid option prevents the evaluation of such access bits and should be used for CDs, DVDs and external data media.

▶ sync

Causes changes to be saved immediately instead of temporarily storing them in RAM for a few seconds and only transferring them to the data medium later. sync minimizes the risk of data loss if you accidentally remove a data medium (e.g., USB flash drive) without using umount or disconnect the cable connection. However, the disadvantage of sync is that writing data is much less efficient depending on the data medium. The speed of cheap USB flash drives and traditional hard disks in particular is reduced by a factor of 10 or more!

▶ user

Allows normal users to unmount the relevant file system using umount and mount it again via mount. Otherwise, only root is able to do this. Data media may only be removed from the file system by the user who initiated mount. This option is only useful in /etc/fstab and is primarily intended for external data media. If you use user, the noexec, nosuid, and nodev options also apply automatically.

▶ users

Has the same meaning as user, but with one small difference: every user may remove data media marked with users from the file system (umount). users therefore allows user A to run mount and user B to call umount later.

btrfs Options

Many mount options depend on the respective file system. The most important file system–specific options are therefore summarized here. We start with btrfs.

▶ compress=zlib|lzo|zstd

Compresses all new or modified files. Existing files remain unchanged as long as they are read only. You have a choice of three compression methods:

– zlib: Good compression

– lzo: Faster than zlib, but worse compression

– zstd: Compression similar to zlib, but slightly faster

Fedora uses zstd by default. This method is currently (in 2025) considered the best compromise between good compression and speed.

▶ compress-force=zlib|lzo

Compresses files even if this doesn't look promising. Usually, the Btrfs file system recognizes files that have already been compressed and doesn't compress them again because this doesn't usually save any space, but costs a lot of CPU time. Using compress-force instead of compress deactivates this recognition.

▶ degraded

Integrates the file system even if some of the RAID devices are missing. This only works if all data is available due to RAID redundancies; that is, a degraded RAID array exists.

▶ noacl

Deactivates the ACL access management.

▶ nobarrier

Deactivates the *write barriers*. A write barrier blocks further write operations until all previous write operations have been physically completed. Without barriers, the file system is faster, but at the same time, the risk of data loss in the event of a crash increases, especially with file systems that are distributed across multiple devices (RAID). Barriers only work if the hard disk actually follows the write request, which is often not the case for performance reasons.

▶ ssd

Optimizes file operations for solid-state drives (SSDs). The option is automatically active if the Btrfs driver can determine that the data medium is an SSD. However, this option has no influence on the trim behavior, which is controlled separately by the discard option.

▶ subvolume=*name* or subvolid=*n*
Uses the specified subvolume (and not the default subvolume). You can determine the volume ID for subvolid using btrfs subvolume list.

CIFS Options

The *Common Internet File System* (CIFS) allows you to integrate Windows network directories into the file system. CIFS replaces SMBFS. To use CIFS, some distributions require you to install a separate package, for example, cifs-utils or smbfs.

Regarding access rights, if the files originate from an ordinary Windows PC, they are assigned to the root user on Linux. All users may read the files, but only root can change them. To give an ordinary user access to the files, you must correctly set the uid, gid, file_mode, and dir_mode options.

However, if the files originate from a file server that supports the CIFS Unix extension (this is the case if Samba is running on a Linux computer), then the UIDs, GIDs, and access bits of the files are forwarded unchanged. The uid, gid, file_mode, and dir_mode options then have no effect unless you deactivate the Unix extensions via the nounix option.

Further details on the CIFS-specific mount options can be found in man mount.cifs.

▶ credentials=*file*
Specifies a file containing the username and password for logging in to the Windows or Samba server. This avoids having to enter passwords in /etc/fstab. The file contains three lines with username=xxx (not user=xxx!), password=xxx, and, if required, domain=xxx.

▶ *dir_mode=n* (dmask for SMBFS)
Specifies which access rights apply for access to directories. dir_mode=0770 gives the user specified by uid and all members of the gid group unrestricted rights (read, open, and change directory).

▶ domain=*workgroup*
Specifies the domain or workgroup.

▶ *file_mode=n* (fmask for SMBFS)
Specifies which access rights apply for file access. *n* must begin with a zero and contains the rights in the octal notation familiar from chmod. file_mode=0444 gives all users read and write permissions. The setting is ignored if the file server supports the CIFS Unix extension.

- ▶ iocharset=*name*
  Specifies the desired character set. If special characters or other non-ASCII characters are displayed incorrectly in file names, iocharset=utf8 usually helps.

- ▶ nodfs
  Deactivates the *distributed file system* (DFS) functions, which allow standardized access to the network directories of different servers. Some old NAS hard disks also use old Samba versions with known DFS errors, which lead to problems especially when the rsync command is used. The nodfs option provides a solution in such cases.

- ▶ nounix
  Deactivates the Unix compatibility mode. At this point, a little background information is essential to understand this option. The CIFS standard has been extended by the Unix extensions to achieve better compatibility with Unix/Linux systems. On some NAS devices, these extensions are also active. In this case, mount takes the user and group information and access rights from the NAS device and ignores the uid, gid, dir_mode, and file_mode options.

  This sounds useful at first glance, but it's not in most cases: The Unix extensions only make sense if the UIDs and GIDs of the local computer and the Samba server have been matched. This is rarely the case, and practically never with NAS devices. For this reason, the compatibility mode must be deactivated so that uid, gid, dir_mode, and file_mode remain effective.

  Sometimes, you can also avoid incompatibilities with nounix, which is expressed in the vague error message, *Operation not supported.*

- ▶ password=*pw*
  Specifies the password for authentication with the Windows or Samba server. Entering the password directly isn't secure! It's better to specify the password in a file and reference it using the credentials keyword.

- ▶ sec=*mode*
  Specifies the authentication mode. This mode controls the authentication with the Samba server. The default setting is ntlmssp. This keyword refers to a Windows-specific procedure with password hashing, whereby the data is packed into an NTLMSSP message. Alternative settings such as ntlmssp or krb5i (Kerberos) are described in man mount.cifs.

- ▶ uid=*u*,gid=*g*
  Specifies the owner and group of the files, as with the VFAT driver. The usernames or group names can also be transferred to the options instead of the UID/GIDs.

- ▶ user=*name* (username for SMBFS)
  Specifies the username for authentication with the Windows or Samba server. The workgroup name and password can also be transferred to the option (user=*work-group/name%password*).

▶ vers=*n*.*n*

Specifies which version of the SMB protocol is to be used. Current mount versions use version 3 by default because older versions have known security problems. To communicate with an old server, you must explicitly specify the desired version, for example, using vers=2.1.

exFAT Options

The exFAT file system (-t exfat) supports the dmask, fmask, gid, noatime, ro (read-only), uid, and umask options. The options have the same effect as the VFAT options of the same name, which are described a few pages further on. Only noatime isn't available for VFAT file systems. This option prevents the access time information from being updated during read accesses, thus reducing the number of write operations.

ext3/ext4 Options

The following options apply to the Linux standard file system ext3 or ext4:

▶ acl

Activates the *access control list* (ACL) access management. This allows the storage of additional access information in the form of *access control lists*. In current kernel versions, acl is automatically active.

▶ barrier=0(only for ext4)

Deactivates the *write barriers*, which ensure that changes are written to the hard disk in the correct order. This increases the speed of the file system but is only safe if the hard disk cache is protected against power failures and crashes by a battery.

▶ commit=*n*

Synchronizes the journal every *n* seconds (every five seconds by default). The laptop-mode package increases this period for notebooks in battery mode to save energy. For ext3 with data=ordered, all file changes are synchronized at the same time as the journal. For ext3 with data=writeback and ext4 with *delayed allocation*, however, the commit interval only applies to the journal.

▶ data=journal/ordered/writeback

Determines the journaling mode (details will follow shortly). By default, the ordered mode applies for ext4.

▶ errors=continue/remount-ro/panic

Determines the behavior if an error occurs during the file system check. You can set the default behavior using the tune2fs command with the -e option.

▶ grpid

Causes new files to automatically adopt the group ID of the directory in which they were created. This makes the file system behave like it's under BSD or the setgid bit is set in all directories. (By default, Linux assigns new files the group ID of the user that created the file.)

► nodelalloc (only for ext4)
Deactivates the *delayed allocation*.

► noload
Ignores the existing journaling file for the mount command. This can be useful if the journaling file is defective.

► sb=*n*
Uses block *n* as a superblock (instead of using block 1 by default). In some cases, this makes it possible to read a damaged file system. Usually, mke2fs creates a copy of the superblock every 8,192 blocks. It's therefore advisable to use the values 8193], 16385, and so on for *n*.

► user_xattr
Activates support for extended file attributes. (This option is the default for current kernel versions.)

ext3/ext4 Journaling Modes

The ext file system has three methods for performing journaling:

► data=ordered
In this mode, only metadata is saved in the journal, that is, *information about* files, but no content. Files are only marked as *committed* in the journal when they have been saved completely. After a crash, the file system can be quickly restored to a consistent state because incompletely saved files are immediately recognized using the journal. However, it's impossible to restore such files.

In the data=ordered mode, the journal is synchronized with the hard disk every five seconds. With ext3, this means that all changes to any files are physically saved to the hard disk within five seconds. Although this standard behavior isn't particularly efficient, it's very secure: even in the event of total crashes and power failures, massive data losses are extremely rare. data=ordered has an unpleasant side effect with ext3, however: each time the fsync function is called, not just a specific file but the entire file system is synchronized. This can result in noticeable delays.

With ext4, the journal is also synchronized every five seconds, but the actual data changes are often not saved until much later due to *delayed allocation*. Only an explicit call of the fsync function ensures the immediate physical storage of a file! Fortunately, fsync for ext4 doesn't require the entire file system to be synchronized, so it can run much faster.

► data=writeback
This mode is similar to the ordered mode. The only difference is that the journal and file operations aren't always fully synchronized. The file system doesn't wait with

the *committed* entries in the journal for the storage operation on the hard disk to be completed. This makes the file system slightly faster than in ordered mode. After a crash, the integrity of the file system is still ensured. However, changed files can contain old data. This problem doesn't occur if application programs—as provided for in the POSIX standard—complete the save process using fsync (see the fsync entry).

▶ data=journal
Unlike in the other two modes, the actual data is now also saved in the journal. This means that all changes must be saved *twice* (first in the journal and then in the relevant file). This is why ext3 is significantly slower in this mode. Files whose changes have already been fully entered in the journal (but not yet in the file) can be restored after a crash.

The journal is physically saved to the hard disk every five seconds. This time span can be changed using the mount option; commit. Internally, the kjournald journaling daemon integrated into the kernel takes care of regularly updating the journaling file.

NFS Options

The NFS file system (-t nfs) is the usual way to share or use network directories on Unix/Linux. Note that -t nfs refers to NFS2 and NFS3. If you want to use NFS4, you must specify -t nfs4 in the mount command.

Most NFS options apply equally to all NFS versions. A comprehensive summary of all options, including a differentiation between NFS2/3 and NFS4, can be found in man nfs.

▶ bg
Causes the mount process to continue in the background if the NFS server isn't accessible. This option is only relevant for /etc/fstab.

▶ hard
Causes a program accessing an NFS file to hang if the NFS server is no longer available. (The alternative to hard is soft, which causes the kernel to give up trying to find the NFS file again after a while. This may sound safer, but in reality, it causes even more problems than hard.)

▶ intr
Makes it possible to stop a program using kill or Ctrl+C even if an open NFS file is no longer available. The option is only valid in combination with hard.

▶ nfsvers=2/3
Specifies the desired NFS version (only for -t nfs).

▶ rsize=*n*,wsize-*n*
Specifies the size of the buffers for read and write operations (in KiB). The default value is 4096 in each case.

235

The right mount options are decisive for the speed that can be achieved with NFS! You can achieve a high speed with the these options: hard, intr, rsize=8192, and wsize=8192. If the NFS speed is still below your expectations, you should ensure that /etc/exports on the NFS server uses the async option.

NTFS Options

The following options apply to the ntfs3g driver, which is used by most distributions to access NTFS file systems

▶ locale=*name*
Specifies the local character set (e.g., in the format locale=en_EN.UTF-8). The option is only necessary in /etc/fstab if the character set isn't yet configured when the file systems are automatically mounted by the Init-V process.

▶ show_sys_files
Also displays NTFS system files.

▶ streams_interface=none|windows|xattr
Specifies how file streams are supposed to be accessed: In the default xattr setting, access occurs via the getfattr or setfattr commands. windows allows access in the usual Windows notation: filename:streamname. The default setting is none.

▶ uid=*u*,gid=*g*,umask=*m*|fmask=*f*,dmask=*d*
Have the same meaning as for the VFAT driver (see the next subsection). The file owner is the user who executed mount. All users have read and write access!

VFAT Options

You can use mount -t vfat to mount Windows VFAT partitions in all variants (FAT12, FAT16, FAT32) in the directory tree. Numerous options control who has which kind of access rights to the files, how the different character sets in the file names are processed, and more.

▶ codepage=*name*
Specifies the code page (a kind of DOS character set) that applies internally to VFAT for short file names (maximum 8 + 3 characters). Long file names are always saved in Unicode format; this option is irrelevant for them. Current Windows versions also save short file names in VFAT format for long file names to maintain the correct uppercase and lowercase notation. For this reason, the option is becoming increasingly less important. The default value of the option can be determined during compilation and is usually cp437.

▶ flush
Causes the driver to immediately start saving cached data to the data medium. This option minimizes the time required to save changed data on external data media

(e.g., a USB flash drive). In contrast to the sync option, the file system doesn't wait for the synchronization to be completed. For this reason, flush has no negative influence on the speed of write operations.

▶ fmask=*f*,dmask=*d*

sets the mask for files (fmask) and directories (dmask) separately. This is particularly useful because you often want to set the x access bit for directories but not for files. This allows the user to go to all directories, but not to run any files or programs.

One possible setting is fmask=177,dmask=077, which allows the owner (UID) to read and write files (rw-------) and change to directories (rwx------). All other users have no access to the files.

fmask=133,dmask=022 is often useful as well. This means that rw-r--r-- applies to files, and rwxr-xr-x applies to directories; that is, anyone can read everything, but only the owner can change anything.

▶ iocharset

Specifies the character set with which Windows file names will be processed on Linux systems. The default setting is iso8859-1. As a rule, this option is only used if you work with an 8-bit character set on Linux. (Within VFAT, the Unicode character set is always used for long file names. The option only determines how the file names are displayed on Linux.)

If you use the UTF-8 character set on Linux, you should use the utf8 option described at the end of this list instead of iocharset. iocharset=utf8 also works at first glance but isn't recommended because the VFAT driver is then case sensitive, which is contrary to Windows conventions.

▶ shortname=lower|win95|winnt|mixed

Specifies in which uppercase and lowercase short file names (maximum 8 + 3 characters) are presented or saved. The behavior in this regard varies on Windows depending on the version.

On Linux, the lower setting applies by default, that is, short file names are presented in lowercase. For new files, the name is saved in VFAT format for long file names (even if the DOS limit of 8 + 3 characters isn't exceeded), so that uppercase and lowercase is retained. The only exceptions to this rule are short file names that consist exclusively of uppercase letters; such names are saved as short file names as is typical for DOS.

The setting has no effect on long file names, which are always case-sensitive once the file has been created.

▶ uid=*u*,gid-*g*,umask=*m*

Determines who (uid) and which group (gid) "owns" the Windows files, that is, who is allowed to read or change the files. You usually enter the desired user or group number via uid and gid. You can pass an octal numerical value with the bit mask of the *inverted* access bits to umask. Thus, umask=0 means that anyone can read, write, and run all files (rwxrwxrwx). umask=022 corresponds to rwxr-xr-x: everyone may read all files, but only the owner may change files.

If these options aren't specified, the following default values apply for the VFAT file system:

- uid: UID of the user running mount. Normally, only root may run the mount command, unless the corresponding line in /etc/fstab contains the user, users, or owner options.

- gid: GID of the user running mount.

- umask: 022, that is, everyone may read everything, but only the owner may change files or directories.

It often happens that fstab contains the setting gid=users. This means that the default group of the user running mount applies to gid. Without this option, it's possible to first activate another group via newgrp, which then also applies to mount.

▶ utf8
Causes the driver to return the file names to Linux as UTF-8 character strings. This option is required for the correct display of non-ASCII characters if you use UTF-8 as the character set on Linux.

XFS Options

Most of the options for the XFS file system concern rarely used functions (see man mount). ACLs and *extended attributes* are available by default on the XFS file system and aren't controlled by options.

▶ logdev=*device*
Specifies the partition in which the journaling data is stored. This option is only necessary if an external journaling partition was specified when the file system was set up.

▶ norecovery
Prevents the journaling data from being evaluated when the file system is mounted. At the same time, you must use the ro option (*read-only access*).

Examples

As a rule, you can integrate Linux partitions into the file system without explicitly specifying the file system (i.e., without the -t option):

```
root# mkdir /windows
root# mount /dev/sda7 /media/backup-partition
```

The following example shows access to the data on a USB flash drive via the /windows directory:

```
root# mkdir /usbstick
root# mount -t vfat /dev/sdc1 /usbstick
```

More options are required so that not only root but also an ordinary user with the user ID 1000 and group ID 1000 has access to the USB flash drive and can write or change files:

```
root# mount -t vfat -o uid=1000,gid=1000,fmask=0022,dmask=0022 \
      /dev/sdc1 /usbstick
```

The following command integrates a network directory of a Windows or Samba server, for example, from a NAS device, into the local directory tree. The user with the UID/GID 1000 can read and change the data.

```
root# mount -t cifs -o username=name,uid=1000,gid=1000,iocharset=utf8,nounix \
      //diskstation/myshare /media/audio-archive
Password for name: ********
```

You can use mount -o remount to change options of a previously mounted file system. For example, the following command activates the noexec option for a USB flash drive and thus prevents programs contained on it from being executed:

```
root# mount /dev/sdc1 -o remount,noexec
```

If problems occur when mounting the system partition during the computer startup, the partition is only mounted read-only. However, in order to fix the cause of the error—such as an incorrect entry in /etc/fstab—it's often necessary to make changes to the file system. To do this, you can run the following command. It remounts the system partition, whereby write access is then also possible.

```
root# mount -o remount,rw /
```

mtr [options] host name

The mtr command periodically sends network packets to the specified host and analyzes the responses. The result list combines data from ping and traceroute[6]. Note that there are two versions of this program: the text command described here and a GTK version with a graphical user interface. Desktop installations of Debian and Ubuntu have the GTK version installed by default. To install the text version instead, you must run apt-get install mtr-tiny.

▶ -c *n*

Performs *n* tests. (By default, the program runs until it's terminated via Ctrl+C).

▶ -n or --no-dns
Omits name resolution and specifies all intermediate stages as IP addresses.

▶ -r or --report
Provides a report in text format after the test has been completed. This option must be combined with -c.

Example

The following lines show a typical mtr result where the connection between a local internet WiFi router and *www.google.com* was tested:

```
user$ mtr -c 10 -r google.com
```

```
HOST: fedora.fritz.box            Loss%   Snt   Last   Avg  Best  Wrst StDev
  1.|-- fritz.box                  0.0%   10    1.4   1.1   0.9   1.5   0.2
  2.|-- dynamic-pd01.res.v6.highw  0.0%   10   37.2  36.0  20.5  42.7   5.8
  3.|-- lg206-9080-3219.as8447.a1  0.0%   10   38.5  40.9  30.0  47.8   5.0
  4.|-- lg23-9080.as8447.a1.net    0.0%   10   41.6  40.0  35.3  45.8   3.8
  5.|-- lg4-9080.as8447.a1.net     0.0%   10   48.4  48.2  46.5  50.0   1.2
  6.|-- 2001:4860:1:1::64e         0.0%   10   49.3  48.9  37.0  56.6   4.9
  7.|-- 2001:4860:0:1::88ab        0.0%   10   47.2  52.4  46.8  58.8   4.6
  8.|-- 2001:4860:0:1::8886        0.0%   10   46.8  49.2  42.0  57.7   4.7
  9.|-- 2001:4860::c:4003:364e     0.0%   10   47.5  47.6  42.5  49.1   2.0
 10.|-- 2001:4860::9:4002:6cfc     0.0%   10   51.5  49.3  26.8  53.6   8.1
 11.|-- 2001:4860:0:1::7651        0.0%   10   53.9  52.5  23.8  60.7  10.4
 12.|-- 2001:4860:0:1::2ea5        0.0%   10   51.7  50.8  22.9  61.9  10.4
 13.|-- bud02s41-in-x0e.1e100.net  0.0%   10   52.4  49.3  22.9  54.9   9.3
```

```
multitail [options] file1 [file2 file3 ...]
```

multitail helps to monitor multiple logging files at the same time. In the simplest form, you should run multitail filei1 filei2 filei3 .... This divides the terminal into multiple areas of equal size, each showing the last lines of the specified files. If you prefix the files with -I, changes to multiple files are output together in one area. However, it's then difficult to see which file has changed.

```
mv source target
mv files target directory
```

mv renames a file or directory or moves (one or more) files to another directory.

▶ -b or --backup
Renames existing files with the same name to backup files (name plus ~ character) instead of overwriting them.

▶ -i or --interactive
Asks before existing files are overwritten.

Example

The following command moves all PDF files in the current directory to the pdf subdirectory:

```
user$ mv *.pdf pdf/
```

mv can't be used to rename multiple files. mv *.xxx *.yyy doesn't work. To perform such operations, you must use for or sed. Corresponding examples can be found in cp.

---

mysql [options] [database name] [< name.sql]

The command interpreter mysql executes SQL commands on a MySQL or MariaDB server, either interactively or in batch mode from the standard input. The commands must end with a semicolon. In the batch version, mysql is often used to import backups that were created using mysqldump. If you don't specify a database name when calling mysql, you must set the desired database via the USE command.

The following options only affect the connection setup and also apply in the same form to the mysqladmin and mysqldump commands described in upcoming entries. In real life, only the -u, -p, and -h options are usually required.

▶ --default-character-set=*name*
Specifies which character set is to be used for the communication with the MySQL or MariaDB server (e.g., latin1 or utf8).

▶ --defaults-extra-file=*filename*
Loads connection options from the specified file. This is useful, for example, if mysql or mysqldump is to be run regularly by cron and interactive password entry is impossible.

▶ -h or --host=*hostname*
Specifies the name or IP number of the computer on which the server is running (localhost by default).

▶ --login-path=*group*
Evaluates the data from [group] in .mylogin.cnf.

▶ -P or --port=*n*
Specifies the port number for the TCP/IP connection (3306 by default).

▶ `-p` or `--password`
Prompts you to enter the password immediately after starting the command. If the option is missing, `mysql` reads the password from the local configuration file `.my.cnf`. If this file doesn't exist or doesn't contain any password data, `mysql` attempts to log in without a password.

▶ `-pxxx` or `--password=xxx`
Transfers the password directly. However, this isn't secure because the password appears in plain text in the process list! Unlike other options, the `-p` option doesn't allow you to specify a space.

▶ `--protocol=tcp/socket`
Specifies which protocol is to be used for the communication between the MySQL client and the database server. By default, `socket` applies if client and server are running on the same computer; otherwise, it's `tcp`.

▶ `-S` or `--socket=name`
Specifies the location of the socket file. This option is only required if `mysql` can't find the socket file on its own. In most distributions, the file has the name `/var/run/mysqld/mysqld.sock`. Its location can be set in configuration file `/etc/my.cnf` or `/etc/mysql/my.cnf`.

▶ `-u` or `--user=name`
Specifies the MySQL or MariaDB username (by default, that's the current username).

There are also some `mysql`-specific options:

▶ `-B` or `--batch`
Separates the columns with tab characters (instead of spaces and line graphics) when outputting the results. In addition, only the results of queries are displayed, but no status information.

▶ `-e` or `--execute='sql command'`
Runs the specified command. You can also enter multiple commands separated by semicolons.

▶ `-h` or `--html`
Formats the query result as HTML code.

▶ `-N` or `--skip-column-names`
Doesn't label the result columns.

▶ `-r` or `--raw`
Outputs the O, tab, newline, and \ characters unchanged in query results. (Usually, these characters are output as \0, \t, \n, and \\.) The option is only effective in combination with `--batch`.

▶ `-s` or `--silent`
Outputs only the result itself.

▶ -x or --xml

Formats the query result as XML code.

Example

The following command imports a database backup on the local MySQL or MariaDB server. The dbname database must already exist.

```
user$ mysql -u name -p dbname < backup.sql
Enter password: ********
```

If the backup file was compressed, you should proceed as follows:

```
user$ gunzip -c backup.sql.gz | mysql -u name -p dbname
```

The following lines show a small script that passes an SQL command to mysql to test whether a database exists or not:

```
#!/bin/bash
dbname="ebooks"
sql="     SELECT schema_name FROM information_schema.schemas "
sql="$sql WHERE schema_name='$dbname';"
result=$(mysql -s -N -e "$sql")
if [[ -n "$result" ]]; then
  echo "Database exists"
else
  echo "unknown database"
fi
```

Alternatives

mysqlsh and mariadb-shell are new shells that have been specifically optimized for MySQL and MariaDB and provide various additional functions. However, the new shells use different options and aren't compatible with the mysql command.

---

mysqladmin [options] command1 command2 ...

mysqladmin helps with various administrative tasks on a MySQL or MariaDB server, such as creating new databases. You can pass multiple commands to mysqladmin, which are then executed in sequence. The names of mysqladmin commands can be abbreviated as long as they are still clearly recognizable (e.g., flush-l instead of flush-logs). Most mysqladmin commands can also be run as SQL commands, such as CREATE DATABASE, DROP DATABASE, FLUSH, and more.

To enable `mysqladmin` to establish a connection to the database server, the connection options described in the `mysql` command must be used, that is, -u, -p, -h, and so on. In addition, `mysqladmin` knows a few more options:

▶ `-f` or `--force`
Runs the command without further inquiry. The execution of multiple commands is continued even if an error occurs with one command.

▶ `-i` or `--sleep=`*n*
Repeats the command every *n* seconds (e.g., for regular status display). `mysqladmin` then runs endlessly and must be terminated via [Ctrl]+[C].

▶ `-r` or `--relative`
In combination with `-i` and the `extended-status` command, shows the change compared to the previous status.

The most important `mysqladmin` commands are described here:

▶ `create` *dbname*
Creates a new database.

▶ `drop` *dbname*
Deletes an existing database irrevocably.

▶ `extended-status`
Shows countless status variables of the server.

▶ `flush-logs`
Closes all logging files and opens them again.

▶ `flush-privileges`
Reloads the database with the relevant access rights.

▶ `kill` *id1, id2 ...*
Closes the specified threads.

▶ `ping`
Tests whether a connection to the database server can be established.

▶ `shutdown`
Terminates the MySQL or MariaDB server.

▶ `status`
Displays various status variables of the database server.

▶ `variables`
Provides a list of the system variables of the database server.

▶ `version`
Determines the version of the MySQL or MariaDB server.

Example

The following command creates a new database on the local MySQL or MariaDB server:

```
user$ mysqladmin -u name -p create newdb
Enter password: ********
```

---

mysqlbinlog [options] loggingfile1 file2 ...

MySQL or MariaDB can log all changes, in particular all INSERT, UPDATE, and DELETE commands, in binary files. These files are used as an incremental update if necessary. mysqlbinlog reads these logging files. The same options apply for establishing the connection as in mysql(-u, -p, etc.).

▶ --database=*name*
Takes only commands into account that affect a specific database.

▶ -j *n*
Evaluates only the first logging file from a certain position.

▶ --no-defaults
Doesn't take configuration files such as .my.cnf into account. The option is required relatively often in practice if .my.cnf contains an option that applies to all MySQL client commands but isn't permitted for mysqlbinlog (e.g., default-character-set=utf8). You'll then avoid the error message *unknown variable xy*.

Example

After importing the last complete backup, the incremental updates from two binlog files are applied:

```
root# cd /var/log/mysql
root# mysqlbinlog -j 120 mariadb-bin.001380 mariadb-bin.001380 | \
      mysql -u root -p
```

---

mysqldump [options] dbname [table1 table2 ...]
mysqldump [options] --databases [dbname1 dbname2 ...]
mysqldump [options] --all-databases

mysqldump is a simple backup tool for the MySQL and MariaDB database systems. The command returns a long list of all SQL commands from which the database can be restored exactly. mysqldump knows three syntax variants, depending on whether one database, several enumerated databases, or all databases managed by MySQL will be saved. Only the first variant allows for the backup to be limited to individual tables.

For `mysqldump` to establish a connection to the database server, the connection options described for the `mysql` command must be used, that is, `-u`, `-p`, `-h`, and so on. A number of other options control the details of the backup.

▶ `--add-locks`
Inserts `LOCK TABLE` before the first `INSERT` command and `UNLOCK` after the last `INSERT` command. This speeds up the reloading of MyISAM tables. However, this option isn't suitable for InnoDB tables.

▶ `--create-options`
Includes MySQL-specific options in the `CREATE-TABLE` command.

▶ `--disable-keys`
Deactivates the index update during the reloading of the data. The indexes aren't updated until the end of the process, which is faster.

▶ `--extended-insert`
Creates longer `INSERT` commands with which several data records are inserted at the same time.

▶ `--lock-all-tables`
Runs `LOCK TABLE READ` for the entire database. This ensures that no table can be changed during the entire backup. This option is only useful for MyISAM tables. For InnoDB tables, it's better to use `--single-transaction`.

▶ `--lock-tables`
Runs `LOCK TABLE READ` for each individual table during the backup before the data is read.

▶ `--no-create-info`
Saves only the data, but not the structure of the database. In the resulting backup, there are `INSERT` commands, but no `CREATE TABLE`.

▶ `--no-data`
Saves only the structure (the schema) of the database, but not the data it contains.

▶ `--no-tablespaces`
Prevents `mysqldump` from including the SQL commands `CREATE LOGFILE GROUP` or `CREATE TABLESPACE` in the backup. This option must be used with current MySQL server versions if you run the command in an account without the `Process` privilege. Otherwise, an *Access denied* error will occur.

▶ `--opt`
Serves as the short notation for the following options: `--add-drop-table`, `--add-locks`, `--create-options`, `--disable-keys`, `--extended-insert`, `--lock-tables`, `--quick`, and `--set-charset`. For historical reasons, `--opt` is active by default. However, the resulting settings are only optimal (hence, the name of this option) if you perform a backup of MyISAM tables.

▶ `--quick`
Outputs the data records to be saved directly without intermediate storage in RAM.

▶ `--routines`
Also saves the code of stored procedures.

▶ `--set-charset`
Causes the active character set to be changed at the beginning of the `mysqldump` output and the previous character set to be restored at the end.

▶ `--single-transaction`
Ensures that the entire backup is performed as part of a single transaction. This is useful for InnoDB tables and prevents individual tables from being changed during the backup. This ensures that the backup is consistent overall.

▶ `--skip-opt`
Deactivates the `--opt` option, which is active by default.

▶ `--triggers`
Also saves the code of triggers.

Examples

Only a few options are required to back up a database with MyISAM tables. If you want to prevent individual tables from being changed during the backup, you need to add the `--lock-all-tables` option to the command. However, note that this blocks all write operations in the database during the backup.

```
user$ mysqldump -u user -p dbname > backup.sql
Enter password: ********
```

If you want to save a compressed backup, you should proceed as follows:

```
user$ mysqldump -u user -p dbname | gzip -c > backup.sql.gz
```

The optimal execution of a backup with InnoDB tables requires considerably more options because the default settings of `mysqldump` aren't optimized for this case:

```
user$ mysqldump -u user -p --skip-opt --single-transaction \
      --create-options --quick --extended-insert \
      --disable-keys --add-drop-table dbname > backup.sql
```

If you run the backup automatically using a cron job, you must replace `-u user -p` with the following option and enter the connection parameters in `root`-only file `/root/.my.cnf`:

```
# in the backup script, --defaults-extra-file must be the first option
mysqldump --defaults-extra-file=/root/.my.cnf ...
```

To restore the backup later, you must run the following commands:

```
user$ mysqladmin  -u user -p create dbname
user$ mysql  -u user -p --default-character-set=utf8 dbname < backup.sql
```

namei [options] file

namei shows the access rights of all directories that form the path of the specified file:

```
user$ namei -l ~/.ssh
f: /home/kofler/.ssh
drwxr-xr-x root    root   /
drwxr-xr-x root    root   home
drwxr-xr-x kofler  kofler kofler
drwx------ kofler  kofler .ssh
```

nc [options] [hostname/ip-address] [port]

The nc command is better known as *Netcat*. It redirects TCP/UDP ports to standard input/output and provides a number of other functions. You can use the command to interactively test network protocols such as HTTP or SMTP in a way similar to telnet. With nc, you can also copy files via any network port, implement a chat, or set up a simple backdoor that receives and executes commands via a port. It's therefore not surprising that nc is extremely popular among hackers and penetration testers.

On some Linux distributions, the nc command is included in the package of the same name; on other distributions, you must install netcat. Note that there are different implementations of Netcat. For example, netcat-traditional is used on Debian and Ubuntu, while RHEL provides a variant from the nmap developers (*https://nmap.org/ncat*). In practice, this doesn't result in any major differences, but individual options may be implemented differently (or not at all) depending on the version.

▶ -4 or -6
Uses only IPv4 or IPv6.

▶ -l
Waits on the specified port for a connection to be established (*listen*).

▶ -p *portnr*
Defines the local port (source port). The port, which is usually specified at the end of the nc command, is the destination port.

▶ -x *proxyadr:portnr*
Uses the specified proxy address and the corresponding port.

A number of other options are described in man nc. A possible alternative to nc is the socat command, which isn't covered in this book. It also supports the SCTP, can work via proxy servers, operate serial interfaces, and encrypt the data for transmission:

www.dest-unreach.org/socat

Examples

To copy a file via any port (here, 1234) from host 1 to host 2, you first start the receiver on host 2 (right column) and then initiate the transfer of the file to host 1 (left column):

```
                              host2$ nc -l 1234 > file
host1$ nc host2 1234 < file
```

It's just as easy to conduct a chat. All you have to do is agree on a port with the other party. The chat is initiated on a computer via nc -l (left column). This way, nc monitors the specified port 1234 and waits for a connection to be established.

On the second host, nc is started without options to connect to the first host. There's no visible confirmation that the connection has been established, but as soon as one of the two parties enters text (standard input) and confirms it by pressing Enter , the text appears in the terminal of the other party (standard output).

```
host1$ nc -l 1234
                              host2$ nc host1 1234

 how are you?
                              how are you?
                              good
good
<Ctrl>+<C>
```

The third example shows the potential danger of Netcat: Here, nc on host 1 is set up to pass all input received on port 1234 to the bash shell. Their outputs are transferred back to the transmitter. Shell commands can now be run on host 1 from a second host. ls therefore shows files that are located on host 1!

```
host1$  nc -l 1234 -e /bin/bash
                              host2$ nc host1 1234
                              ls
                              file1
                              file2
                              file3
```

However, the -e option for executing a command isn't available with all Netcat versions. It's especially missing in the netcat-traditional implementation that's common on Debian and Ubuntu. A solution to that is to install the nmap package and run the ncat command included there.

ncdu [options] [directory]

ncdu is an interactive variant of the du command. Usually, it's simply called without options and parameters. It then displays the subdirectories in the current directory that take up the most space.

You can now navigate through the subdirectories using the cursor keys and ⟨Enter⟩. Various other functions are accessible via keyboard shortcuts. An overview of the most important shortcuts can be obtained by entering ⟨?⟩ But be careful—⟨D⟩ deletes the current file or directory after a query.

▶ -r

Runs the command in read-only mode. Accidental deletion of files or directories is therefore impossible.

▶ -x

Remains in the current file system, that is, doesn't take any mount directories into account.

Example

The space required by the directories in the home directory is also symbolized by bars:

```
user$ ncdu
--- /home/kofler ---
    1.5 GiB [##########] /Nextcloud
    1.0 GiB [######    ] /Downloads
  528.5 MiB [###       ] /Images
  209.4 MiB [#         ] /Documents
    ...
```

needs-restarting [options]

The needs-restarting command available on Fedora, Red Hat, and related distributions specifies whether the computer or individual programs need to be restarted due to an update of the kernel, a basic library, or a firmware file.

The command isn't available for distributions based on Debian or Ubuntu. There, the /var/run/reboot-required file indicates that a reboot is required.

▶ -r

Indicates whether and why a computer restart is required.

▶ -u

Lists processes of the current user that need to be restarted. This option applies by default.

Example

The following commands show the status of a computer that should be restarted:

```
root# needs-restarting -r
Core libraries or services have been updated since boot-up:
  * glibc
Reboot is required to fully utilize these updates.
More information: https://access.redhat.com/solutions/27943
root# needs-restarting | sort -n
782 : /usr/lib/polkit-1/polkitd --no-debug
783 : /usr/libexec/power-profiles-daemon
784 : /usr/bin/qemu-ga --method=virtio-serial ...
787 : /usr/libexec/accounts-daemon
...
```

netplan [command]

Netplan (*https://netplan.io*) is a framework developed by Ubuntu that configures and integrates other network backends such as the NetworkManager and networkd (a systemd component, see networkctl). It's used in Ubuntu and evaluates the configuration files located in /etc/netplan (see /etc/netplan/netplan.yaml).

▶ apply
Runs netplan generate and then activates the changed configuration.

▶ generate
Reads the Netplan configuration from the YAML files in /etc/netplan, /lib/netplan and /var/netplan, and then generates the corresponding configuration files for the network backends. The new files end up in /run/NetworkManager or in /run/systemd/network.

The new configuration isn't activated! The changed settings only take effect when the respective backend is requested to evaluate the files (or at the next restart).

▶ ip leases *interface*
Outputs the current DHCP lease data for the relevant interface. This only works if the networkd backend and a DHCP configuration are in use. Otherwise, the command will return misleading error messages.

netstat [options]

netstat provides information about the network activity on the local computer. If the command is called without options, it returns a list of all open internet connections and sockets.

▶ -a
Also takes into account inactive sockets, that is, services that are waiting for a connection on a network port (LISTEN state).

▶ -e
Also specifies the user for internet connections.

▶ -n
Provides numerical addresses and port numbers instead of host and network service names.

▶ -p
Also specifies the process number (PID) and the process name of the program responsible for the connection. netstat requires root permissions so that it can provide information about nonnative processes.

▶ -t / -u / -w / -X
Limits the output to connections that use the TCP, UDP, Raw, or Unix protocol.

Example

The following command lists all active connections (established) or monitored ports (LISTEN). The output is heavily abridged due to space limitations.

```
root# netstat -atupe
Active Internet connections (servers and established)
Proto Local Address        Foreign Addr   State    User    PID/Prog name
tcp    *:nfs               *:*            LISTEN   root    -
tcp    *:ldap              *:*            LISTEN   root    5842/slapd
tcp    localhost:mysql     *:*            LISTEN   mysql   5785/mysqld
tcp6   [::]:ssh            [::]:*         LISTEN   root    5559/sshd
udp    *:nfs               *:*                     root    -
```

networkctl [command]

networkctl is one of the administration commands from the systemd family. If networkd is used as the network backend, networkctl helps to analyze the network status.

Although networkd is installed by default on many distributions, it's only active on a few of them. One of the exceptions is Ubuntu Server, where networkd serves as a backend for Ubuntu's own Netplan system (see netplan). networkd evaluates configuration files from the /etc/systemd/network, /lib/systemd/network, and /run/systemd/network directories (see /etc/systemd/network/networkd.network).

▶ list
Lists all network interfaces. The SETUP column of the result shows whether the interface is controlled by networkd (configured) or not (unmanaged).

▶ lldp

Shows other devices in the network that were discovered via the *Link Layer Discovery Protocol* (LLDP).

▶ status [*interface*]

Provides detailed information about the status of a network interface. If no interface is specified, the command attempts to summarize the entire network status.

Examples

The following outputs of networkctl were created on an Ubuntu server in a virtual machine:

```
root# networkctl list
IDX  LINK      TYPE       OPERATIONAL   SETUP
 1   lo        loopback   carrier       unmanaged
 2   ens3      ether      routable      configured
root# networkctl status ens3
      Link File: /lib/systemd/network/99-default.link
   Network File: /run/systemd/network/10-netplan-ens3.network
           Type: ether
          State: routable (configured)
                 ...
     HW Address: 52:54:00:0a:8d:fc
        Address: 138.201.20.182
                 fe80::5054:ff:fe0a:8dfc
        Gateway: 138.201.20.176 (Fujitsu Technology Solutions GmbH)
            DNS: 213.133.100.100
                 213.133.98.98
 Search Domains: ubuntu-buch.info
```

The second listing shows the status summary of another server installation with IPv6 configuration:

```
root# networkctl status
         State: routable
  Online state: online
       Address: 168.119.33.110 on eth0
                172.17.0.1 on docker0
                2a01:4f8:242:1f88::4 on eth0
                fe80::5054:ff:fe4a:7321 on eth0
       Gateway: 168.119.33.119 on eth0
                2a01:4f8:242:1f88::2 on eth0
           DNS: 213.133.100.100
```

```
213.133.99.99
2a01:4f8:0:1::add:1010
2a01:4f8:0:1::add:9999
```

newaliases

The /etc/aliases file contains an alias list for the email server, which ensures, for example, that all emails addressed to postmaster are forwarded to root. For changes to this file to be taken into account by the mail server, you must run newaliases.

newgrp [group name]

The newgrp command determines the currently active group of a user who belongs to several groups. The active group determines which group newly created files belong to. The groups available for selection can be determined via groups. If no group name is specified, the primary group will be used. This group is also automatically considered an active group after a login.

Example

In the following example, the newgroup command makes docuteam the active group of the user kofler. As a result, the newly created newfile file is assigned to the docuteam group and can be edited by other members of the documentation team.

```
user$ groups
kofler docuteam wheel
user$ newgroups docuteam
user$ touch newfile
```

newusers file

newusers reads a text file and creates a new user for each line. The text file basically has the same format as /etc/passwd. However, the passwords must be entered unencrypted in the second column. Most of the other parameters (UID, GID, etc.) are optional. newusers creates a new account for each specified user, whereby the associated groups are also created if required. For missing parameters, newusers selects suitable default values, taking into account the settings in /etc/login.defs.

Note that newusers creates new home directories if their location is specified in the sixth column. However, the command doesn't take care of copying the contents of /etc/skel there.

Example

The following lines show the minimalist text file `users.txt`, which corresponds to the requirements of `newusers`. As `users.txt` contains passwords in plain text, you must make sure that no one other than `root` can read the file or that you delete the file again after running `newusers`.

```
hawks:secret1:::Howard Hawks:/home/hawks:/bin/bash
moore:secret2:::Grace Moore:/home/moore:/bin/bash
smith:secret3:::Peter Smith:/home/smith:/bin/bash
```

`newusers` now creates the three new users `hawks`, `moore`, and `smith`, as well as primary groups of the same name. `newusers` itself decides on suitable UIDs and GIDs.

```
root# newusers users.txt
```

nft command [options]

Most current Linux distributions now use the new *nftables* firewall system, which replaces its predecessor, *Netfilter*. However, the old `iptables` command is often still used to configure the firewall, which continues to work thanks to a compatibility layer.

However, new or nftables-specific functions can only be controlled using the `nft` command. The following description provides a rough overview of the most important subcommands and options of this command. You can find a lot more details about the functionality of nftables and the syntax of `nft` in `man nft` and the following wiki:

https://wiki.nftables.org

Before you call `nft`, you should make sure that a firewall isn't already running on your distribution. On Fedora, RHEL, and SUSE, for example, *Firewalld* is active (see `firewall-cmd`). Custom-defined firewall rules then often result in conflicts with the firewall of your distribution.

nft Terminology

A separate terminology applies to `nft`: Firewall *rules* are always part of a *chain*. *Tables*, in turn, consist of multiple chains. Unlike `iptables`, there are neither predefined tables nor predefined chains. Thus, you can compose as many custom tables as you like from chains also defined by yourself.

Tables, chains, and rules apply to different types of network packets, which are assigned to the following *families*:

▶ `ip`
Rules for IPv4 only.

▶ `ip6`
Rules for IPv6 only.

▶ `inet`
Rules that apply equally to IPv4 and IPv6.

▶ `arp`
Level 2 rules, which are analyzed before Level 3 rules.

▶ `bridge`
Switching rules.

▶ `netdev`
Low-level rules that are analyzed before all other rules and enable, for example, a particularly efficient defense against DDOS attacks.

nft Options

▶ `-a`
Integrates *handles* into the output of `nft list`. These are numbers that uniquely identify each object within a group (e.g., a rule within a chain). Handles make it possible to delete or replace individual objects or to specify the position where new objects are to be inserted.

▶ `-f` *fname*
Reads the commands to be executed from the specified file. Two syntax variants are permitted: the file can either contain ordinary `nft` commands line by line or a hierarchically structured sequence of tables, chains, and rules. The structure of the file for the second variant corresponds to the output of `nft list ruleset`.

▶ `-S`
Uses the names of network services specified in `/etc/services` instead of port numbers (e.g., `ssh` instead of 22).

▶ `-v`
Shows the version number.

nft Commands

▶ `add chain` *family tname cname*
Creates a new chain named *cname* in the specified table.

▶ `add rule` *family tname cname matches statements*
Adds a new rule to the specified *cname* chain. Here, *matches* specifies the packets to which the rule applies. *statements* describes what should happen to the respective packets.

Instead of add, the insert and replace keywords are also permitted to insert a rule at a specific position in the rule list or to replace an existing rule with a new one.

▶ add table *family tname*
Creates a new table for the specified family.

▶ delete/flush chain *family tname cname*
Removes the specified chain from the table. flush deletes all rules of the chain, but not the chain itself.

▶ delete rule *family tname cname* handle *came*
Deletes the rule specified by *handle* from the *cname* chain.

▶ delete/flush table *family tname*
Deletes the specified table. With flush, all chains and rules in the table are deleted, but the now empty table is retained.

▶ flush ruleset
Deletes all tables, including all chains and rules contained therein. This corresponds to a complete reset of the firewall. The firewall now accepts every packet.

▶ list tables
list table family name
list chain *family tname cname*
Lists tables, chains, and rules.

▶ list ruleset
Lists all tables, chains, and rules. The output structured by curly brackets fulfills the syntax rules for nft -f and can therefore be used as the basis for a new rule file. With the additional -j option, the output is in JSON format.

▶ monitor [filter criteria]
Displays nftables events (e.g., as debugging help).

Examples

The first two commands show which firewall tables are defined on the test computer (Fedora 32) and which chains and rules apply to the firewalld table within the ip family. (nftables allows the use of matching table names for different families. For this reason, the family must also be specified in the second list command.)

```
root# nft list tables
table bridge filter
table bridge nat
table inet firewalld
table ip firewalld
table ip6 firewalld
root# nft list table ip firewalld
table ip firewalld {
```

```
chain nat_PREROUTING {
    type nat hook prerouting priority dstnat + 10; policy accept;
    jump nat_PREROUTING_ZONES
}
chain nat_PREROUTING_ZONES {
    iifname "enp1s0" goto nat_PRE_FedoraWorkstation
    goto nat_PRE_FedoraWorkstation
}
...
```

`systemctl` now deactivates the default Fedora firewall daemon:

root# **systemctl disable --now firewalld**

Now, you can create your own firewall using nft. It's not usual to define the rules by repeatedly calling nft. It's much more elegant to formulate the rules in a text file and run this file like a script. To do this, you must enter nft -f as a shell interpreter in the line.

```
#!/sbin/nft -f
# delete existing firewall
flush ruleset
# new table for IPv4 and IPv6
table inet myfilter {
  chain input {
      type filter hook input priority 0; policy drop;
      # block faulty packets
      ct state invalid drop
      # always accept packets from self-generated connections
      ct state {established, related} accept
      # accept internal network traffic
      iif lo accept
      # Block packets to loopback but from external addresses
      iif != lo ip daddr 127.0.0.1/8 drop
      iif != lo ip6 daddr ::1/128 drop
      # accept ICMP packets
      ip protocol icmp accept
      ip6 nexthdr icmpv6 accept
      # accept SSH connection from outside
      tcp dport 22 accept
      # for server configuration: more rules for
      # HTTP + HTTPS (Port 80 + 443), Samba etc.
      # ...
```

```
    }
    # no forwarding
    chain forward {
        type filter hook forward priority 0; policy drop;
    }
}
```

Now you can run the script:

```
root# chmod +x myfirewall
root# ./myfirewall
```

nft -f first performs a syntax check. If it detects errors, it aborts the process with an error message. In this case, the previous firewall is retained.

The preceding script has been slightly modified from the Gentoo wiki. You can find more examples there and in the ArchLinux Wiki:

https://wiki.gentoo.org/wiki/Nftables/Examples

https://wiki.archlinux.org/title/Nftables

ngrep [options] [grep-search-expression] [pcap-filter-expression]

ngrep is referred to as a *packet sniffer*. The command reads the network traffic of a port and filters it. ngrep provides similar functions to the tcpdump command and, like this, uses the pcap library to read the network packets. Unlike tcpdump, however, ngrep also takes the contents of the packages into account. Naturally, this only works for nonencrypted protocols, such as HTTP or FTP.

The search expression must be formulated as a regular pattern as in grep. The syntax summarized in tcpdump applies to the pcap filter expression.

▶ -d *interface*|any
Determines the network interface.

▶ -i
Ignores uppercase and lowercase in the grep search expression.

▶ -v
Inverts the search. ngrep therefore only returns the packets in which the grep search pattern was *not* recognized.

▶ -w
Interprets the grep search expression as a word.

▶ -W byline
Takes line breaks into account during output, resulting in more legible output.

Example

The following example listens on all interfaces for HTTP packets containing the keywords user, pass, and so on:

```
root# ngrep -d any -i 'user|pass|pwd|mail|login' port 80
interface: any
filter: (ip or ip6) and ( port 80 )
match: user|pass|pwd|mail|login
T 10.0.0.87:58480 -> 91.229.57.14:80 [AP] POST /index.php
  HTTP/1.1..Host: ... user=name&pass=geheim&login=Login
...
```

nice [options] program

nice starts the specified program with a reduced or increased priority. The command can be used to start non-time-critical programs with low priority so as not to affect the rest of the system too much.

▶ -n +/-n

Specifies the nice value. By default (i.e., without nice), programs are started with the nice value of 0. A value of -20 means highest priority, and a value of +19 means lowest priority. Values less than 0 may only be specified by root; that is, most users can only start programs with reduced priority with nice. If this option is omitted, nice starts the program with the nice value of +10.

Note that nice only controls the CPU load. If you want to reduce the I/O load of a command, it's better to use ionice.

Example

The following command starts the mybackup.sh script with lower priority:

```
user$ nice -n 10 mybackup.sh
```

nl [options] file

nl numbers all nonempty lines of the specified text file and writes the result to the standard output. By setting the numerous options, you can have page-by-page numbering, the numbering of headers and footers, and more.

nmap [options] hostname/ip-address/ip-address-range

The nmap command (*network mapper*) from the package of the same name performs a port scan and attempts to determine which network services are active on the specified computer or in the specified network. nmap should only be used to analyze your own computers or after consultation with the respective administrator. A port scan of other computers can be interpreted as an attempted attack!

▶ -A

Performs an extensive ("aggressive") scan, corresponding to -sV -O -sC --traceroute.

▶ -F

Considers only the 100 most important ports from /usr/share/nmap/nmap-services (quick scan).

▶ -iL *file*

Scans the IP addresses specified in the file.

▶ -oN *file* / -oG *file* / -oX *file*.*xml*

Writes the results to an ordinary text file, to a text file that can be easily processed with grep, or to an XML file. Without the option, nmap uses the standard output and the ordinary text format.

▶ -O

Attempts to recognize the operating system; this option must be combined with a scan option, such as -sS, -sT, or -sF.

▶ -p1-10,22,80

Takes only the specified ports into account.

▶ -Pn

Doesn't perform a ping test. This means that nmap considers all hosts to be online and performs a scan in any case (slowly!).

▶ -sL

Lists all ports and specifies host names assigned in the past. This is particularly fast, but provides outdated data even from devices that are currently no longer online.

▶ -sP

Performs only a ping test (fast).

▶ -sS

Performs a TCP SYN scan (applies by default).

▶ -sU

Takes UDP into account as well. This option can be used together with another -s option.

▶ -sV

Tries to find out which service is provided on open ports (*service version detection*) or which program is responsible for the service and in which version.

▶ -T0 to -T5

Selects a timing schema:

  – -T5 is the fastest.

- – -T3 is the default.

- – -T0 and -T1 are extremely slow, but minimize the risk of the scan being detected.

▶ -v

Outputs detailed information (*verbose*).

You must choose an -s option when calling. Only -sU may be combined with other -s options. In general, the right choice of options is a trade-off between thoroughness and speed.

In many cases, nmap -v -A name is sufficient to get an initial overview of the network services of the specified computer. Advanced nmap users can find more details on the man page and on the following websites:

https://insecure.org/nmap

https://nmap.org/book

Graphical user interfaces also exist for nmap, such as nmap-frontend or zenmap.

Examples

The following command performs a quick network scan on the local network (256 IP addresses). Thanks to the focus on the most important 100 ports, the job is done within about two seconds. The nmap outputs have been shortened for reasons of space and only show the results of two of the devices found:

```
root# nmap -F  -T4 10.0.0.0/24
Nmap scan report for imac (10.0.0.2)
  Host is up (0.00019s latency).
  PORT    STATE SERVICE
  22/tcp  open  ssh
  88/tcp  open  kerberos-sec
  445/tcp open  microsoft-ds
  548/tcp open  afp
  MAC Address: AC:87:A3:1E:4A:87 (Apple)
Nmap scan report for raspberrypi (10.0.0.22)
  Host is up (0.00038s latency).
  Not shown: 99 closed ports
  PORT    STATE SERVICE
  22/tcp open  ssh
  MAC Address: B8:27:EB:11:44:2E (Raspberry Pi Foundation)
...
Nmap done: 256 IP addresses (6 hosts up) scanned in 2.42 seconds
```

The second example shows the nmap result for a NAS device that is located in a different local network (again, it has been heavily shortened):

```
root# nmap -v -A 192.168.178.28
Nmap scan report for DiskStation.fritz.box (192.168.178.28)
Host is up (0.0023s latency).
...
Discovered open port 445/tcp on 192.168.178.28
Discovered open port 139/tcp on 192.168.178.28
Discovered open port 80/tcp on 192.168.178.28
Discovered open port 443/tcp on 192.168.178.28
Discovered open port 22/tcp on 192.168.178.28
Discovered open port 5001/tcp on 192.168.178.28
Discovered open port 5000/tcp on 192.168.178.28
Discovered open port 548/tcp on 192.168.178.28

PORT    STATE SERVICE VERSION
22/tcp  open  ssh      OpenSSH 7.4 (protocol 2.0)
  ssh-hostkey:
    2048 5a:e7:3a:66:f4:99:9f:0a:0a:...  (RSA)
    256 06:1a:bf:9f:e9:d0:64:3a:92:49:... (ECDSA)
    256 ad:b7:7d:ab:ae:70:0a:c9:a6:0c:... (ED25519)
    ...
80/tcp  open  http         nginx
139/tcp open  netbios-ssn Samba smbd 4.6.2
445/tcp open  netbios-ssn Samba smbd 4.6.2
...
Nmap done: 1 IP address (1 host up) scanned in 64.19 seconds
```

nmblookup [options] workgroupname

nmblookup from the samba-common-bin (Debian/Ubuntu) or samba-client (Fedora/RHEL) package determines which devices or servers in the local network provide SMB shares. Previously, smbtree could be used for this, but this command requires the browsing functions of SMB1 and is obsolete in modern networks.

▶ -S

Performs a status query for each server found and then displays the detailed results.

▶ -T

Provides an ordered list of server names and IP addresses. In my experience, the results of nmblookup are more reliable if you run nmblookup -S first or repeat the command again after a few seconds. The first result is often incomplete.

Example

There are two SMB servers, a NAS device, and an appropriately configured Raspberry Pi in the local network:

```
user$ nmblookup -T WORKGROUP
DiskStation.fritz.box, 192.168.178.43  WORKGROUP<00>
pi5.fritz.box,         192.168.178.123 WORKGROUP<00>
```

> **nmcli** [options] con|dev|nm command

Network Manager is usually controlled via a menu in the Gnome or KDE panel. You can also use the nmcli command to control network connections via the command line or a script.

The commands available for selection depend on which object they refer to. Options -t, -p, and so on can be used to control the form of the nmcli output—depending on whether the output is to be formatted properly or processed further by a script (for details, see man nmcli).

▶ con down *name* or con down uuid *n*
  Deactivates the specified connection.

▶ con show
  Lists all configured connections, specifying the name and UUID of each connection.

▶ con up *name* or con up uuid *n*
  Activates the specified connection.

▶ dev disconnect *name*
  Terminates the connection for the specified interface.

▶ dev
  Lists all network interfaces known to Network Manager and specifies their properties.

▶ dev show
  Displays detailed information on all network interfaces.

▶ dev wifi hotspot
  Configures the Wi-Fi adapter as a router.

▶ dev wifi list
  Provides a list of all Wi-Fi networks within range, including the signal strength. The command requires the wpa_supplicant service to be running.

▶ general [status]
  Displays the status of Network Manager.

▶ networking [on|off]

Indicates whether there's a network connection or not. All network connections can be interrupted or restored by selecting off or on.

▶ radio [wifi|wwan|wimax|all] [on|off]

Displays the status of the wireless networks or changes their status. wifi refers to Wi-Fi connections, wwan to mobile connections, and wimax to WiMAX technology.

Examples

The first command lists all connections known to Network Manager. The following two commands deactivate and reactivate the connection named *Wired connection 1*:

```
root# nmcli con show
NAME                      UUID          TYP           DEVICE
Wired connection 1        effd094d-...  802-3-ethernet enpOs3
virbr0                    207d9f4f-...  bridge        virbr0
root# nmcli con down   id 'Wired connection 1'
root# nmcli con up     id 'Wired connection 1'
```

The second example filters out the IP addresses of the name server from the detailed interface information. This is particularly useful on Ubuntu, where these addresses aren't listed in /etc/resolv.conf. Rather, resolv.conf references a local name server in this distribution, that is, the address 127.0.0.1.

```
root# nmcli dev show | grep DNS
IP4.DNS[1]:  10.0.0.138
IP4.DNS[2]:  4.4.8.8
```

The third command sets up a hotspot on a Raspberry Pi connected to the LAN via ethernet cable:

```
root# device wifi hotspot ssid 'wlan-name' password 'wlan-pw'
```

nohup command

If you start a command as a background process in a shell window and then close the window, or if you start the command in a text console and then log out, the background process is automatically terminated. As a rule, this is reasonable behavior.

However, sometimes you want to start a process that continues to run after you log out—and that's exactly what nohup is there for. The command must be specified with its full path. It can't write text outputs to the standard output. If necessary, such outputs are redirected to the nohup.out file in the local directory.

Example

In the following example, an administrator logs in to a server via ssh, starts a backup script in the background, and then logs out again. The backup script continues to run.

```
someone@localhost$ ssh user@remotehost
user@remotehost$ nohup backup-script &
user@remotehost$ exit
```

nproc

nproc indicates how many CPU cores are available. For CPUs with hyperthreading, the virtual cores are also included in the calculation. The lscpu command and the contents of the /proc/cpuinfo file provide more detailed information about the CPU.

nvidia-xconfig [options]

nvidia-xconfig helps to configure the proprietary NVIDIA graphics driver. The command is only available if this driver is installed.

If the command is executed without parameters, it creates or changes the /etc/X11/xorg.conf file so that the NVIDIA graphics driver will be used in the future. If the graphics system is working after restarting the computer, the nvidia-settings graphical user interface can be used for further configuration.

▶ -c file
   Uses the specified file instead of /etc/X11/xorg.conf.

▶ --mode=WxH
   Adds a graphics mode for WxH pixels to xorg.conf.

▶ --query-gpu-infos
   Displays details of all available graphics cards and connected monitors.

▶ -t
   Reads xorg.conf and displays the settings it contains in a clear tree view.

openssl command

openssl from the package of the same name helps with the creation and further administration of certificates as well as private and public keys. The command uses the OpenSSL library. New certificates or keys are written to the standard output if you don't specify a target file via -out.

▶ dhparam [options] [n]
   Generates or manipulates parameter files for the Diffie-Hellman method. The n

parameter specifies the key size in bits. In the simplest case, `openssl dhparam -out keyfile 512` generates a Diffie-Hellman key with a length of 512 bits.

▶ `enc [options]`
Reads data from the standard input, encodes or decodes it symmetrically (option `-d`), and writes the result to the standard output. You need to specify the algorithm to be used and the key with options, for example, in the format `-aes-256-cbc -pass file:mykeyfile`. Note that `openssl enc` is only suitable for relatively small amounts of data. To encrypt large files, it's better to use the `gpg` command.

▶ `genpkey [options]`
Generates a private key. To write a 2,048-bit RSA key to a file, the following options are required:

`openssl genpkey -algorithm RSA -pkeyopt rsa_keygen_bits:2048 -out server.key`

If the key itself is to be encrypted, you must specify the name of the encryption algorithm with an additional option, such as `-aes256`. During the execution of the command, `openssl` asks for an encryption password twice.

▶ `list-standard-commands | list-message-digest-commands |`
`list-cipher-commands | list-cipher-algorithms |`
`list-message-digest-algorithms | list-public-key-algorithms`
Lists the commands, algorithms, and so on that are supported by `openssl`.

▶ `rand [options]` *n*
Writes *n* bytes of pseudo-random data to the standard output. Note that the output is binary data, not ASCII text! The `-base64` option causes `openssl` to perform Base64 encoding in ASCII format. This increases the length of the resulting character string by around a third.

▶ `req [options]`
Generates a request to sign a certificate (i.e., a *certificate signing request* [CSR] file). With `-new`, you can specify that a new certificate must be set up. `-key` specifies the key file to be used. Without this option, `openssl` generates a new RSA key for this purpose, which you can save in a separate file using `-keyout`. `openssl req -new -sha256 -key server.key -out server.csr openssl` then interactively asks for the key data of the certificate, that is, the country code, location, your name, host name (common name), and more. With the additional `-x509 -days` *n* options, `openssl` generates a self-signed certificate with a validity of *n* days instead of a certificate request.

▶ `rsa [options]`
Processes RSA keys and converts them between different forms and formats. `-in file` and `-out file` specify the file from which the original key is read and where the new key is written. If you don't specify any further options, `openssl` removes the encryption (*pass phrase*) of the key. The form of the key can be specified with `-inform` and `-outform` (by default, it's PEM, alternatively DER or NET). With `-in file -text`, you

can determine the properties of an existing key, such as the key length (the number of bits).

▶ sclient [options]

Acts as an SSL/TLS client to test connections to an encrypted server. You can find an application example at the following address:

https://www.misterpki.com/openssl-s-client/

▶ speed

Performs multiple benchmark tests, each of which lasts three seconds. The result shows how many bytes were processed for the respective algorithm (the more there are, the faster it runs or the better optimized the hardware on which the test is performed).

▶ x509 [options]

Displays certificate data, signs certificates, and converts certificates between different forms. Again, -in and -out specify the file names of the source and target certificates. With -req, openssl expects a certificate request as input, not the certificate itself. -CA specifies which CA certificate is to be used for the signature. -CAkey specifies the corresponding private key.

There are separate man pages available for most openssl commands. For example, man req describes the countless options of openssl req, man rsa describes the options of openssl rsa, and so on.

Examples

If you want to have a certificate signed for your web server by an official certification authority (e.g., Thawte), you should first generate a key (if you don't already have one) and then a certificate signing request. The key isn't encrypted here because otherwise the web server would have to ask for the encryption password every time it's started.

```
root# genpkey -algorithm RSA -pkeyopt rsa_keygen_bits:2048 -out server.key
root# chmod 400 server.key
root# openssl req -new -key server.key -out server.csr
...
Common Name (eg server FQDN or YOUR name) []: company-abc.com
```

You can take a look at the CSR file to check this:

```
root# openssl req -in server.csr -noout -text
```

Then, you need to send the CSR file to the certification authority. You will then receive the signed certificate from there for a fee. The usual file identifier is .pem or .crt. You must enter this file, your own key, and a CA certificate from the certification authority in the Apache configuration.

```
# Apache configuration file
SSLCertificateFile      /etc/apache2/server.pem
SSLCertificateKeyFile   /etc/apache2/server.key
SSLCertificateChainFile /etc/apache2/sub.class1.server.ca.pem
SSLCACertificateFile    /etc/apache2/ca.pem
```

If you don't want to pay for an official signature, you can sign your certificate yourself:

```
root# openssl x509 -req -days 1900 -in server.csr \
        -signkey server.key -sha256 -out server.pem
Signature ok
subject=/C=DE/L=Boston/O=John Doe/CN=www.company-abc.com/
  emailAddress=webmaster@company-abc.com
Getting Private key
```

To take a look at the certificate file, you can run openssl x509 -text:

```
root# openssl x509 -text -in server.pem
Certificate:
    Data:
        Version: 1 (0x0)
        Serial Number: 12669601459972319941 (0xafd37766c36baac5)
    Signature Algorithm: sha256WithRSAEncryption
        Issuer: C=DE, L=Boston, O=John Doe,
                CN=www.company-abc.com/emailAddress=webmaster@company-abc.com
        Validity
            Not Before: Sep 28 14:48:03 2025 GMT
            Not After : Dec 10 14:48:03 2029 GMT    ...
```

The generation of a key, a certificate request, and self-signing can also be carried out in a single command: The following command generates a self-signed certificate for a mail server that is valid for 10 years. It's important that you enter the host name of your server as the common name when running openssl. Because the certificate itself is signed, your mail client will indicate during configuration that the certificate isn't trustworthy.

```
root# openssl req -new -x509 -days 3650 -nodes \
        -out    /etc/ssl/certs/postfix.pem \
        -keyout /etc/ssl/private/postfix.key
...
Common Name (eg server FQDN or YOUR name) []: company-abc.com
root# chmod 400 /etc/ssl/private/postfix.key
```

pacman [options]

pacman stands for *package manager* and is the package management command for Arch Linux and compatible distributions.

▶ -Q
Provides a list of all installed packages (*query*).

▶ -Qi *package name*
Displays a description of the package.

▶ -Ql *package name*
Lists the files of the package.

▶ -Qo *filename*
Indicates to which package the specified file belongs.

▶ -R *package name*
Removes the specified package (*remove*). If you use the -Rs variant instead of -R, dependent packages that are no longer required will also be uninstalled.

▶ -S *package name* or -S *reponame/package name*
Installs or updates the specified package, whereby in the second variant only the specified repository is taken into account. If the package has already been installed, the installation will be repeated and the package will be updated if necessary. The additional --needed option prevents unnecessary work and only carries out the installation if the package isn't yet installed or is installed in an old version. The ubiquitous -S option stands for *sync*, by the way.

▶ -Sy
Reloads the package sources.

▶ -Su
Displays all installed packages.

▶ -Syu
First updates the package sources and then all installed packages, that is, performs a system update.

▶ -Ss *pattern*
Searches the names and short descriptions of all packages, including those not installed, for the specified search pattern.

Examples

The following commands first perform a complete update of all package sources and installed packages and then install the emacs editor, including all required libraries. --needed prevents already installed packages from being installed again.

pacman -Qo determines which package provides the /etc/resolv.conf file. pacman -Qi then provides more detailed information about this package.

```
root# pacman -Syu
root# pacman -S --needed emacs
root# pacman -Qo /etc/resolv.conf
/etc/resolv.conf is owned by filesystem 2025.01.19-1
root# pacman -Qi filesystem
  Name            : filesystem
  Description     : Base Arch Linux files
  Install Reason  : Installed as a dependency for another package
  ...
```

Alternatives

pacman only works with "ordinary" packages. However, *Arch User Repository* (AUR) packages are also very often used in Arch Linux. AUR is a collection of build scripts on GitHub. Because the manual installation of this type of packages is tedious, AUR helpers take over this task, such as aura, aurman, pacaur, or yay. These commands integrate the functions of pacman and treat AUR packages like ordinary packages. Thus, yay -S *name* installs the specified package, regardless of whether it's an ordinary package or an AUR package.

---

pactl command [parameter]

pactl controls the PulseAudio system. This network-enabled sound server is installed by default on most Linux distributions and usually uses ALSA as its basis. PulseAudio makes it possible for several programs to use the audio system in parallel without any problem. pactl includes approximately 30 commands, of which only the most important are presented here:

▶ info|stat
  Displays the key data or the memory consumption of the PulseAudio system.

▶ list [short] [modules|sinks|sources|clients|cards|...]
  Provides a detailed description of all modules, audio sources and outputs, streams, and so on that are managed by PulseAudio. The additional short keyword reduces the output to one line per element. The specification of modules or sinks, etc., causes pactl to list only the relevant elements.

▶ set-default-sink *n*
  Defines the default audio output. Instead of the number *n* of the audio system (see /proc/asound/cards), you can also enter its name. You can determine the permitted names using pactl list short cards.

- ► `set-sink-mute` *n* `0|1|toggle`
  Deactivates, activates, or changes the mute function for audio output *n*.

- ► `set-sink-volume` *n vol*
  Sets the volume for audio output *n*. The volume can be specified as an integer, a percentage, or as a decibel value (e.g., 10% or 20dB).

Example

The following two commands result in the first audio device being used as the standard output device at a volume of 20%:

```
user$ pactl set-default-sink 0
user$ pactl set-sink-volume 0 20%
```

---

pandoc `[options]` `in1` `[in2 in3 ...]` `[> out]`

The primary task of pandoc from the package of the same name is to convert Markdown text files into other formats, such as HTML files, PDF documents, or Microsoft Office files. With some limitations, pandoc can also process other input formats instead of Markdown files, including HTML and LaTeX.

- ► `-c cssfile`
  Embeds the specified CSS file in the HTML document. The option can be specified multiple times if several CSS files are required.

- ► `-f format`
  Specifies the source format (*from*). Important source formats are asciidoc, docbook, latex, and markdown. pandoc attempts to recognize the desired source and target formats itself based on the file ID of the input and output files. The -f and -t options are only required if that isn't successful.

- ► `-H header file`
  Inserts the specified header file at the start of the HTML or LaTeX source text. The option can be specified multiple times if multiple header files are required.

- ► `-o output.identifier`
  Writes the result to the specified file instead of to standard output. This option is mandatory for binary output formats (DOCX, ODT, PDF).

- ► `-s`
  Creates an independent document (*standalone*). This option is particularly relevant for HTML, LaTeX, and RTF output formats. Without this option, pandoc creates files without any header information; such files can't be used on their own, but must be embedded in another document.

- -t *format*

  Specifies the target format (*to*). Important target formats are docx, latex, html, html5, man, odt, and rtf. Note that pdf isn't included in the list of supported formats. To create PDF documents, you must specify the -o out.pdf option. pandoc first converts the source text into LaTeX format and then creates the PDF file using pdflatex.

- --toc

  Embeds a table of contents in the resulting document.

Examples

The first command creates an independent HTML document from the input.text Markdown file. The second command creates an EPUB file from a book whose chapters are saved in separate files. The third command creates a PDF file, using LaTeX in the background.

```
user$ pandoc -s -c my.css input.text > output.html
user$ pandoc chap1.text chap2.text chap3.text -t epub3 -o book.epub
user$ pandoc preface.text chap01.text chap02.text -o out.pdf
```

paplay [options] [file]
parecord [options] [file]

paplay plays the specified audio file via the PulseAudio system. parecord makes a recording and saves the audio stream as a file.

- -d *id*

  Indicates the output device (paplay) or the audio source (parecord). Without this option, the respective default device is used.

- --file-format=*name*

  Specifies the desired audio format, such as raw or wav.

- --list-file-formats

  Lists all supported audio formats.

- --volume=*n*

  Indicates the desired playback volume. The value range is from 0 to 65535 for maximum volume.

Example

The following command plays one of the audio files that were provided with Libre-Office:

```
user$ paplay /usr/lib/libreoffice/share/gallery/sounds/train.wav
```

paps [options] text files > out.ps

paps converts the specified UTF-8 text files into PostScript format and writes the result to the standard output. The UTF-8 characters aren't displayed as letters, but as lines. For this reason, the display of the PostScript file in a PostScript viewer (Evince, Okular, etc.) may look pixelated. It's not possible to select a text. However, the print quality is good.

▶ --columns=*n*
Defines the number of text columns (by default, it's one).

▶ --cpi=*n*
Controls the number of characters per *inch* and thus the font size.

▶ --font=*name*
Specifies the desired character set (the default setting is monospace 12).

▶ --landscape
Formats the text in landscape format.

Example

The following command produces the two-column PostScript file text.ps:

```
user$ paps --landscape --columns 2 text.txt > print.ps
```

parallel [options] ['command']

parallel from the package of the same name helps to run multiple commands in parallel, possibly even on different computers. parallel supports some of the same options as xargs.

▶ -eta
Provides information on the status of parallel processing and the time at which parallel will probably be finished, that is, the *estimated time of arrival*. This time estimate assumes that each job takes approximately the same amount of time.

▶ -j *n*
Runs a maximum of *n* instances in parallel.

▶ -j +*n* or -j -*n*
The same as -j *n*, but the specification is relative to the number of CPUs/cores. On a CPU with eight cores, -j +0 results in a maximum of eight instances of the command that are run in parallel. Accordingly, -j -1 limits the number of parallel calls to seven.

▶ -S *host1,host2,host3*
Distributes the jobs to the hosts: *host1*, *host2*, *host3*, and so on. If the host list ends with a comma, this means that the local computer will also be used.

-S assumes that communication with the hosts via ssh and scp is possible without a password; you must therefore set up the corresponding key files beforehand using ssh-copy-id. In addition, the command to be executed must of course be available on all hosts. Finally, the command should be installed in parallel on all hosts. It's required to determine the number of CPUs (cores).

-S is usually combined with the --transfer, --return {.}.result, and --cleanup options to copy the local file to the external host, transfer back the new file with the *.result identifier, and finally delete the temporary data on the host. A short notation for these three options is -trc {.}.result.

In theory, -S promises a simple distribution of jobs across multiple hosts that can be accessed via SSH. In my tests, however, the option proved to be extremely error-prone. If you want to parallelize tasks via SSH and, for example, run a command on dozens of hosts or cloud instances, you can use the pssh (*parallel SSH*) or clusterssh commands instead of parallel, although these aren't covered in this book.

Examples

A simple application of parallel is to compress, decompress, or otherwise process a number of files. Instead of compressing multiple files one after the other using gunzip *.gz, you can start a separate instance of gunzip for each file in parallel. On a computer with several CPUs/cores, this will be significantly faster than the serial processing of the files.

```
user$ ls *.gz | parallel gunzip
```

In the second example, a number of PNG images are supposed to be converted to EPS format using convert. However, it's not sufficient to simply transfer the file names. Instead, convert expects the original file name and the name minus the identifier *.png, but plus the new identifier *.eps. For this purpose, parallel replaces {} in the command string with the file name, and it replaces {.} with the name without the identifier.

```
user$ ls *.png | parallel -eta -j +0 \
        'convert -density 90 -flatten -background white {} {.}.eps'
```

parted [options] [device [command [options]]]

parted supports the partitioning of hard disks and SSDs. In contrast to the outdated fdisk command, parted also uses GUID Partition Tables (GPTs).

If you don't pass a command to parted, you can use the program interactively and enter multiple commands in sequence. The commands can be abbreviated as long as the input is unique (i.e., q instead of quit).

You should note, however, that unlike with fdisk, the commands are executed immediately and can't be undone! Instead of the less convenient parted program, you can also use its graphical user interface called gparted.

▶ -a or --align
Specifies how new partitions are to be aligned. Permissible settings are none, cylinder (alignment to cylinder limits), minimal (alignment to block limits), or optimal (alignment to multiples of 1 MiB). You should always use the -a optimal option to use the data medium at maximum speed.

▶ -l or --list
Lists all partitions.

▶ -s or --script
Runs all commands passed as parameters without queries.

The most important parted commands are briefly described here:

▶ align-check min/opt *nr*
Checks whether the specified partition starts at an optimal position. parted align-check min only tests whether the partition starts at a physical block boundary. With parted align-check opt, parted attempts to calculate the optimal start position of partitions from the key data of the data medium. If this isn't possible, multiples of 1 MiB are considered optimal.

▶ mklabel msdos/gpt
Sets up a new partition table in MBR or GPT format. Caution: the entire contents of the hard disk will be lost! parted also supports various other partition formats, but these are rarely important in the Linux environment (see man parted).

▶ mkpart primary/logical/extended/*name* [*file system*] *start end*
Creates a new partition. Note that you must specify the start and end positions, not the start position and size! The end position can also be specified as a negative number and is then calculated from the end of the data medium.

Specifying the partition type (primary, logical, or extended) is only useful for data media with MBR partition tables. If there is a GPT on the data medium, you should give the partition a name with the first mkpart parameter.

The optional specification of the file system is only used to define the partition type. However, no file system is set up on the new partition. Permitted types are fat16, fat32, ext2, HFS, linux-swap, NTFS, and ufs.

▶ print
Lists all partitions.

▶ quit
Terminates parted.

▶ resize *nr start end*
Changes the size of the specified partition. Unfortunately, the command can only be used for the extended partition of an MBR disk. To change the size of other partitions, you must delete them and then set them up again with the exact same starting point. This is of course an error-prone operation, which also requires that there is free space behind the partition.

▶ rm *nr*
Deletes the specified partition.

▶ set *nr flag* on/off
Changes the flags (additional attributes) of the partition. parted can use the boot, root, swap, hidden, raid, lvm, lba, and bios_grub flags, among others.

▶ unit *bios grub*
Defines the unit for position and size specifications. The following options are available:

- s (sectors)
- B (bytes)
- kB, MB, GB, and TB (10^3, 10^6, 10^9, and 10^{12} bytes)
- KiB, MiB, GiB, and TiB (2^{10}, 2^{20}, 2^{30}, and 2^{40} bytes)
- % (percentage, relative to the total size of the data medium)
- compact (decimal megabytes for input, reader-friendly output)

The compact setting applies by default. If parted was started with the --align option, the position and size specifications are adjusted accordingly when new partitions are created.

Examples

The following lines show how you can set up a new LVM partition on an SSD with GPT using parted:

```
root# parted -a optimal /dev/sda
(parted) print
Model: ATA SAMSUNG SSD 830 (scsi)
Hard disk  /dev/sda:  512GB
Sector size (logical/physical): 512B/512B
Partition table: gpt
Number  Start   End     Size    Filesys  Name             Flags
 1      17.4kB  1000MB  1000MB  fat32    EFI System Part...  boot, hidden
 2      1000MB  21.0GB  20.0GB  ext4                         hidden
```

3	21.0GB	21.1GB	134MB		Microsoft reser...	hidden, msftres
4	21.1GB	61.2GB	40.1GB	ntfs	Basic data part...	hidden
5	61.2GB	82.2GB	21.0GB	ext4		hidden
6	82.2GB	103GB	21.0GB	ext4		boot, hidden
7	103GB	124GB	21.0GB	ext4		boot
8	124GB	281GB	157GB	ext4		
9	281GB	386GB	105GB			lvm
10	386GB	407GB	21.0GB	ext4		

```
(parted) mkpart lvm-partition 407GB 450GB
(parted) set 11 lvm on
(parted) print
...
```

11	407GB	450GB	42.7GB		lvm-partition	lvm

```
(parted) quit
```

parted can also process commands directly without manual interaction. The following commands assume that /dev/sdb is the device of a USB flash drive or SD card. The first parted command creates a new MBR partition table there and deletes all existing data. The second command creates a primary partition, leaving 1 MiB free at the beginning and end of the disk. mkfs.vfat sets up a VFAT file system in this partition.

```
root# parted /dev/sdb mklabel msdos
  Warning: The existing partition table and all data
  on /dev/sdb will be deleted. Do you want to continue?
  Yes/No? yes
root# parted /dev/sdb 'mkpart primary fat32 1mib -1mib'
root# mkfs.vfat -F 32 -n FOTOS /dev/sdb1
```

The last example shows the use of parted in script mode. The command sets up a GPT on the specified data medium. Caution: all previously saved data will be lost!

```
root# mkpart /dev/sdb -s mklabel gpt
```

partprobe [options] [devices]

partprobe is usually run without any options or other parameters after changing the partitioning of local hard disks or SSDs, for example, after fdisk has been completed. The command then informs the kernel about the changes made so that the new or modified partitions can be used without restarting the computer.

partx [options] [partition] [disk]

partx is a low-level command that determines information about the partitioning of data media or passes it on to the kernel. partx is equally suitable for the MBR partition tables and GPTs.

The partition to be edited is usually specified directly by a device file. Alternatively, you can specify the data medium using its device file and the partition number via the -n option, for example, with partx -n 3 /dev/sda. If an individual partition itself contains subpartitions and is therefore to be treated as an independent data medium, you want to precede the partition device with a hyphen, such as partx - /dev/sda4.

Unlike fdisk and parted, partx doesn't change the data medium, so it's *not* suitable for setting up or deleting physical partitions! The -a, -d, and -u options only change the partition table of the Linux kernel, but leave the disk unchanged. The sole purpose of these options is to inform the kernel about a partitioning changed by another program. This is usually easier with the partprobe command. In practice, partx is often used for processing virtual data media, for example, for loop devices or image files of virtual machines.

► -a

Adds new partitions. The partition must already physically exist on the data medium. -a only updates the partition table of the kernel!

► -b

Processes size specifications in bytes.

► -d

Deletes partitions from the kernel's partition table.

► -n M:N

Specifies numerically the partitions of the data medium to be processed. Different kinds of notation are possible here, as the following examples illustrate:

- -n 4 denotes the fourth partition.
- -n -2 denotes the penultimate partition.
- -n 3:6 refers to partitions 3 to 6.
- -n 3: refers to all partitions, starting with partition 3.
- -n :5 includes partitions 1 to 5.

► -o columns or --output columns

Displays selected information about the selected disk or partition. columns can contain the following keywords, among others: NR (partition number), START (start sector), END (end sector), SIZE (size in bytes), TYPE (partition type as hex code or UUID), and FLAGS (additional information). partx -o NR,START,END thus provides a list of partitions with their start and end sectors. The sector size is generally 512 bytes, regardless of the actual sector size of the data medium.

▶ -s or --show

Displays information on the selected data medium or partition. The data displayed with -s or -o is determined from the physical disk and may not match the partition table of the Linux kernel.

▶ -u

Changes the size or other data of a partition in the kernel's partition table. The changes must first be made physically on the data medium.

Example

The following command provides detailed information on a hard disk with an MBR partition table:

```
root# partx -o NR,TYPE,FLAGS,START,END,SECTORS,SIZE /dev/sda
NR TYPE FLAGS  START      END  SECTORS  SIZE
 1 0x83 0x80    2048   411647   409600  200M
 2 0x8e 0x0   411648 31457279 31045632 14,8G
```

---

passwd [options] [username]

---

passwd without parameters allows you to change the password of the current user. To do this, you must first enter the old password and then the new password twice in succession. The new password is entered encrypted in the /etc/shadow file. However, in some distributions, the new password must comply with certain security rules to be accepted.

root can also change the password of other users via passwd name. The old password doesn't need to be entered; that is, root can change the password even if the user has forgotten their password. The preceding password restrictions don't apply to root, so it can also define a password consisting of just a single character. However, not even root is permitted to enter no password at all (i.e., simply pressing Enter).

By specifying options, passwd can also lock individual accounts, reactivate them, and set expiry times for the account or its password (see also the description of the chage command, which offers even more setting options in this respect).

▶ -g

Sets the password of a group. This function is rarely used. It allows you to secure access to nonprimary groups via newgrp using a password. Because the group password must be known to multiple users, the procedure is inherently insecure.

▶ -l

Deactivates the account (*lock*).

- -u

  Reactivates a deactivated account (*unlock*).

- -x *n*

  Specifies the maximum number of days a password remains valid. In other words, -x 180 means that a new password must be set approximately every six months.

passwd can only be used interactively. It's not possible to pass the password as a parameter or option. To set passwords automatically, you must use the chpasswd command.

Example

In the following example, root sets a new password for the user hawks. In the future, they will have to change their password once a year.

```
root# passwd hawks
change password for user hawks.
Enter a new password: *********
Enter the new password again: *********
passwd: all authentication features successfully updated.
root# passwd -x 365 hawks
```

paste file1 file2 . . .

paste combines the lines of the specified files into new (longer) lines and displays the result on the screen. The first line of the resulting text therefore results from the first line of the first file plus the first line of the second file, and so on. Tab characters are inserted between the components of the new line. The result can be saved in a file using > target file.

- -d 'separator'

  Specifies a separator string. With paste -d ':' file1 file2, a colon is placed in each line between the contents of file1 and file2, that is:

  ```
  line1 from file1:line1 from file2
  line2 from file1:line2 from file2
  line3 from file1:line3 from file2
  ...
  ```

patch [options] < patch file

patch applies the changes summarized in a diff file. The command is generally used to apply code changes (e.g., for the kernel code).

▶ -b
  Creates backup files for all changed files.

▶ --dry-run
  Tests the patch, but doesn't make any changes. In general, you should make sure that no problems occur before applying any patch with the --dry-run option. Nothing is more annoying than an incorrectly or only partially applied patch!

▶ -p*n*
  Removes *n* directory levels from the file names of the patch file. If the original file name is /a/b/name.c, -p1 makes it a/b/name.c, -p2 returns b/name.c. The correct value for *n* depends on the directory in which patch is run.

▶ -R or --reverse
  Applies the patch inversely. This way, a patch that has already been applied will be undone.

Example

The following commands show how to patch the kernel code from version 6.7.5 to 6.7.6. To do this, the patch from version 6.7 to 6.7.5 must first be undone before the patch for version 6.7.6 can be applied. Kernel patches usually refer to the latest major version, in this case, version 6.7.

```
root# cd /usr/src/linux-6.7.5
root# bunzip2 -c patch-6.7.5.bz2 | patch -R -p1 --dry-run (test inverse patch)
... no error messages
root# bunzip2 -c patch-6.7.5.bz2 | patch -R -p1          (6.7.5 -> 6.7)
root# bunzip2 -c patch-6.7.6.bz2 | patch -p1 --dry-run   (test patch)
... no error messages
root# bunzip2 -c patch-6.7.6.bz2 | patch -p1             (6.7 -> 6.7.6)
root# cd /usr/src root# mv linux-6.7.5 linux-6.7.6
```

pdf2ps source.pdf [target.ps]

pdf2ps creates a PostScript file from a PDF document. If you don't specify target.ps, the new PDF file will be named source.pdf. An alternative to pdf2ps is the pdftops command from the Poppler library. To convert PostScript files into PDF documents, you can use ps2pdf.

▶ -dLanguageLevel=*n*
  Specifies which PostScript level is to be used for the output (by default, this is PostScript level 2; alternatively levels 1 and 3 are also supported).

```
pdfimages [options] file.pdf [image name]
```

pdfimages attempts to extract images from PDF files. However, this doesn't always work perfectly because often countless tiny images are recognized. The best way to recognize the relevant images is by their file size.

You pass a PDF file and an image name to the command. The command then creates images, which it names using the `imagename-nnn.type` format. By default, the command writes images in the unusual PBM and PPM formats, unless you select a different format by using `-f`.

▸ `-all`
  Saves embedded images in the specified format if possible. Where this isn't possible, PNG images will be generated.

▸ `-f n` and `-t n`
  Specifies from which page to which page the PDF file should be analyzed. Without these options, the command takes the entire document into account.

▸ `-j`
  Keeps the format of images that are embedded in the PDF file in JPEG format.

▸ `-list`
  Lists the images found, but doesn't create any image files.

▸ `-png` or `-tiff`
  Saves all images in PNG or TIFF format.

Example

`pdfimages -list` provides comprehensive information on all images found. The output has been shortened here for reasons of space.

```
user$ pdfimages -list book.pdf
page   type   width height color comp bpc  enc interp  object  ...  size ratio
   1 stencil     1      1   -      1   1  image  no   [inline]       1B   -
   4 stencil     1      1   -      1   1  image  no   [inline]       1B   -
   4 stencil     1      1   -      1   1  image  no   [inline]       1B   -
   4 stencil     1      1   -      1   1  image  no   [inline]       1B   -
   4 image     415    324  rgb     3   8  image  no        71    11.5K 2.9%
   5 image     626    441  rgb     3   8  image  no        81    19.3K 2.4%
   7 image    1173    635  rgb     3   8   jpeg  no       134    59.6K 2.7%
...
```

In the following example, pdfimages extracts 11 bitmaps from the PDF file:

```
user$ pdfimages -all book.pdf tst
user$ ls tst*
tst-000.png  tst-001.png  tst-002.png  tst-003.png  tst-004.png  tst-005.png
tst-006.jpg  tst-007.png  tst-008.png  tst-009.png  tst-010.png
```

pdftk file1.pdf file2.pdf . . . command

pdftk manipulates PDF files. You can use it to extract pages, merge multiple PDF documents, encrypt PDF documents or remove the encryption (if you know the password), and more. For reasons of space, the syntax is simplified here. The complete syntax is provided by pdftk. You can combine multiple commands, but you must adhere to the order in which the commands are described here:

▶ input_pw *password*
  Specifies the password for file1.pdf.

▶ cat
  Merges all PDF files.

▶ cat *page list*
  Extracts the specified pages. Multiple page ranges are separated by spaces, for example, 1-5 7-10. end denotes the last page. To merge the pages from multiple PDF files, you must give each input file an abbreviation and refer to it in the page details (e.g., pdftk A=file1.pdf B=file2.pdf cat A2-4 B5 B7). If you only want pages with even or odd page numbers, you need to add even or odd to the page number, such as 1-10odd.

▶ background *watermark.pdf*
  Stores each output page with a page from watermark.pdf. If watermark.pdf has fewer pages than the output, the last page from watermark.pdf will be repeatedly used as the watermark. The source PDF file must be transparent; otherwise, the watermark won't be visible.

▶ stamp *stamp.pdf*
  Writes a page from stamp.pdf over each output page. If stamp.pdf has fewer pages than the output, the last page from stamp.pdf will be repeated. stamp.pdf must be transparent; otherwise, it will cover the original PDF file.

▶ burst
  Creates a separate file for each page of the PDF document and names it *page_n*.pdf, where *n* is the page number.

▶ output *result.pdf*
  Saves the resulting PDF file as result.pdf.

▶ owner_pw *password* or user_pw *password*
  Encrypts result.pdf with the specified password. The owner_pw password applies to printing and other PDF operations, while the user_pw password applies to opening the file.

Examples

The following command reads pages 10 to 20 and 30 to 40 from in.pdf and writes them to the new out.pdf file:

```
user$ pdftk in.pdf cat 10-20 30-40 output out.pdf
```

You can also use the cat command to join multiple PDF files together:

```
user$ pdftk in1.pdf in2.pdf in3.pdf cat output out.pdf
```

The following example creates a separate PDF file named *pg_n* for each individual page in in.pdf, where *n* is the page number:

```
user$ pdftk in.pdf burst
```

The next example creates an encrypted PDF file. The file can be read without the xxx password, but can't be printed or edited in any other way. If you want to protect the reading of the file yourself, you must use the user_pw command instead of owner_pw.

```
user$ pdftk in.pdf output encrpyted.pdf owner_pw xxx
```

Finally, pdftk is used to mark a PDF file with a watermark:

```
user$ pdftk in.pdf background watermark.pdf output out.pdf
```

pdftops [options] source.pdf [target.ps]

pdftops from the Poppler library, which is provided in the poppler-utils package on Debian and Ubuntu, generates a PostScript file from a PDF document. Unlike pdf2ps, it supports numerous options for influencing the PostScript file.

▶ -eps

Creates an EPS file. For multipage PDF documents, a page must be selected with -f and -l.

▶ -f *n* and -l *n*

Specifies the first and last page (*first* and *last*).

▶ -level*n*

Specifies the desired PostScript level (1 to 3).

▶ -level*n*sep

Also performs a color separation. For this purpose, all colors are converted to CMYK format.

▶ -opw *password* or -upw *password*

Specifies the owner or user password to process password-protected PDF documents.

▶ -paper *format*
Specifies the desired paper format (A3, A4, letter, or legal).

▶ -paperw *n* and -paperh *n*
Indicates the paper size in dots.

pdftotext [options] source.pdf [target.txt]

pdftotext extracts the text from a PDF file and saves it in a plain text file (source.txt by default). The entire formatting and all images will be lost. pdftotext is also part of the Poppler library.

The -f, -l, -opw, and -upw options have the same meaning as in pdftops (see there).

▶ -layout
Tries to preserve the page layout.

▶ -nopgbrk
Doesn't mark the end of the page with a special character.

pdfunite in1.pdf in2.pdf . . . out.pdf

pdfunite from the Poppler library merges multiple PDF files into a new document (last file name). You can also do this with pdftk, but pdfunite is easier to use.

pg_dump [options] [> backup.sql]

pg_dump creates a backup of a PostgreSQL database. The options for establishing a connection correspond to those in the psql command (see the "psql" entry). To automate backups, you can save the password in a .pgpass file (see *www.postgresql.org/docs/current/libpq-pgpass.html*).

▶ -a
Creates a pure data backup (i.e., without CREATE DATABASE).

▶ -C
Includes the CREATE DATABASE command in the backup.

▶ -F plain|directory|tar
Defines the database format. Usually, a SQL file is generated (plain), which can be reloaded via psql -f. -F directory creates binary database files that can be restored later using pg_restore. -F tar creates the same files, but packs them into a TAR archive.

▶ -f *backupfile/backupdirectory*
Defines the storage location of the backup. If necessary, the directory will be created.

▶ -s

Creates only a backup of the schema. The backup then only contains CREATE commands, but no data.

▶ -t *pattern*

Takes only those tables into account for the backup that have been specified in the sample. You can specify the option multiple times (-T table1 -T table2).

▶ -T *pattern*

Ignores the specified tables.

Example

The following command establishes a network connection to the PostgreSQL server running on the local computer and then creates a backup of the test database:

```
user$ pg_dump -h localhost -U name -d test > test.sql
Password: ********
```

pidof program name

pidof determines the process numbers of all instances of a named process.

▶ -o %PPID

Ignores the parent process, that is, in the case of shell scripts, the shell in which the script is currently being run.

▶ -s

Returns only the first matching process (*single shot*).

Example

The following command returns the PIDs of all running bash shells:

```
root# pidof /bin/bash
32329 21636 21600 3351 1739 922
```

pinctrl command

The command determines or changes the state of the GPIO pins of the Raspberry Pi. It replaces the obsolete raspi-gpio command and is compatible with the RP1-I/O chip of Raspberry Pi 5.

▶ funcs [*nr*]

Indicates which functions the specified GPIO can perform. If *nr* is missing, raspi-gpio func provides this data for all GPIOs.

nr refers to the numbering of the GPIOs according to the manufacturer Broadcom's documentation. These aren't the pin numbers of the J8 header!

▶ get*[nr]*
Displays the status of all GPIOs or the GPIO specified by *nr*.

▶ help
Displays a relatively detailed help text. Unfortunately, there is no man page available in this context.

▶ set *nr state*
Activates the desired state for the GPIO. Possible values are listed here:

  – ip = input

  – op = output

  – no = deactivate (*no function*)

  – dl = state 0 (*drive low*)

  – dh = state 1 (*drive high*)

  – pu = pull-up resistor active

  – pd = pull-down resistor active

  – pn = pull-up/down resistor deactivated (*no pull*)

  – a0 to a5 = alternative function 0 to 5

As far as meaningful combinations arise, multiple keywords from the preceding list may be used at once, each separated by spaces. You can use pinctrl func to check which alternative functions a GPIO supports and how they are numbered.

Example

The first command determines which functions are supported by the GPIO with the number 23. On the Raspberry Pi, this GPIO is connected to pin 16 of the J8 header. The second command uses this GPIO as a simple signal output, whereby the level is set to high (3.3 V). An LED connected to pin 16 via a series resistor would now light up. The third command verifies the status of GPIO 23.

```
user$ pinctrl funcs 23
23, GPIO23, SDO_CMD, DPI_D19, I2SO_SDO1, SCL3, I2S1_SDO1,
    SYS_RIO023, PROC_RIO023, PIO23, -
user$ pinctrl set 23 op dh
user$ pinctrl get 23
23: op dh pd | hi // GPIO23 = output
```

```
ping [options] address
ping6 [options] address
```

ping sends a network packet (ECHO_REQUEST according to the Internet Control Message Protocol [ICMP]) to the specified address once per second. If the address is reachable and echo packets aren't blocked by a firewall, ping receives response packets and indicates how long the communication took there and back. By default, ping runs indefinitely until it's terminated by `Ctrl`+`C`.

▶ -4 or -6

Uses only IPv4 or IPv6. In some distributions, you can alternatively call the ping4 or ping6 commands.

▶ -c *n*

Sends only *n* packets and terminates afterwards.

▶ -i *n*

Specifies the interval time between two packets in seconds. *n* is a float; that is, ping -i 0.1 is also possible.

▶ -n

Displays only the IP address, but not the host name of the recipient.

pip or pip3 command

The command from the python3-pip package supports the installation of Python packages from the *Python package index* (PyPI). In many distributions, the command must be run as pip3 to differentiate it from the previously used pip command for Python 2.

In contrast to Linux package management commands, pip can usually be run without root permissions or without sudo. The packages are still available to all users (unless you use the --user option described previously).

The local installation of PIP packages sometimes results in conflicts with packages from Linux package management. In current Python versions, pip therefore displays an error message. If possible, you should install the desired module as a Linux package (e.g., using apt install python3-*name*). If there is no suitable package in the distribution, you should organize your project in a Python environment (see the following example). Alternatively, you can force the installation using the --break-system-packages option, but this should really only be an emergency solution.

▶ install [--upgrade] [--user] [--break-system-packages] *package name*

Installs or updates the specified package. With the --user option, the package gets installed in such a way that it's only available to the current user.

With --break-system-packages, you force the installation of a package, even if the protection mechanism of new Python versions advises against it. The option has a pretty daunting name: problems can occur, but only if there are a lot of packages in play or if an update of the Python version is performed—so ultimately, this happens relatively rarely. Nevertheless, the use of a virtual environment is almost always the better solution.

▶ `list [--outdated]`
Provides a list of all installed packages or all packages that can be updated.

▶ `search` *sudor term*
Searches PyPI for packages with the specified search term.

▶ `show [--files]` *package name*
Displays information on an installed package or lists its files.

▶ `uninstall` *package name*
Removes the specified package.

Example

Using the following commands for Debian/Ubuntu, `pip3` itself is installed first. `python3 -m venv` sets up a virtual environment in the project directory. The environment is activated via the script provided for this purpose (see also `source`). This will change the prompt. Finally, the Python module for the OpenAI API gets installed in the environment.

```
user$ sudo apt install python3-pip python3-venv
user$ mkdir my-project
user$ cd my-project
user$ python3  -m venv .
user$ source bin/activate
(my-project) user$ pip3 install openai
```

pkcon command

PackageKit is a general interface to the various Linux package management systems. PackageKit is mainly used by graphical user interfaces such as *Gnome Software*, but it can also be controlled via the `pkcon` command. Depending on the backend used by PackageKit (APT, DNF, YUM, etc.), some of the following commands may not be available.

▶ `get-groups`
Determines package groups.

▶ `get-updates`
Provides a list of all packages that can be updated.

▶ `install` *package name*
Installs a package.

▶ install-local *file*
Installs a local file.

▶ refresh [force] [options]
Updates the cache with the package information. If you use pkcon refresh force -c -1, the permitted age for the cache files will be set to less than zero. This causes the entire cache to be deleted, which can save a lot of space.

▶ remove *package name*
Uninstalls a package.

▶ search [name|details|group|file] *search term*
Searches for packages.

▶ update [*package name*]
Updates the specified package or all installed packages.

popd

The bash command popd returns to a directory that was previously saved using pushd. The directory is removed from the directory list. popd and pushd are usually only used in shell scripts.

postconf [options] [parameter[=value]]

postconf helps to read the configuration of the Postfix mail server and to change it during operation. If the command is called without options, it simply displays the currently valid setting of the approximately 1,000 Postfix parameters.

▶ *parametername*
Shows the current status of the parameter.

▶ *parametername=new value*
Sets the parameter again and also makes the change in the /etc/postfix/main.cf configuration file.

▶ -d or -df
Displays the default value of all parameters (even if the parameters are currently set differently). The -df option (*fold*) instead of -d causes long entries to be wrapped over several lines.

▶ -d *parametername*
Displays the default state of the parameter (possibly different from the current setting).

▶ -n or -nf
Lists all parameters that don't have the default value. -nf wraps long lines.

`postqueue` [options]

`postqueue` provides information about emails from the Postfix mail server that haven't yet been delivered or initiates a new transmission attempt.

▶ `-f`
  Attempts to resend all emails in the queue immediately (*flush*). Postfix takes care of this automatically but maintains wait times between multiple transmission attempts.

▶ `-i` *id*
  Tries to resend the specified email immediately.

▶ `-p` or `-j`
  Shows all emails in the queue. With `-j`, the output is in JSON format.

`powertop`

The interactive `powertop` program from the package of the same name analyzes which processes consume the most energy or wake up the CPU or hard disk most frequently from sleep mode. While the program is running, you can use the [Tab] key to switch between different result pages.

The most interesting page from the point of view of energy-saving functions is called **Tunables**. There, `powertop` shows a list of settings whose current state can be **Bad** or **Good**. You can then select individual points with the cursor keys and change them by pressing [Enter]. `powertop` shows which command it's executing.

Then, you can try changing individual settings step by step and test what effect this has: Does the energy consumption of the previously charged notebook, recorded with a power meter, actually drop noticeably? Do the disabled functions cause problems? Can USB devices still be used? Does switching the Wi-Fi adapter on and off still work? Does the audio system work without interference, and so on?

The changes made using `powertop` are only valid until the next restart of the computer. To activate the power saving measures permanently, you enter the commands displayed by `powertop` (e.g., `echo '1' > /sys/xxx`) into a file that is executed at every system start. In most distributions, `/etc/rc.d/rc.local` is suitable for this. If necessary, you must create this file and mark it as executable by using `chmod a+x`. Depending on the distribution, you may also need to run `systemctl daemon-reload` so that systemd includes the file.

A radical solution would be to enter `powertop --auto-tune` in `rc.local` instead of individual tuning commands. Then, `powertop` simply performs all known optimization measures. Unfortunately, `powertop` often overshoots the mark with this: What good is it if the notebook runs for an hour longer than before, but the network connection is unreliable?

You can get a list of all possible tuning commands by running powertop with the --html option. After a measurement time of approximately 20 seconds, the command then generates the HTML file powertop.html, which contains various statistical data as well as a summary of all tuning parameters.

ppa-purge ppa:ppaowner[/ppaname]

The Ubuntu-specific ppa-purge command from the command of the same name deactivates the specified private package repository (PPA, see also the description of the add-apt-repository command). If the PPA provides alternative versions of official packages, these packages will be removed and replaced by the original packages.

Example

The following command deactivates the libreoffice package source and replaces the packages originating from this package source with standard Ubuntu packages:

```
root# ppa-purge ppa:libreoffice/ppa
```

printenv [variable]

Outputs the content of the specified environment variable or all environment variables line by line.

printf format para1 para2 para3 ...

printf allows you to format output in the syntax of the C command; printf. You can obtain detailed information on the formatting options via man 3 printf.

Example

The following command outputs an integer, a floating-point number with two decimal places, and an end-of-line character.

```
root# printf "%d %.2f\n" 123 3.1415927
123 3.14
```

pro command

The pro command is available on all current Ubuntu versions. Although it's primarily used to control the paid Ubuntu Pro offering, some of the functions can also be used without a Pro subscription.

- attach *token*

  Activates the Ubuntu Pro functions. You can obtain valid tokens via the Ubuntu Pro dashboard.

- security status

  Provides an overview of how many packages are being serviced and for how long. The five-year update guarantee for Ubuntu LTS versions only applies to the package sources *main* and *restricted*. A typical Ubuntu installation contains a mix of packages from different package sources.

- status

  Shows the status of the Ubuntu Pro services.

Example

The following results were obtained on a server with Ubuntu 22.04, on which the free version (with a maximum of five installations) of Ubuntu Pro was activated:

```
user$ pro status
SERVICE         ENTITLED   STATUS     DESCRIPTION
anbox-cloud     yes        disabled   Scalable Android in the cloud
esm-apps        yes        enabled    Expanded Security Maintenance for Applica-
tions
esm-infra       yes        enabled    Expanded Security Maint. for Infrastructure
livepatch       yes        enabled    Canonical Livepatch service
Subscription: Ubuntu Pro - free personal subscription
user$ pro security-status
1186 packages installed:
 938 packages from Ubuntu Main/Restricted repository
 243 packages from Ubuntu Universe/Multiverse repository
   5 packages no longer available for download
 This machine is attached to an Ubuntu Pro subscription.
Main/Restricted packages are receiving security updates from
Ubuntu Pro with 'esm-infra' enabled until 2032.
Universe/Multiverse packages are receiving security updates from
Ubuntu Pro with 'esm-apps' enabled until 2032.
```

ps [options]

ps displays the list of running processes (programs) and kernel threads. The command is particularly useful in conjunction with kill to forcibly terminate programs that have hung themselves up. ps is equipped with countless options, which are described in detail in the online manual (man ps). There, you'll also find explanations of what the numerous pieces of information output by ps mean. A variant of ps is pstree, which displays the process tree and thus makes the hierarchy of the processes clear at a glance.

In contrast to other commands, ps has options with and without a preceding hyphen. Some of these even have different meanings. (ps -a and ps a aren't equivalent!) Both option types can be mixed in groups, such as ps -A ul.

▶ a

Also displays processes of other users (not just your own).

▶ -A

Displays all processes.

▶ f

Displays the process tree.

▶ -f -l or l

Displays various additional information (memory requirements, priority, and more).

▶ -p n or -p n1,n2,n3

Displays only processes with the specified process numbers (PIDs).

▶ --ppid n

Displays only processes that have n as their parent process ID, that is, that were started by process n.

▶ u

Also displays the names of the users of the respective processes. The option can't be combined with l.

▶ x

Also displays processes to which no terminal is assigned. These include internal Linux processes for managing the system (*daemons*).

▶ Z

Also displays the SELinux context of the processes.

Examples

The following examples show basic applications of the ps command:

```
root# ps ax          (shows all processes)
root# ps ax | grep ssh  (shows processes whose description contains ssh)
root# ps axu         (shows all processes including owner/account)
```

On Linux, the kthreadd kernel thread daemon always has process number 2. All other kernel threads are started by kthreadd. So, if you only want to list the kernel threads, you can use ps -ppid 2:

```
root# ps --ppid 2  (only shows kernel threads)
```

ps2pdf [options] source.ps [target.pdf]

ps2pdf creates a PDF file from any PostScript or EPS file. If you don't specify target.pdf, the PDF file will have the same name as the PostScript file (but the identifier .pdf, i.e., *source*.pdf).

▶ -dEncodeColorImages=false
Prevents any compression of images. The resulting PDF files are now very large. In combination with the /prepress setting, ps2pdf now delivers PDFs in optimum print and exposure quality.

▶ -dPDFSETTINGS=/default | /screen | /printer | /prepress
Specifies whether the PDF document is to be optimized for any use, for screen display, for a regular printout, or for exposure (e.g., book printing, etc.). The price for the higher quality is the increasing file size. These four default settings save you having to set numerous options individually.

▶ -r*n*
Specifies the resolution for bitmap fonts (unit DPI, i.e., *dots per inch*).

Countless other options are described at the following address:

https://web.mit.edu/ghostscript/www/Ps2pdf.htm

Example

The following command creates the PDF document book.pdf from the PostScript file book.ps:

user$ **ps2pdf book.ps**

psql [options]

psql enables the execution of SQL commands for PostgreSQL databases. psql is part of the postgresql package. If you only need the client without the associated database server (e.g., because the database is running on another server or in a Docker container), you can install the postgresql-client (Debian, Ubuntu) or libpq (Fedora, RHEL) package for some distributions.

▶ -c 'SQL command'
Runs the specified SQL command and then closes the connection.

▶ -d *dbname*
Specifies the database to which the connection is to be established. If a connection already exists, you can change the active database using \c dbname.

▶ -f *file.sql*

Runs the SQL commands contained in the specified file.

▶ -h *hostname*

Specifies the host on which the PostgreSQL server is running. The option can be omitted for local installations if communication via a socket file is possible.

▶ -l

Lists all databases and other objects (users, roles, etc.). If a connection already exists, \l lists the database objects. In the active database, \d provides a list of all tables.

▶ -U *username*

Defines the username. Without this option, psql uses the currently active Linux account name.

Example

The following listing shows the creation of a database connection, the creation of a test database with a simple table, the insertion of two data records, and a simple query:

```
user$ psql -U name -h localhost
Password for user name: ********
postgres~# CREATE DATABASE test;
postgres~# \c test;              (activate database)
postgres~# CREATE TABLE mytable (id SERIAL PRIMARY KEY, data TEXT);
postgres~# INSERT INTO mytable (data) VALUES ('abc'), ('efg');
postgres~# SELECT * FROM mytable;
 id | data
----+------
  1 | abc
  2 | efg
```

> **pssh** [options] command
> **pscp** [options] local_file remote_file
> **pnuke** [options] pattern

The Python script pssh from the package of the same name runs a command on multiple servers at the same time. pscp copies files from all or to all specified hosts. pnuke terminates a program whose name or pattern has been specified on all hosts via kill -9. All three commands require that authentication with keys is possible or that the ssh-agent background program takes care of authentication.

▶ -h hosts.txt

Specifies the location of the text file that contains the host names of the computers on which the command is executed line by line.

▶ -o *directory*

Specifies a directory in which pssh saves the output of the commands and pscp saves the files to be copied. pssh and pscp set up separate output files for each host, the names of which indicate the host.

Example

The example assumes that all hosts are RHEL systems. Then, a software update is performed on these computers using yum.

```
root# pssh -h hosts.txt yum update
```

pstree [options] [pid]

pstree from the psmisc or pstree package displays a tree with all processes on the screen. The tree makes it clear which process was started by which other process. If a process number is specified, the tree starts at this point; otherwise, it starts with the first process that is run at system startup (usually systemd).

▶ -h

Identifies the current process and all its parent processes.

▶ -p

Also shows the process number (PID) for each process.

▶ -u

Shows the username or account name for all processes whose UID (user ID) differs from the UID of the parent process.

pushd directory

The bash command pushd saves the current directory and then changes to the specified directory. popd takes you back to the original directory. dirs displays the list of saved directories. pushd and popd are mainly used in shell scripts.

pvcreate [options] device

The LVM command pvcreate declares a partition or a device as a physical volume (PV) for later use in a volume group (VG; see also vgcreate and vgextend).

pvcreate requires that the partition or device has previously been marked as an LVM partition. In fdisk, you must use the [T] shortcut command and code 8e for this purpose. In parted, the required command is set *partition number* lvm on.

Example

The following command turns the /dev/sdc1 partition into a physical volume for LVM:

```
root# pvcreate /dev/sdc1
  Physical volume "/dev/sdc1" successfully created
```

pvdisplay device

pvdisplay shows detailed information on the specified PV.

pvremove device

pvremove removes the PV label of an unused PV.

pvscan

pvscan lists all PVs.

```
pw-cat audio file
pw-cli [command]
pw-mon [options]
pw-top [options]
```

Most current Linux distributions use PipeWire as an audio and video server. There are various commands for controlling PipeWire, each with a name beginning with pw-. Here, I'll focus on the most frequently used commands:

▶ pw-cat transfers raw files from or to the audio system. The command can also be run as pw-play, pw-record, pw-midiplayrecord, or pw-midirecord.

▶ You can change the audio setup using pw-cli. There are countless subcommands, of which only the most important are mentioned here:

  – connect or disconnect establishes or terminates a connection to an external PipeWire server.

  – create-link connects two PipeWire nodes.

  – create-node and export-node manage PipeWire nodes, that is, objects for signal management or forwarding.

  – info *id* displays detailed information on a specific object.

  – list-objects lists all PipeWire objects.

  – list-remotes lists all external PipeWire instances.

  – load- and unload-module loads or removes PipeWire modules.

- – `send-command` sends a command to a PipeWire object.
- – `switch-remote` defines the active remote instance.

`pw-cli` is a low-level command. Its use is extremely cumbersome. Because PipeWire has a compatibility layer with PulseAudio, in many cases, it's easier to control the setup via `pactl`.

▶ `pw-mon` lists all audio and video objects known to the PipeWire system, including countless properties.

▶ `pw-top` provides an overview of the currently active PipeWire components, including the latencies occurring during audio processing.

pwd

The `bash` command `pwd` specifies the current directory.

In addition to the version included in the `bash`, `pwd` is also available as an independent command, often under the file name of `/bin/pwd`. Note that there may be cases in which the two variants of `pwd` don't produce the same result. This is because the `bash` version of `pwd` works on a relative basis, whereas the external command works with absolute values. Symbolic links to directories can trick the `bash` version of `pwd` to a certain extent.

Example

In the following example, `ln` creates a symbolic link to the `/tmp` directory. `cd` changes to this directory. The `bash` believes that this directory is the current directory, while `/bin/pwd` recognizes that `/tmp` is actually the current directory.

```
user$ ln -s /tmp symlink
user$ cd symlink
user$ pwd          (bash version of pwd)
/home/kofler/symlink
user$ /bin/pwd     (standalone pwd command)
/tmp
```

pwgen [options] [n]

`pwgen` from the package of the same name generates an entire list of random but easy-to-remember passwords. The idea is that the user selects one of the passwords and then clears the screen using `Ctrl`+`L` before anyone can look over their shoulder and read the password. Compared to the passwords generated by `makepasswd`, the `pwgen` passwords are less random and therefore less secure, but they are still good enough for many use cases.

By default, passwords are eight characters long. You can generate longer passwords by specifying the optional parameter *n*.

▶ -0

Doesn't use numbers in the passwords.

▶ -1

Provides only 1 password and not 100.

▶ -B

Avoids letters or numbers that are easy to confuse depending on the font, that is, O and 0 as well as l and 1. This makes passwords less secure.

▶ -y

Also includes special characters in the passwords.

Example

As the example shows, even the passwords generated in the default settings are by no means trivial:

```
user pwgen
Ohshu3yo Ea1wedoe OhCh2Zua Aili8ooc Xu4iiyix eug3Chee Gaesh2pu Eeth6mah
eMeeOjio xieL6oob ob8uYah9 shaifOEd uep8Eive lang3Eho thaiS7xa Sah3See4
...
```

The following command generates an 80-character password that also contains special characters:

```
root# pwgen -1 -y 80
AhB9eiv]oo~sOgo2ievaF7eePe6ooyoo7Cim%oh8eishiloosaeSeex+eeza9Iem7Ahchuhi6ahch[
i1
```

qalc [expression]

qalc from the package of the same name calculates the transferred mathematical expression and returns the result. To avoid special characters being processed by the bash, you must enclose most expressions in quotation marks:

```
user$ qalc "(2+3)*4^5"
(2 + 3) * (4^5) = 5120
```

If no expression is transferred, qalc activates an interactive mode. It then performs calculations until the command is ended via [Ctrl]+[D].

```
user$ qalc
>2^64
 2^64 = approx. 1.8446744E19
```

```
>sin(0.2)
 sin(0.2 * radian) = approx. 0.19866933
>sin(pi)
 sin(pi * radian) = 0
>sqrt(2)
 sqrt(2) = approx. 1.4142136
>10 EUR to USD
 10 * euro = $10.651
```

qemu [options] [imagefile]

Quick Emulator (QEMU) emulates various CPUs. There is a special variant of the command for each important CPU architecture: qemu-aarch64 for 64-bit ARM systems, qemu-mips64 for 64-bit MIPS systems, and more. For x86 systems, qemu-x86_64 is the most important command. In some distributions, the names are qemu-system-aarch64, qemu-system-x86, and more. In the following, I'll simply stick to qemu for the sake of brevity. Depending on the distribution and CPU architecture, you must call the appropriate variant of the command.

If possible, qemu is combined with *Kernel-Based Virtual Machines* (KVM), a Linux kernel extension. This means that virtual machines with the same architecture as the host system can be run much faster. KVM is activated with the -enable-kvm option or with -accel kvm. Depending on the distribution, there are tiny scripts such as kvm or qemu-kvm that start qemu with the KVM option. On RHEL systems, the qemu-kvm command is located in the /usr/libexec directory.

In the simplest case, you only transfer the name of an image file to qemu. QEMU then runs the virtual machine with default settings, including an IDE hard disk. If you want to use other settings or multiple data media, you can add the options -drive or -hda, -hdb, and so on. In this case, the direct specification of the image file in the KVM command can be omitted.

▶ -accel kvm

Activates KVM. QEMU can also cope with other acceleration systems, which is why the option also supports the xen, hax, or tcg settings.

▶ -boot order=xxx,once=xxx,menu=on/off

Specifies the order in which the data media are to be considered for the boot process. Here, xxx is a sequence of letters that expresses the order of the data media (e.g., adc: first the floppy disk drive, then the CD/DVD drive, then the first hard disk). The letters a to d correspond to the Windows drive letters.

once=xxx specifies the boot sequence for the *first* boot process only. For example, if the virtual machine is supposed to boot from the CD/DVD drive on the first attempt, but from the hard disk on subsequent restarts, you enter -boot order=c,once=d or simply -boot once=d. (The virtual DVD drive is usually connected to the ISO image for the installation system.)

menu=on displays the message **Press F12 for boot menu** at the start of the boot menu. The boot disk can then be selected interactively by pressing the F12 key.

▶ -cdrom *iso file*
Uses the specified ISO file as the data source for the virtual CD/DVD drive. The option corresponds to -drive file=iso-file,index=2,media=cdrom.

▶ -cpu *host*
Transfers all properties of the host CPU to the guest. This isn't the case by default: only one subset is passed on to maximize the compatibility of virtual machines between different CPUs.

▶ -device *cup*
Adds an additional device to the virtual machine. A list of all supported devices is provided by qemu -device ? The device name is case-sensitive! You can determine the options available for a specific device using qemu -device *cup*,?, for example, qemu -device isa-serial,?.

Note that you can define most components of a virtual machine in two ways: with the very universal -device option described here or with device-specific options (e.g., -drive, -soundhw, -usb-device, or -vga).

▶ -drive *details*
Defines the properties of a virtual hard disk. The detailed parameters are separated from each other by commas only (without spaces!). The option can be used multiple times if the virtual machine is to be equipped with multiple data media.

boot=on/off specifies whether the data medium should be taken into account when booting.

cache=writethrough/writeback/none specifies whether and how write accesses are cached. By default, writethrough is used: in the guest system, a write access doesn't appear as completed until the host system has acknowledged the save operation.

file=fname specifies the file name of the image, ISO file, or the device name of a logical volume.

if=ide/scsi/virtio specifies which interface the virtual machine should use to access the data medium (the default value is ide). virtio is more efficient with Linux guests.

index=*n* determines the numbering of the data media of an interface. The parameter is only required if the data media aren't specified in sequence.

media=disk/cdrom specifies whether it should be a hard disk (default setting) or a CD/DVD drive.

▶ `-enable-kvm`
Corresponds to `-accel kvm`.

▶ `-hda/-hdb/-hdc/-hdd` *details*
Specifies a virtual IDE hard disk.
`-hda fname` corresponds to `-drive file=fname,index=0,media=disk`.
`-hdb fname` corresponds to `-drive file=fname,index=1,media=disk`, and more.

▶ `-k` *language identifier*
Uses the specified keyboard layout. Permitted language identifiers include `de` (German) and `en-us` (US English). This option is only required if the virtual machine is operated by an external VNC client. The VNC clients of the virtual machine manager or the `virt-viewer` command automatically recognize the keyboard setting.

▶ `-localtime`
Initializes the virtual CMOS clock of the guest system with the local time (instead of the default UTC time).

▶ `-m` *n*
Sets the memory size of the virtual machine (in MiB). The default setting varies depending on the distribution.

▶ `-machine` *name[,para1=value1,para2=value2 ...]*
Specifies which hardware is supposed to be emulated. A list of permitted type names is provided by `-machine help`. The parameters can be used to activate or deactivate additional properties that differ from the basic settings of the respective type.

▶ `-monitor` *device*
Redirects the input and output of the QEMU monitor to the specified device. If you want to operate the monitor via the current console, you must enter `stdio` as the device. With `pty`, qemu creates a new pseudo TTY device at startup and uses it for communication.

▶ `-net nic,`*details*
Configures a virtual network adapter. If this option isn't specified, QEMU emulates an RTL8139-compatible network card by default.

`model=ne2k_pci/i82551/i82557b/i82559er/rtl8139/e1000/pcnet/virtio` defines the network adapter QEMU is supposed to emulate. For Linux guests, you'll achieve the best results with `model=virtio`. You use `macaddr=52:54:00:nn:nn:nn` to specify the desired MAC address.

▶ `-net user,`*details*
Uses user mode networking (this is the default setting). Although the guest system can use the internet connection of the host system thanks to NAT and masquerading, a direct network connection between guest and host isn't possible.

- ▶ `-rtc base=utc/localtime`

  Specifies which start time the virtual machine clock should have. `utc` is the correct setting for Linux guests, while `localtime` is suitable for Windows guests. By default, the clock is always synchronized with that of the host computer. If you don't want this to happen, you can specify the additional `clock=vm` parameter.

- ▶ `-smp` n or `-smp cores=`c`, threads=`t`, sockets=`s

  Short notation to specify how many CPUs or cores are to be assigned to the virtual machine (only one core by default). For host systems with multiple CPUs, c specifies how many cores are to be used per CPU. t specifies the desired number of threads per core; the value 2 is usually useful here for Intel CPUs that support hyperthreading. s finally specifies how many CPUs (sockets) are to be used. $c*t*s$ provides the number of CPUs the virtual machine is able to see.

- ▶ `-soundhw ac97/es1370/hda/sb16/all`

  Adds one of the specified audio devices to the virtual machine (or all if you use `all`). `ac97` stands for Intel 82801AA AC97, `es1370` for Ensoniq AudioPCI ES1370, `hda` for Intel High Definition Audio, and `sb16` for Creative Sound Blaster 16.

- ▶ `-spice port=`n`[,options]`

  Activates the Spice graphics system. For this purpose, you must specify at least the desired port.

- ▶ `-usb`

  Activates the USB driver.

- ▶ `-usbdevice mouse/tablet/disk/host...`

  Adds a USB device to the virtual machine. Most of the time you'll need the `-usbdevice tablet` option. It replaces the standard emulated PS/2 mouse with a virtual USB pointing device that understands absolute coordinates and thus enables synchronization of the guest's mouse position with the VNC or Spice client.

  `-usbdevice disk` makes it possible to pass an image file from the host to the guest in such a way that the guest sees a USB data medium.

  `-usbdevice host:bus.addr` or `-usbdevice host:vendorid:productid` redirects a USB device from the host to the guest. The USB device must not be used by the host. The easiest way to determine the bus and device number or the vendor and product IDs on the host server is to use `lsusb`.

- ▶ `-vga cirrus/qxl/std/virtio/vmware`

  Specifies the desired type of virtual graphics card. By default, QEMU emulates a Cirrus-compatible graphics card with a resolution of up to 1,024 × 768 pixels. This graphics system is recognized correctly by almost all guest systems and runs at an acceptable speed. The `qxl` graphics card can only be used in combination with `-spice`.

▶ -vnc *n.n.n.n:n*[,options]
   Runs a VNC server via which clients can display the contents of the virtual graphics card. You can use *n.n.n.n* to specify the IP address from which the connection to the VNC server can be established (e.g., 127.0.0.1 for connections from localhost). :*n* indicates the display number. The port for the VNC server is *n*+5900. If you only enter the display number without the IP address (e.g., :0), the connection can be established by any computer.

Only in exceptional cases is qemu run directly. It's more common to use the virsh command from the libvirt package to set up and start virtual machines, to use an interface such as virt-manager, or to use a cloud framework such as OpenStack.

Example

In the following example, an image file with a maximum size of 20 GiB is first created, which is then used as a virtual hard disk to install a Debian system from an ISO file:

```
user$ qemu-img create -f qcow2 disk.img 20G
user$ kvm -accel kvm -m 2048 -smp 2 -boot once=d -cdrom debian.iso \
        -drive file=disk.img,if=virtio,format=qcow2 \
        -net user -net nic,macaddr=52:54:00:12:e4:4e,model=virtio \
        -vga cirrus -vnc 127.0.0.1:0 -k de -usb -usbdevice tablet
```

To operate the virtual machine, you must now start a VNC client, such as the vncviewer program:

```
user$ vncviewer localhost:0
```

qemu-img command

qemu-img supports the creation and management of image files for virtual hard disks. The first parameter you specify must be a command, for example, create or convert. This is followed by further options, the file name of the image file, and more, depending on the command.

▶ convert [-f *source format*] -O *target format source file target file*
   Converts an image file from one format to another. The original file is retained. However, the image file must not be used by a virtual machine during the conversion. The target file is compressed with the additional option -c.

▶ convert -f qcow2 -s *snapshotname* -O *target format source file target file*
   Transfers only the snapshot of a QCOW2 image selected with -s to the new image file.

▶ create [-f raw/qcow2/qed] [-o opt1=val1,opt2=val2,...] *file size*
Creates a new image file of the specified type and size. The size is specified in bytes by default. Optionally, the suffixes k or K, M, G, or T can be used (for KiB, MiB, GiB, or TiB). Depending on the image format, various additional options can be specified. You can determine the parameters available for selection via -o '?', whereby this option must be specified in a complete, syntactically correct command:

qemu-img create -f qcow2 -o '?' test.img 1G

▶ create -f qcow2 -o backing_file=basis.img overlay.img
Creates an overlay file. When the virtual machine is run, all changes are saved in the overlay file; the base file remains unchanged.

▶ info *file*
Provides information about the specified image file. Information about snapshots in QCOW2 images can be obtained using qemu-img snapshot (see that entry at the end of this list).

▶ resize *file* +/-*size*
Enlarges or reduces the image file by the specified size. The command is only suitable for raw and QCOW2 images.

▶ snapshot [options] *snapshotname file*
Edits snapshots. This command is only available for QCOW2 images. -c creates a new snapshot, -a applies a snapshot to the image (thus revoking all changes that have been made since the snapshot was created), -d deletes the snapshot, and -l lists all snapshots. However, with the exception of -l, you may not use these commands while the image file is being used by a running virtual machine!

Examples

The first command creates a QCOW2 image file. A major advantage of this image format is that the file initially takes up hardly any memory on the data medium and grows only gradually.

root# **qemu-img create -f qcow2 disk.img 10G**

The second command creates an equivalent raw image from a QCOW2 file:

user$ **qemu-img convert -f qcow2 image.qcow2 -O raw image.raw**

rclone command

The rclone command from the package of the same name synchronizes files or entire directories with a storage location in the cloud. The files can also be encrypted or decrypted when they are transported back. rclone is compatible with more than 70 cloud providers, including AWS S3.

- copy *remote name:bucketname localdir*
  Copies files from a remote directory to a local directory. (A *bucket* is a storage location with the cloud provider.)

- config
  Guides you through an interactive setup program in which you select your cloud provider and enter the authentication data. This data is saved under the name you specify in a section in .config/rclone/rclone.conf. Each section of this type describes a remote directory.

  To configure automatic encryption, you first need to configure a normal setup for the cloud provider, and then configure a second setup (a second remote directory) of type crypt that refers to the original setup. Note that you must enter two passwords. Alternatively, rclone config generates random passwords, which you should make a note of.

- lsd *remote name:*
  Lists all accessible buckets with the specified cloud provider.

- ls *remote name:bucketname*
  Lists the files in the bucket.

- mkdir *remote name:mynewbucket*
  Creates a new bucket.

- sync *localdir remotename:bucketname*
  Synchronizes the files from a local directory with those in the bucket. Note that bucket files which don't exist locally will be deleted! You should initially use the --interactive option so that queries prevent possible errors.

Example

Before you can use rclone, you need to set up your cloud provider once. You can then use rclone sync to synchronize the content of a local directory with a bucket on the cloud provider's side.

```
user$ rclone config
...
user$ rclone sync mydata/ myremote:mybucket
```

rdfind [options] directory1 directory2

rdfind from the package of the same name searches two directories for duplicates (i.e., identical files). The command supports various methods for determining whether two files are identical and how it assesses which file is the original and which file is the duplicate. Depending on the options, the duplicates are listed, replaced by links, or even deleted.

The following options affect the search behavior:

- -checksum md5|sha1|sha256
  Calculates checksums for two files of the same size using the specified method. By default, sha1 is used.

- -followsymlinks
  Takes symbolic links into account in the comparison.

- -ignoreempty
  Ignores empty files (with a file length of 0).

- -minsize *n*
  Ignores files that are smaller than *n* bytes.

You can use other options to specify what should happen with duplicates. By default, their names are written to the results.txt file.

- -deleteduplicates true|false
  Deletes duplicates (default: false).

- -dryrun true|false
  Runs in test mode, doesn't make any changes (default: false).

- -makesymlinks true|false
  Replaces the duplicate with a symbolic link to the original (default: false).

- -makehardlinks true|false
  Replaces the duplicate with a fixed link (default: false).

- -makesresultfile true|false
  Writes the file names of matching files to results.txt (default: true). The result file not only contains the file names of both versions of each duplicate but also keywords that indicate which file rdfind considers to be the original:

  - DUPTYPE_FIRST_OCCURRENCE: Original
  - DUPTYPE_WITHIN_SAME_TREE: Duplicate in directory1
  - DUPTYPE_OUTSIDE_TREE: Duplicate in directory2

Examples

The following example determines all identical files in two directories:

```
user$ rdfind dir1 dir2
Now scanning "dir1", found 211 files.
Now scanning "dir2", found 215 files.
Now have 426 files in total.
Removed 0 files due to nonunique device and inode.
Total size is 85754591 bytes or 82 MiB
Removed 103 files due to unique sizes from list. 323 files left.
```

```
Now eliminating candidates based on first bytes:
   removed 4 files from list. 319 files left.
Now eliminating candidates based on last bytes:
   removed 0 files from list. 319 files left.
Now eliminating candidates based on sha1 checksum:
   removed 2 files from list. 317 files left.
It seems like you have 317 files that are not unique
Totally, 39 MiB can be reduced. Now making results file results.txt.
```

The second command searches two directories that contain photos. Duplicates that are at least 100 bytes in size are replaced by fixed links:

user$ **rdfind -minsize 100 -makehardlinks true photos1 photos2**

---

rdiff-backup [options] source directory target directory

rdiff-backup synchronizes the target directory with the source directory in a way that's similar to rsync. In contrast to rsync, rdiff-backup also archives old versions of changed or deleted files when executed repeatedly, whereby only the changes are saved in compressed form, that is, in the form of incremental backups, to save space. This makes it possible to reconstruct files that have been deleted or overwritten by mistake.

The source and target directories can be located on network servers. In this case, communication takes place by default via ssh. The rdiff-backup command must also be installed on the external computers. When specifying external directories, the same syntax applies as for rsync. The only difference is that *two* colons must be entered after the host name, for example, in the following way: user@backupserver::directory.

▶ -r *time* or --restore-as-of *time*
Reconstructs the data as it was at the given time. However, restoring old files requires a considerable amount of CPU as the number of versions increases and is correspondingly slow!

▶ --remove-older-than *time*
Deletes incremental backup files that are older than the specified time. You can set the time as an absolute value (e.g., 2020-12-31) or as a relative value in hours (h), days (D), weeks (W), and so on (see also man rdiff-backup in the TIME FORMATS section). Instead of a specific point in time, you can also use *n*B to specify the maximum number of backup versions that should remain archived. By default, rdiff-backup only deletes *one* backup version at a time. If you want to delete multiple backup versions at once, you must also specify the --force option.

Examples

The following command creates a backup of /home in the /home-backup directory. If this command is run regularly (e.g., daily), all changes will automatically be archived. The required space in the backup directory will then increase accordingly.

```
root# rdiff-backup /home /home-backup
```

The following command reconstructs the backed-up data in a temporary directory:

```
root# rdiff-backup -r now /home-backhup /tmp/home-current
```

The following command restores the state of the /home directory to the way it was 10 days previously:

```
root# rdiff-backup -r 10D /home-backup/ /tmp/home-historical
```

read [var1 var2 var3 ...]

read loads a line of text into the specified variables in bash scripts. read expects the data from the standard input. If no variable is specified, read writes the input to the REPLY variable. If exactly one variable is specified, read writes the entire input to this one variable. If multiple variables are specified, read writes the first word to the first variable, the second word to the second variable, and so on until the remainder of the input is written to the last variable. Words are separated by spaces or tab characters.

The read command doesn't provide the option of outputting an info text as a prompt. It's therefore advisable to inform the user of the purpose of the input via echo -n before running read commands.

readonly

The bash command readonly displays the read-only variables of the shell. Variables can be protected against changes via declare -r.

reboot [options]

reboot terminates all running processes and restarts the computer. reboot is equivalent to shutdown -r now.

recode character set1..character set2 file
recode character set1..character set2 < source > target

recode performs a character set conversion from character set 1 to character set 2. recode -l provides a comprehensive list of all supported character sets.

Examples

The following command converts the DOS file `dosdat` into a Linux file with the Latin-1 character set:

```
user$ recode ibmpc..latin1 < dosdat > linuxdat
```

The second `recode` example replaces all line endings (CR plus LF, i.e., *carriage return* and *line feed*) in the `windowsdat` file with the line ending that is common on Linux systems (LF only). The actual character set doesn't get changed. The resulting file is saved in `linuxdat`.

```
user$ recode latin1/cr-lf..latin1 < windowsdat > linuxdat
```

`recode` loads the text file `latin1dat` encoded in the Latin-1 character set and saves it as a UTF-8 file (Unicode):

```
user$ recode latin1..u8 < latin1dat > utf8dat
```

rename

Most distributions have a preinstalled `rename` command that helps with renaming multiple files. However, different variants with different syntax are used depending on the distribution. In the following sections, I'll describe the two most common variants—first the Perl script from the `rename` package, which is common on Debian, Raspberry Pi OS, and Ubuntu, and then the `rename` command from the `util-linux` package, which is common on Fedora, RHEL, and SUSE.

rename on Debian, Raspberry Pi OS, and Ubuntu

The syntax of this `rename` variant looks as follows:

```
rename [options] rename_expression files
```

In a syntax similar to `sed`, *rename_expression* specifies how to proceed with the files named in the other parameters. You can read the details of the syntax using `man perlexpr`. There are options available to you to influence the renaming process:

▶ `-f` or `-force`
Overwrites existing files.

▶ `-n`
Shows which files would be renamed, but doesn't make any changes.

The following example renames all `*.jpg` files to `*.jpeg` files:

```
user$ rename -n 's/.jpg/.jpeg/' *.jpg
```

The second command replaces all uppercase letters with lowercase letters:

user$ `rename -n y/A-Z/a-z/ *`

You can find many more examples at the following site:

www.howtogeek.com/423214/how-to-use-the-rename-command-on-linux

rename on Fedora, RHEL, and SUSE

The rename syntax on Fedora, RHEL, and SUSE is simpler:

rename [options] *find replace files*

The command simply *replaces* the *find* expression in the file name with *replace*. There is no way to test the effect of rename in advance.

▶ -s

 Also changes the names of files that are referenced by symbolic links.

▶ -v

 Displays the name changes that have been made.

To rename .jpg files back to .jpeg files, you must run rename as follows:

user$ `rename jpg jpeg *.jpg`

renice n pid

renice changes the priority of the process with the specified PID. *n* is either the new priority (a value between -20 and 20) or a delta value with a positive or negative sign (e.g., +3 or -2). Only root is allowed to increase the priority of processes. The highest priority is -20, the lowest is 20.

reset

reset restores the font in the text terminal if it has been corrupted by the output of special characters. reset also resets the terminal settings to the default settings previously saved using setterm -store.

resize2fs [options] device [size]

resize2fs changes the size of an ext2/ext3/ext4 file system. For the size specification, the notations *n*K, *n*M, and *n*G for KiB, MiB, or GiB are permissible. If you don't specify a size, resize2fs adapts the file system to the size of the underlying partition or logical volume (ideal for enlargements).

The file system can be enlarged during operation. To reduce the size, you must unmount the file system and run `fsck` first. Note that you must *first* enlarge the underlying partition or LV if you're enlarging it, but may only reduce the partition or LV *afterwards* if you're reducing it!

Example

In the following example, a logical volume is first enlarged using `lvextend`. The file system it contains is then enlarged using `resize2fs`.

```
root# lvextend -L 40G /dev/mapper/vg1-test
    Extending logical volume test to 40,00 GB
    Logical volume test successfully resized
root# resize2fs /dev/mapper/vg1-test
The file system on /dev/mapper/vg1-test is mounted on /test;
    online resizing required
Perform an online resizing of /dev/mapper/vg1-test
    to 10485760 (4k) blocks.
The size of the file system on /dev/mapper/vg1-test is now 10485760 blocks.
```

`resolvectl` [options] command [name]

In modern distributions, you can use `resolvectl` to read the local DNS configuration and, under certain circumstances, change it. `resolvectl` is a systemd component connected to `systemd-resolved` or `systemd-networkd`. `resolvectl` is particularly useful if your distribution runs a local caching name server. The `/etc/resolv.conf` file then only contains a reference to this local name server, but no information about the underlying configuration. `resolvectl status` provides information about the DNS server used by the local cache in such cases.

▶ dns *network interface ipaddr1 ipaddr2* ...
  Temporarily changes the IP addresses for the DNS server for the specified network interface. The setting isn't saved permanently and is lost if the network connection changes (at the latest on the next restart).

▶ monitor
  Displays all incoming DNS queries until you end the monitoring via `Ctrl`+`C`.

▶ query *hostname*
  Returns the IPv4 and IPv6 addresses assigned to the specified host name.

▶ query *ipaddr*
  Returns the host name assigned to the IP address (if known).

▶ statistics
  Shows the number of DNS operations and the status of the DNS cache.

▸ status

Displays the current DNS configuration.

Example

The test system, an Ubuntu desktop installation, uses Fritz!Box with the IP address 192.168.178.1 as DNS server. The Rheinwerk Publishing server was using the IP address 46.235.24.168 at the time of the test.

```
user$ resolvectl status
Link 2 (enp0s1)
  Current Scopes: DNS
        Protocols: +DefaultRoute +LLMNR -mDNS -DNSOverTLS DNSSEC=no/unsupported
  Current DNS Server: 192.168.178.1
          DNS Servers: 192.168.178.1
          DNS Domain: fritz.box
        ...
user$ resolvectl query rheinwerkverlag.de
rheinwerkverlag.de: 46.235.24.168    -- link: enp0s1
-- Information acquired via protocol DNS in 46.3ms.
```

restorecon [options] files

restorecon restores the SELinux context intended for a directory. This is necessary if SELinux was temporarily deactivated or after you've moved or copied files (e.g., using cp -a) in such a way that an automatic setting of the SELinux context wasn't possible. To define the SELinux context independently of the directory, you can use the chcon command.

▸ -0

Expects the file names to be separated by 0 bytes and not by tab or space characters. This helps in combination with find -print0 when processing files whose names contain spaces.

▸ -e *directory*

Skips this directory. The option can be repeated to exclude multiple directories from the context changes.

▸ -r or -R

Also takes into account all subdirectories (*recursive*).

▸ -v

Shows the changes made (*verbose*).

Example

The following command sets the correct SELinux context for all files in the DocumentRoot directory of the Apache web server:

```
root# restorecon -R -v /var/www/html/*
```

rfkill command

rfkill from the package of the same name makes it possible to switch Bluetooth, Wi-Fi, and mobile phone adapters on and off, for example, to save power on a notebook for functions that aren't currently required.

▶ block *index/type*
Deactivates the device specified by the index number or all devices of a certain type. Permitted types are all, wifi = wlan, bluetooth, uwb = ultrawideband, wimax, wwan, gps, and fm.

▶ event
Waits for rfkill events and outputs them.

▶ list
Lists all wireless adapters and their status. soft blocked means that the device can be switched on and off by rfkill. hard blocked means that rfkill can't control the device, for example, because it's connected to a mechanical switch.

▶ unblock *index/type*
Activates the specified device(s), which are specified in the same way as with block.

Example

Two Bluetooth and one Wi-Fi adapter are available on the test computer. One of the Bluetooth adapters gets activated:

```
root# rfkill list
0: tpacpi_bluetooth_sw: Bluetooth
        Soft blocked: yes
        Hard blocked: yes
1: hci0: Bluetooth
        Soft blocked: yes
        Hard blocked: no
2: phy0: Wireless LAN
        Soft blocked: yes
        Hard blocked: no
root# rfkill unblock 1
```

rkhunter [options]

The rkhunter shell script (*Rootkit Hunter*) attempts to detect rootkits installed on the computer. Compared to chkrootkit, rkhunter performs even more tests. As is the case with chkrootkit, however, the program isn't infallible: you must expect both *false positives* and that the program won't detect a modern rootkit at all.

In some distributions, a cron script gets set up when the rkhunter package is installed, which calls rkhunter once a day and sends an email with the results. On Debian and Ubuntu, this feature is also provided, but it's disabled in the /etc/default/rkhunter configuration file.

▶ -c or --check
   Performs a rootkit search and displays the results.

▶ --update
   Updates the pattern files for detecting rootkits.

Example

The shortened version of the program reads as follows:

```
root# rkhunter -check
[ Rootkit Hunter version 1.4.6 ]
Checking system commands
...
Performing file properties checks
  Checking for prerequisites             [ OK ]
  /usr/sbin/adduser                      [ OK ]
  /usr/sbin/chroot                       [ OK ]
  ...
Performing check of known rootkit files and directories
  55808 Trojan - Variant A               [ Not found ]
  ADM Worm                               [ Not found ]
  AjaKit Rootkit                         [ Not found ]
  ...
Performing system configuration file checks
  Checking for an SSH configuration file [ Found ]
  Checking if SSH root access is allowed  [ Warning ]
  Checking if SSH protocol v1 is allowed  [ Not allowed ]

System checks summary:
File properties checks:  Files checked: 151,   Suspect files: 1
Rootkit checks:         Rootkits checked: 475, Possible rootkits: 0
...
All results have been written to the log file: /var/log/rkhunter.log
```

rm [options] files

rm deletes the specified files. Directories aren't deleted unless the -r option is used. The rmdir command is provided for deleting individual directories. If files with special characters are to be deleted, the file names must be enclosed in single apostrophes. The most important options of the rm are listed here:

▶ -f
Deletes without further inquiry (including directories). Caution!

▶ -i or --interactive or -v or --verbose
Displays a confirmation prompt before deleting each individual file.

▶ -r or -R or --recursive
Also deletes files in all subdirectories (watch out!). If the entire content of the subdirectory is deleted, the subdirectory itself is also eliminated.

Examples

The following command deletes all backup files (files ending with the ~ character) in the current directory:

user$ **rm *\char126**

rm deletes the backup file or directory. If backup is a directory, all subdirectories and files contained in it will be deleted as well!

user$ **rm -r backup**

rm deletes all files that start with the hash character (#). The apostrophes are required so that the shell doesn't interpret # as a comment.

user$ **rm '#'***

rpicam-still [options]

The rpicam-still command available on the Raspberry Pi OS takes a photo with a camera connected to the Raspberry Pi.

▶ --camera 0|1
Selects the camera. The option needs to be specified only if two cameras are connected at the same time.

▶ -e jpg|bmp|gif|png
Specifies the desired image format (by default, this is jpg).

▶ --ev *n*

Sets the *exposure compensation*. With --ev 0.5 the resulting image becomes half a light level brighter, with --ev -0.5 darker.

▶ --height *n*

Specifies the image height (by default, that's the maximum resolution of the camera).

▶ -n

Doesn't display a preview window during recording.

▶ -o *file*

Specifies the name of the image file.

▶ -q *n*

Indicates the JPEG image quality (between 0 = minimum and 100 = maximum).

▶ --rotation *n*

Indicates the desired rotation of the image in degrees. Only the values 0 and 180 are permitted.

▶ -t *n*

Specifies the time span in milliseconds after which the photo is to be created (by default, the span after 5,000 ms, i.e., after 5 seconds). To take the picture as quickly as possible, you must enter -t 1.

▶ --timelapse *n*

Specifies the number of milliseconds after which additional photos should be taken. In this case, -t controls the entire runtime of the command (e.g., -t 60000 means one minute).

▶ --width *n*

Specifies the image width (by default the maximum resolution of the camera).

Example

The following command creates a photo in a resolution of 800 × 600 pixels with minimum wait time and without displaying a preview window:

```
user$ rpicam-still -n -t 1 -w 800 -h 600 -o foto.jpg
```

The second example creates numbered photos that can later be combined into a time-lapse movie. Over a period of two minutes, a photo is taken every 500 ms (resulting in 240 images).

```
user$ rpicam-still -n -t 120000 --timelapse 500  -o foto-%d.jpg
```

Variants

The command can also be run under the old name of libcamera-still. Instead of rpicam-still, you can also run rpicam-jpeg. The command has slightly fewer options and doesn't support any image formats except JPEG. Finally, you can use rpicam-hello to check the function of the connected camera. This command displays a preview window for five seconds, but doesn't save any images.

rpicam-vid [options]

The rpicam-vid command records a video in H264 format with the camera, which is available as an extension for the Raspberry Pi. The default resolution depends on the camera model.

Many control options correspond to the rpicam-still command described previously, including --height, -n, -o, --rotation, and --width. At this point, I'll focus on two rpicam-vid-specific options:

▶ --framerate *n*
Specifies how many frames per second are to be recorded (30 by default). Current Raspberry Pi models can manage up to 120 frames per second at a reduced resolution, but various additional options must then be used. Tips for recording videos with as many frames per second as possible can be found here:

https://github.com/raspberrypi/documentation/blob/develop/documentation/asciidoc/computers/camera/rpicam_vid.adoc

▶ -t *n*
Specifies the desired length of the video in milliseconds.

Example

The following command records a video with a length of 10 seconds at a resolution of 1,024 × 768 pixels:

```
user$ rpicam-vid --width 1024 --height 768 -t 10000 -o test.mp4
```

rmdir [options] directory

rmdir deletes the specified directory. rmdir can only be run if the directory is empty. Any existing files must first be deleted using rm. The most important option is the following:

▶ -p or --parents
Also deletes subdirectories in the specified directory if the directories—with the exception of subdirectories—are empty.

rmmod [options] module name

rmmod removes the specified module from the kernel. This only works if the module is no longer required.

route [options]
route add/del [-net/-host] target address [options]

route displays the routing table for the IP network protocol, adds a new entry (add), or removes an entry (del). The target address is specified either in numerical format or as a network name. The default keyword is also permitted as a target address to define a standard rule (e.g., for the gateway).

The route packet is considered obsolete. If possible, you should use the ip route command instead.

▶ -host
  Addresses a single computer.

▶ -n
  Displays only numerical IP addresses and doesn't contact a name server to resolve the assigned names.

▶ -net
  Addresses a network.

▶ dev *interface*
  Specifies the desired network interface (e.g., eth0). If dev is the last option, the keyword can be omitted (i.e., route add -net 192.34.34.34 netmask 255.255.255.0 eth1).

▶ gw *routing address*
  Redirects the packets to the specified router.

▶ netmask *mask*
  Specifies the desired network mask (e.g., 255.255.255.0).

Example

The first route command defines the IP address 192.168.0.1 as the default gateway. The second command displays the new routing table.

```
root# route add default gw 192.168.0.1
root# route -n
Kernel IP routing table
Destination Gateway     Genmask         Flags Metric Ref Use Iface
192.168.0.0 0.0.0.0     255.255.255.0   U     0      0   0 eth0
0.0.0.0     192.168.0.1 0.0.0.0         UG    0      0   0 eth0
```

`rpcinfo` [options]

rpcinfo runs a remote procedure call (RPC) and returns the result returned by the RPC server. rpcinfo is used, among other things, to check the NFS server configuration.

▶ -p [hostname]
Determines a list of all active NFS and RPC services on the local computer or on an external computer.

`rpm` options [filename/package name]

rpm installs, updates, or removes RPM packages or displays information about these packages. rpm is available on all distributions that are based on RPM packages, such as Fedora, Red Hat, and SUSE.

▶ -e or --erase
Removes an existing package. If you also use the --nodeps option, the package will be removed even if other packages are dependent on it.

▶ -i or --install
Installs the specified package file(s). The installation location can be changed using the additional --root directory option. If the specified package is a source package, the program code and the configuration files will be installed in /usr/src. In this case, the installation location can be set by changing /etc/rpmrc.

With the additional --test option, rpm doesn't make any changes, but outputs only what would happen during an actual installation.

rpm attempts to automatically check the integrity of signed packages. However, the public key of the signature must be available for this (see the --import option). If you want rpm to omit the check, you must use the --nosignature option.

With the additional --nodeps option, you can install a package even if rpm believes that various dependencies aren't fulfilled. (If two packages are mutually dependent, there's no need to use --nodeps. You can then simply install both packages at the same time!)

The additional --ignorearch option enables an installation even if rpm recognizes that the CPU architecture isn't correct. This is sometimes necessary to install 32-bit packages on a 64-bit distribution.

The --force option forces an installation in any case, even if rpm believes that the package has already been installed.

The additional --noscripts option makes sure that rpm doesn't automatically run installation programs. However, the program you've just installed may not work properly.

▶ `--import`

Installs the specified GPG key. `rpm` can use this key to check the integrity of signed packages. The required key files can be found on all websites that provide signed RPM packages.

▶ `-q` or `--query`

Provides information about installed packages or the contents of a package file. In its simplest form, `rpm -q package name` outputs the exact package name, including the version number of the specified package.

You can use `rpm -qf file` to determine which package a specific file on your system belongs to. This is particularly useful for configuration files. For example, `rpm -qf` /sbin/hwclock returns the package name; *util-linux-n.*

`rpm -qp package.rpm` provides information about a package that has *not* yet been installed. In this case, the file name of a *.rpm file must be used as a parameter.

`rpm -q --whatprovides attribute` or `rpm -q --whatrequires attribute` searches for all installed packages that provide or require the specified dependency attribute. These two options aren't suitable for searching packages that haven't yet been installed!

`rpm -q --provides package` or `rpm -q --requires` creates a list of all attributes provided or required by the specified package. The options can also be used for package files that haven't yet been installed (additional `-p` option).

For all query variants that return a package as a result, you can use additional options to control the amount of information about the package found:

— `-i` : Short description of the package, compressed size

— `-l` : List of all files in the package

— `-lv` : As above, but with file size, access rights, and more

— `-c` : List of all configuration files of the package

— `-d` : List of all files with online documentation for the package

— `--scripts` : List of all installation programs

If you simply need an unsorted list of all installed packages, you should run `rpm -qa` without any additional parameters. `rpm -qa --last` sorts all packages by installation date (the most recently installed package first).

▶ `-U` or `--upgrade`

Updates an existing package. Backup files are automatically created from the modified configuration files of the previous package; all other files of the previous package are replaced or deleted by the new version. With the additional `--oldpackage` option, you can replace a newer package with an older one. You can also use `--nodeps` to ensure that `rpm` ignores package dependencies when upgrading.

▶ -V or --verify

Checks whether any files in a package have changed since installation. The command returns a list of all changed files. Documentation files aren't checked.

Examples

The following command installs the specified package:

```
root# rpm -i abc-2.0.7-1.x86_64.rpm
```

The second rpm example provides a list of all installed packages, whereby the packages are sorted by installation date (the most recently installed package is at the top of the list):

```
root# rpm -qa --last
```

rpm -q --whatprovides tells you which previously installed package provides the mysqli.so attribute:

```
root# rpm -q --whatprovides mysqli.so php-mysql
```

If you don't know which package the /etc/magic file belongs to, you can determine this information via rpm -qf. Then, rpm -qi provides a description of the package:

```
root# rpm -qf /etc/magic
    file-n.n
```

```
root# rpm -qi file
    ...
    The file command is used to identify a particular file according to the type
    of data contained by the file.
```

> **rpm2archive** name.rpm
> **rpm2cpio** name.rpm

rpm2archive creates a compressed TAR archive (name.rpm.tgz) with all files of the package from an RPM package available as a file. rpm2archive is useful for extracting individual files from a package without installing the package.

> **rsync** [options] source
> **rsync** [options] source target

rsync copies files from the source directory to the target directory or synchronizes the two directories. In the first syntax variant (no target directory), the files are copied to the local directory. rsync transfers the files either locally or encrypted through the network. The network variant requires an SSH or RSYNC server to be running on the second computer. The password must be entered before the copy command starts. The following syntax variants are available for specifying the source and target directories:

file1 file2	Local files
directory	Local directory
host:dir	Directory on the host computer
user@host:dir	Same as host:dir, but SSH login under the name of user
rsync://user@host/dir	Communication with RSYNC server
rsync://user@host:port/dir	RSYNC server on the specified port

The developers of the SSH project recommend using rsync instead of the scp command since the implementation of scp is inadequate and a solution to the scp issues isn't foreseeable. If rsync is also installed on the external host and no special scp options are involved, you can replace scp one-to-one with rsync. Corresponding examples can be found in scp.

▶ -4 or -6
Forces the IPv4 or IPv6 protocol if both variants are available for selection.

▶ -a or --archive
Copies the directory contents recursively and retains all file information. The option is a short version for -rlptgoD.

▶ --delete
Deletes files or subdirectories in the target directory that no longer exist in the source directory (caution!).

▶ -D or --devices
Also copies character and block device information.

▶ -e command or --rsh=command
Specifies which external shell program is used to run rsync on the target computer. As a rule, this will be ssh. If you want to pass additional options to the command, you must put it in quotation marks, such as -e "ssh -p 1234".

▶ --exclude=pattern
Specifies a pattern for files that shouldn't be copied. For example, --exclude="*.o" excludes *.o files. (To formulate more complex patterns, see man rsync.)

▶ -g or --group
Gives the files in the target directory the same group assignment as for source files.

- ▶ -l or --links
  Duplicates symbolic links from the source directory in the target directory.

- ▶ -o or --owner
  Gives the files in the target directory the same owner as the files in the source directory.

- ▶ -p or --perms
  Gives the files in the target directory the same access rights as the source files.

- ▶ -r or --recursive
  Recursively copies all subdirectories as well.

- ▶ -t or --times
  Gives the files in the target directory the same modification time as the source files. In this way, rsync can clearly recognize which files have changed when it's called again.

- ▶ -u or --update
  Ignores files that already exist in the target directory and are newer than the source file.

- ▶ -v or --verbose
  Provides detailed information about what is currently going on.

- ▶ -W or --whole-file
  Copies the entire file when changes have been made. With fast network connections, this can increase the synchronization speed a little.

  By default, rsync transfers only the changes and thus minimizes the amount of data transferred. The -W option increases the amount of data to be transferred, but reduces the effort required to detect the changes. -W applies automatically if the source and target directories are local directories. In this case, you can use --no-whole-file to ensure that rsync still uses the incremental change mode.

Examples

The following command synchronizes all files from the mydata directory with the backup directory of the same name. Files that have been deleted in mydata will also be deleted in the backup directory.

```
user$ rsync -av --delete mydata/* /backup/mydata/
```

The second example synchronizes the local dir1 directory with the dir2 directory on the mars.sol computer. The password entry is the responsibility of ssh. (The login password of the user username on the mars.sol computer must therefore be entered.)

```
user@saturn.sol$ rsync -e ssh -az dir1/ mars.sol:dir2/
user@mars.sol's password: ********
```

rsync can also be used to resume an aborted scp transfer of a large file. (scp isn't able to do this).

```
user$ rsync --partial --progress --rsh=ssh local_file user@host:remote_file
```

scp [options] source target

The widely used scp command copies a file between two computers in a network. The transfer is encrypted. scp requires an SSH server to be running on the second computer. The source and target files or the target directory must be specified as follows:

file	Local file
directory	Local target directory
host:	Target computer (copies files to the home directory)
host:file	File on the specified computer
host:directory	Target directory on the specified computer
user@host:file	File of user host: directory on the specified computer
user@host:directory	Target directory of user user@host

Examples

The following command copies the local abc.txt file into the ~/efg directory of the mars computer:

```
user@uranus$ scp abc.txt mars:efg/
user@mars's password: ******
```

-r enables you to transfer an entire directory. The --links option makes sure that rsync also takes symbolic links into account, which is usually not the case (but is with scp).

```
user$ scp -r local-dir/ mars:other-dir
user@mars's password: ******
```

With scp, you can use the -P option to pass a port number if the SSH server on the other side doesn't use the standard port 22. This option isn't available with rsync. Instead, you must use -e to specify how the ssh command called in the background should be composed:

```
user$ scp -P 2222 file mars:
user@mars's password: ******
```

screen [options] [tty-device]

screen from the package of the same name is a screen manager that connects multiple processes (sessions) to a terminal. For example, you can use it to run multiple commands in parallel within an SSH connection and switch back and forth between the sessions. screen can also be used to "redirect" serial interfaces (TTY devices) to a terminal.

screen is controlled by numerous keyboard shortcuts, all of which start with `Ctrl`+`A`. `Ctrl`+`A`, `?` displays a summary of the most important shortcuts, `Ctrl`+`A`, `\` terminates screen. Further shortcuts are summarized in Table 15 in the "Keyboard Shortcuts" section.

You can detach longer running processes from screen using `Ctrl`+`A`, `D` (*detach*). The process then continues to run in the background. If you restart screen with the -r option at a later time, screen reestablishes the connection to the session.

► -ls
Lists all sessions separated from screen.

► -L
Logs all commands executed in screen as well as their output in the files screenlog.0 (first session), screenlog.1 (second session), and so on. Logging can also be switched on and off using `Ctrl`+`A`, `⇧`+`H`. If you want to read the logging files via less, you must use the -r option so that the logged control characters are displayed correctly.

► -r [*session-id*]
Reestablishes the connection to a disconnected session. If there's only one session, you can omit the session ID.

Example

It's difficult to explain how screen works in text form. If you've never used screen before, you should watch a short video. For example, you could search for "screen command video" or "screen command video tutorial"!

scrot [options]

scrot from the package of the same name creates a screenshot if X is used as the graphics system. (If Wayland is active instead, you must use the grim command.) Without additional parameters, the entire screen content is saved as <time>_<resolution>_scrot.png in the local directory.

► -
Forwards the screenshot to the standard output instead of saving it to a file.

- -d *n*

  Takes the screenshot after a delay of *n* seconds.

- -f *name*.*png*

  Saves the screenshot in the specified file.

- -s

  Allows you to select the window whose content is to be displayed before creating the screenshot. This option is often combined with -d so that there's a sufficient amount of time after the window selection to bring this window to the foreground.

- -u

  Creates a screenshot of the active window (not of the entire screen).

Example

In the following example, scrot first waits 10 seconds, then takes a screenshot of the window in the foreground, and finally saves it as myscreenshot.png:

```
user$ scrot -d 10 -u -f myscreenshot.png
```

sdptool [options] command

You can use sdptool from the bluez package to send requests to Bluetooth devices in accordance with the *service discovery protocol* (SDP). The browse command is used most frequently.

- browse *bt-mac*

  Provides information about the external Bluetooth device, which is specified by its MAC address. You can use hcitool scan to determine the MAC address.

- browse local

  Provides detailed information about the local Bluetooth adapter.

Example

sdptool determines that the specified Bluetooth device is an input device (human interface device) that supports the *Logical Link Control and Adaptation Protocol* (L2CAP) and *Human Interface Device Protocol* (HIDP).

```
user$ sdptool browse 60:FB:42:FC:BB:8C
Service Name: Wireless Mouse
Service Class ID List:
  "Human Interface Device" (0x1124)
Protocol Descriptor List:
  "L2CAP" (0x0100)
```

```
    PSM: 17
  "HIDP" (0x0011)
...
```

sealert [options]

sealert helps with the diagnosis of SELinux rule violations. The command requires the settroubleshootd background service to be running. This service logs to the systemd journal:

root# **journalctl | grep setrouble**
```
...
setroubleshoot[2685]: SELinux is preventing httpd from read access
on the file test.html. For complete SELinux messages run
sealert -l ca111078-1aba-4481-bbfc-fbd59d8ca6b9

...
```

If you now run sealert on a non-English system, you'll receive a text that is a motley mix of local language and English terms. The text will be easier to read if you change the localization to English:

root# **LANG= sealert -l ca111078-1aba-4481-bbfc-fbd59d8ca6b9**
```
SELinux is preventing httpd from read access on the file test.html.

**  Plugin catchall_boolean (89.3 confidence) suggests

If you want to allow httpd to read user content, then you must tell  SELinux
about this by enabling the 'httpd_read_user_content' boolean. You can read
'httpd_selinux' man page for more details. Do

setsebool -P httpd_read_user_content 1

**  Plugin catchall (11.6 confidence) suggests

If you believe that httpd should be allowed read access on the test.html  file
by default, then you should report this as a bug ...
```

Unfortunately, the result is rarely of real help.

▶ -a *logfile*
Analyzes the specified logging file and provides information on all SELinux events logged in it.

▶ -l *id*
Specifies the SELinux event for which sealert should provide information. -l '*' handles all logged SELinux rule violations.

---

sed [options] command [< source > target]

sed is a *stream editor*. Usually, the command is used as a text filter to find and edit certain characters or character combinations in the source text (delete, replace with other characters, etc.).

sed is controlled by commands that are either applied to all lines of the text or only to those lines that meet certain conditions.

▶ -E

Activates *Extended Regular Expressions* (EREs). POSIX Basic Regular Expressions are used by default. sed isn't PCRE-compatible.

▶ -i *file*

Changes a file directly (*inline*). Usually, sed reads from standard input and writes to standard output. Then, it's impossible to change a file directly.

▶ -i.backup-identifier *file*

Like the previous list item, but the original file is saved as a backup with the specified ID.

Unfortunately, dealing with sed isn't easy. Even a description of the basic syntax options would go far beyond the scope of this reference. That is why I'll limit myself to a few simple examples in this book. You can find an excellent introduction to sed at the following site:

www.grymoire.com/Unix/Sed.html

Example

In the first example, sed deletes the first three lines of the test file and forwards the rest of the file to the standard output, where d stands for sed command *delete*. 1,3 specifies the address range in which this command is effective.

```
user$ sed 1,3d < test
```

In the second example, sed replaces all occurrences of "a" with "A", where s stands for the *regular find and replace* command. The texts enclosed in / are the search pattern and the text to be inserted. As the command isn't preceded by an address, it will be applied to all lines.

```
user$ sed s/a/A/ < test
```

The third example deletes all LaTeX statements starting with \newpage from a Markdown file. A backup of the original file is saved as test.md.bak.

```
user$ sed -i.bak '/\\newpage.*/d' test.md
```

The way in which you can use sed to copy or rename files is shown in another example in the description of the cp command.

```
seq [options] end
seq [options] start end
seq [options] start delta end
```

seq returns a sequence of digits. The following examples demonstrate how it works. Note that in the case of floating-point numbers, the final value may not be reached due to rounding errors, as it's demonstrated in the final example.

▶ -s *string*

Specifies which character is used to separate the results. By default, this is done using \n, that is, a line break.

Examples

The following examples show the use of seq. For space reasons, a space is used as a separator. Usually, each number is output on a separate line (i.e., without the -s option).

```
user$ seq -s ' ' 5
1 2 3 4 5
user$ seq -s ' ' 4 9
4 5 6 7 8 9
user$ seq -s ' ' 10 5 30
10 15 20 25 30
user$ seq -s ' ' 1 0.1 2
1 1.1 1.2 1.3 1.4 1.5 1.6 1.7 1.8 1.9
```

If you formulate for loops in bash scripts, you should use the notation {*start..end*} or {*start..ende..delta*}, as in the following command, for example:

```
user$ {\bfs for i in {100..150..10}; do echo $i; done}
100 110 120 130 140 150
```

```
sestatus [options]
```

sestatus shows the status of the SELinux system. If the command is used without additional options, it provides a summary of the current status.

▶ -b

Also shows the status of all Boolean parameters of the SELinux rules. These parameters can be changed using setsebool.

▶ -v

Also displays context information on the files and processes listed in /etc/sestatus.conf.

Example

SELinux is active on the test computer, and compliance with the rules is ensured (Current mode = enforcing). The targeted rule set is used.

```
root# sestatus -b
SELinux status:              enabled
SELinuxfs mount:             /sys/fs/selinux
SELinux root directory:      /etc/selinux
Loaded policy name:          targeted
Current mode:                enforcing
Mode from config file:       enforcing
Policy MLS status:           enabled
Policy deny_unknown status:  allowed
Memory protection checking:  actual (secure)
Max kernel policy version:   32

Policy booleans:
abrt_anon_write              off
abrt_handle_event            off
abrt_upload_watch_anon_write on
antivirus_can_scan_system    off
...
```

set

The bash command set displays all variables known to the shell, including the environment variables, which can also be displayed via printenv. set -x causes the bash to display what the internal command line looks like after alias abbreviations and file name expansion have been taken into account before running each command.

setcap [options] capability file name

setcap from the libcap-ng-utils package specifies which operations are permitted for the program (*capabilities*) for executable files. If the -r option is passed instead of capability, all previously defined capabilities will be deleted. Capabilities requires a file system with extended attributes. For this reason, you must use the mount option user_xattr for ext file systems.

Example

The `ping` network command used to be equipped with the `setuid` bit in many distributions so that it could be used by ordinary users. As soon as you delete this bit, only `root` is able to use `ping`:

```
user$ ls -l /usr/bin/ping
-rwsr-xr-x 1 root root 72776 Jan 31  2025 /usr/bin/ping
user$ sudo chmod u-s /bin/ping
user$ ping yahoo.de
ping: icmp open socket: The operation is not permitted
```

Instead of setting the insecure `setuid` bit again, it's also sufficient to give the `ping` command access to the kernel's network functions by using `setcap`. You can use `getcap` to see which capabilities a command has:

```
user$ sudo setcap cap_net_raw=ep /bin/ping
user$ getcap
/bin/ping /bin/ping = cap_net_raw+ep
user$ ping yahoo.de
PING yahoo.de (212.82.102.24) 56(84) bytes of data.
...
rtt min/avg/max/mdev = 58.054/58.054/58.054/0.000 ms
```

In most current distributions, `ping` is equipped with capabilities by default, for example, on Debian, Fedora, openSUSE, and Ubuntu.

setenforce 0|1|Enforcing|Permissive

`setenforce` switches SELinux between the `Enforcing` (1) and `Permissive` (0) modes. In `Permissive` mode, rule violations are logged, but the affected program can continue to work unhindered. In `Enforcing` mode, however, SELinux prevents the program from executing processes prohibited by SELinux rules.

If you want to completely deactivate SELinux, you must make the `SELINUX=disabled` setting in `/etc/selinux/config` and restart the computer. Note that reactivation at a later date is difficult because you then have to correct the SELinux context of all new files. This is why the `Permissive` mode is generally preferable.

setfacl [options] [aclaction] file name

`setfacl` changes the extended access rights of the specified files or directories. This only works if the file system supports *Access Control Lists* (ACLs). For ext3/ext4 file systems, the `mount` option `acl` must be used.

The command is usually used to perform one of the following four actions:

▶ -m *aclrule*
Adds another rule (an *Access Control Entity* [ACE]) to the existing ACLs. (m stands for *modify*.)

▶ -M *aclrule file*
Like -m, but reads the rule from a file. Rule files must look like the output of getfacl.

▶ -x *aclrule*
Deletes the specified ACL rule.

▶ -X *aclrule file*
Like -x, but reads the rule from a file.

The simplified structure of an ACL rule (an ACE) is shown in the following table. The complete syntax is documented in man setfacl.

[u:]uid [:rights]	Changes rights for a user
g:gid [:rights]	Changes rights for a group
o[:] [:rights]	Changes rights for all other users
m[:] [:rights]	Sets the ACL mask (*effective right mask*)

The rights consist of up to three letters: r for *read*, w for *write*, and x for *execute*. To remove all rights from a user or group, you must specify -.

If you prefix the entire rule with d:, it applies to the standard ACLs. Instead of the letters u, g, o, and m, you can also use the user, group, other, and mask keywords.

setfacl can be controlled by further options:

▶ -B or --remove-all
Removes all ACL rules.

▶ -d
Applies the transferred rule to the standard ACL.

▶ -k or --remove-default
Removes the standard ACL rules.

▶ -n
Dispenses with the automatic recalculation of the ACL mask for every ACL rule change.

▶ --restore=*file*
Applies the ACL rules specified in the file to the files in the current directory. You can create an ACL backup file using getfacl -R.

▶ -R

Applies the specified rule recursively to all files and subdirectories.

Examples

In a file system with ACLs, the default access rights normally apply, which are often also referred to as the *minimum ACL*. getfacl displays these rights in ACL form:

```
user$ touch file1
user$ getfacl file1
# file: file1
# owner: kofler
# group: kofler
user::rw-
group::r--
other::r--
user$ ls -l file1
-rw-r--r-- 1 kofler kofler ...  file2
```

Using setfacl, you can now define additional access rules. The following commands give the user grace and all members of the docuteam group write and read access to the file, but deny the user katherine any access:

```
user$ setfacl -m grace:rw file1
user$ setfacl -m g:docuteam:rw file1
user$ setfacl -m katherine:- file1
```

The getfacl rights list is now somewhat longer. With ls -l, the usual access letters are followed by the + sign to indicate that there are ACL rules.

```
user$ getfacl file1
# file: file1
# owner: kofler
# group: kofler
user::rw-
user:grace:rw-
user:katherine:---
group::r--
group:docuteam:rw-
mask::rw- other::r--

user$ ls -l file1
-rw-rw-r--+ 1 kofler kofler ... file1
```

setfattr [options] file name

setfattr changes the extended attributes of the selected files or directories. This only works if the file system supports *extended attributes* (EAs). For ext3/ext4 file systems, the mount option user_xattr must be used.

▶ -n *attribute name* or --name=*attribute name*
Specifies the name of the attribute to be changed. The actual name must be preceded by user. (i.e., -n user.myattribute).

▶ -v *value* or --value=*value*
Specifies the value to be saved in the attribute.

▶ -x *attribute name*
Deletes the specified attribute.

▶ --restore=*file*
Applies the EA definitions specified in the file to the files in the current directory. You can create an EA backup file via getfattr -R -d.

Examples

The following examples show how you can use setfattr to store attributes and how to read them using getfattr: the number of attributes per file is limited in ext file systems.

```
user$ touch file2
user$ setfattr -n
user.language -v en file2
user$ setfattr --name=user.charset --value=utf8 file2
user$ getfattr -d file2
 # file: file2
user.charset="utf8"
user.language="en"
```

getfattr usually returns only attributes whose name starts with user.. If you want to see other attributes, you must specify their names using -n or their patterns by using -m:

```
user$ getfattr -n security.selinux -d tst
# file: tst security.selinux="user_u:object_r:user_home_t:s0^000"
```

`setsebool` [options] bool1=value1 bool2=value2 ...

`setsebool` changes Boolean parameters at runtime. Boolean parameters can be used to change the behavior of some SELinux rules. You can determine which Boolean parameters are available via `sestatus -b`.

▶ `-N`

Saves the change, but doesn't activate it. The change will therefore only take effect the next time the computer is restarted.

▶ `-P`

Performs the change and saves the setting permanently. The change applies immediately and will also apply in the future, that is, after the next restart.

Example

The following command allows the Squid program to act as a transparent proxy:

`root# setsebool -P squid_use_tproxy 1`

`setterm` [option]

`setterm` changes various settings of the terminal. If the command is executed without specifying an option, it displays a list of all possible options. Useful options for shell programming are listed here:

▶ `--background` *color*

Sets the background color. The colors `black`, `red`, `green`, `yellow`, `blue`, `magenta`, `cyan`, and `white` are permitted. To ensure that the setting remains valid after color output by `ls` or `grep`, the option must be combined with `--store`.

▶ `--blank` *n*

Activates the screen saver after *n* minutes without input (only for text consoles).

▶ `--bold` on|off

Activates or deactivates the bold font. In text consoles, the text doesn't appear in bold, but at least in a different color than the other text.

▶ `--clear`

Deletes the contents of the terminal.

▶ `--default`

Restores colors and text attributes to the default setting.

▶ `--foreground` *color*

Sets the text color.

▶ `--half-bright` on|off

Switches highlighted font on/off.

▶ `--inversescreen on|off`
Inverts the display in the text console (black text on a white background) or restores the normal state.

▶ `--reverse on|off`
Sets inverse font on/off.

▶ `--store`
Remembers the color settings (`--foreground` and `--background` options) so that they are retained even if commands such as `ls` or `grep` temporarily use different colors. `--store` is only valid in the Linux text consoles, but not within terminal windows. The settings aren't saved in a file there either, but are lost when you log out.

▶ `--underline on|off`
Switches underlined font on/off.

Example

If the following line is inserted in `.bashrc`, text consoles use a white background and black font after login:

```
# insert into .bashrc
setterm -background white -foreground black -store
```

sfdisk [options]

`sfdisk` lists the partitions of a hard disk or repartitions the hard disks. The command is particularly suitable for automated partitioning (e.g., in a script) or for transferring an existing partitioning to a second hard disk, for example, when setting up a RAID array.

`sfdisk` only works with MBR partition tables and is unsuitable for disks with a GUID partition table (GPT)!

▶ `-d [device]`
Lsts all partitions of all hard disks or the specified hard disk. The output format is suitable for further processing by `sfdisk`.

▶ `--force`
Performs partitioning even if `sfdisk` believes that the partition or hard disk is in use, or if partition boundaries don't match cylinder boundaries.

▶ `-l [device]`
Lists all partitions of all hard disks or the specified hard disk in a readable format. With the additional `-uS`, `-uB`, `-uC`, or `-uM` option, `sfdisk` uses sectors (512 bytes), blocks (1,024 bytes), cylinders, or MiB as the unit of measurement.

- ▶ -s *device*
  Returns the number of blocks (1,024 bytes each) of the specified hard disk or partition.

- ▶ -uS|-uB|-uC|-uM
  Uses sectors (512 bytes), blocks (1,024 bytes), cylinders, or MiB as the unit of measurement.

- ▶ *device* < *partitioning table*
  Repartitions the hard disk using the transferred table. The partitioning table must have the same format as the output of sfdisk -d.

Example

The following command partitions the second hard disk (/dev/sdb) exactly like the first hard disk (/dev/sda). The second hard disk must be at least as large as the first one. All partitions and data on the second hard disk will be lost.

```
root# sfdisk -d /dev/sda | sfdisk /dev/sdb
```

```
sftp [options] sshserver
sftp [optionen] user@sshserver:filename
```

sftp is an alternative to the insecure ftp command. In the first syntax variant, sftp establishes a connection to the specified SSH server. After logging in, you can interactively transfer files between the local computer and the SSH server. In the second syntax variant, sftp transfers the specified file directly to the local computer after the password has been entered.

- ▶ -b *batch file*
  Runs the commands specified in the batch file.

```
sgdisk [option] device
```

sgdisk helps to edit GUID partition tables on hard disks and SSDs. Unlike parted, the command is controlled exclusively by options and isn't intended for interactive use. Instead, you can use sgdisk to create or manipulate partition tables under script control. Despite the similar sounding name, sgdisk expects completely different options than sfdisk. sgdisk uses MiB and GiB in IEC notation.

- ▶ -a *n*
  Specifies the multiples of sectors to which new partitions are aligned. By default, sgdisk uses the value 2048. This corresponds to 1 MiB.

▶ -b *file*

Writes a backup of the partition table to a binary file.

▶ -d *nr*

Deletes the specified partition.

▶ -G

Assigns new, random GUIDs to all partitions. This is required after cloning a hard disk.

▶ -l *file*

Partitions the data medium according to the partition table stored in the file. The file must first be created using the -b option. All existing partitions on the data medium will be deleted. The combination of the sgdisk -b and sgdisk -l commands can be used to transfer the partition layout of a hard disk to a second disk of the same size, for example, when setting up a RAID system.

▶ -n *nr*:*start*:*end*

Creates the new partition with the number *nr*, which extends from the specified start position to the end position. By default, start and end are specified in sectors of 512 bytes. Alternatively, you can also specify the position in MiB and GiB. sgdisk -n 3:10G:35G creates a new partition that is 25 GiB in size.

The start and end positions can be preceded by the + and - characters. + enables entries that are relative to the default start sector; - enables specifications that are calculated from the end of the hard disk. sgdisk -n 4:+1M:-20G creates a partition that starts 1 MiB after the last partition and leaves 20 GiB free at the end of the data medium.

Example

The following commands first create a backup of the current partition table on the /dev/sdb disk and then set up the new partition 3:

```
root# sgdisk -l partbackup /dev/sdb
root# sgdisk -n 3:70G:430G /dev/sdb
```

shansum files

sha1sum, sha224sum, sha256sum, sha384sum, and sha512sum calculate checksums for all specified files. The number in the command indicates the bit length of the checksum (160 bits for sha1sum). The checksums are used to ensure that the file is unchanged after it has been transferred.

Example

The following command calculates a 256-bit checksum for an ISO file:

```
root# sha256sum image.iso
6ad62fe91eefa4315f852eb1bc8732bc341d95b47f6b4d1f5e362d3629fc981b  image.iso
```

shift [n]

The bash command shift shifts the parameter list passed to the shell script through the predefined variables $1 to $9. If shift is used without parameters, the parameters are shifted by one position; otherwise, they are shifted by *n* positions.

shift is only used for script programming. The command is a valuable aid if more than nine parameters are to be addressed. Once parameters have been shifted out of the variables with shift, they can no longer be addressed. They are also removed from the $* variable.

showmount [options]

Without additional options, the showmount command, when executed on an NFS server, returns a list of all clients using the shared directories via NFS 3. Note that NFS 4 directories aren't included in the result and are only taken into account with the showmount -e variant.

▶ -a
  Also provides the IP address for each NFS 3 client.

▶ -e [hostname]
  Provides a list of all directories that can be used via NFS, regardless of whether this service is currently in use. If you specify a host name, the list of directories available on this computer will be determined. showmount -e also works for NFS 4.

Example

```
root# showmount -e
/nfsexport       *.lan,10.0.0.0/24
/nfsexport/fotos *.lan,10.0.0.0/24
/nfsexport/iso   *
```

shutdown [options] time [message]

shutdown shuts down the computer. A time (hh:mm) or the number of minutes (calculated from the current time, +m) or the now keyword (i.e., immediately) must be specified as the time. shutdown can only be executed by root. Linux is often configured in such a way that users without root permissions can restart the computer via [Alt]+[Ctrl]+ [Del].

shutdown informs all other users that the system will be shut down shortly and will no longer allow new logins. All processes are then warned that they will be stopped shortly. Some programs (e.g., emacs or vi) use this warning and save all open files in backup copies.

▶ -c

Attempts to cancel a shutdown process that has already been initiated. This is possible if the time that precedes the start of the shutdown process hasn't yet expired.

▶ -f

Initiates a restart like -r, but faster.

▶ -F

Causes the file systems to be checked at the next reboot. shutdown creates the /forcefsck file for this purpose. If this file exists, most distributions trigger a file system check when restarting.

▶ -h

Stops the system after shutdown. The computer then no longer responds to inputs, and the *system halted* message appears on the screen. Then, the computer gets switched off by an ACPI signal.

▶ -n

Performs the shutdown particularly quickly, bypassing the init system.

▶ -r

Initiates a restart after the system has been shut down.

▶ -t *seconds*

Defines the wait time between the warning message and the kill signal for the processes (usually, 20 seconds).

sleep time

sleep puts the running program into sleep mode for the specified time. During this time, the program consumes virtually no compute time. The time is usually given in seconds. Optionally, the letters m, h, or d can be added to specify the time in minutes, hours, or days.

slurp

The command from the package of the same name enables the selection of a rectangle on the screen on Wayland. It then returns the coordinates of the top-left corner and the size of the rectangle as text. The command can be used in combination with `grim`, for example, to select the area for a screenshot.

Unfortunately, `slurp` isn't compatible with every Wayland compositor. `slurp` works excellently on Raspberry Pi OS, but fails with distributions that use Gnome or KDE as the desktop system.

`smartctl` options device

`smartctl` controls the SMART functions of the hard disk. This allows you to detect possible hard disk problems at an early stage.

▶ `-a` or `--all`
Provides all available SMART information.

▶ `-d ata/scsi/3ware,`*n*
Indicates the hard disk type. This option is only required if `smartctl` can't recognize the type. This applies in particular to SATA hard disks, which `smartctl` often believes to be SCSI hard disks. The `3ware,`*n* setting enables the use of SMART functions for hard disks that are connected to a 3ware RAID controller (for details, see `man smartctl`).

▶ `-H` or `--health`
Indicates whether the hard disk is currently OK and will probably continue to function for the next 24 hours.

▶ `-i` or `--info`
Provides basic information about the hard disk.

▶ `-l error/selftest` or `--log=error/selftest`
Provides information on the past five errors or the results of the last 21 self-tests.

▶ `-s on/off` or `--smart=on/off`
Activates or deactivates the SMART functions.

▶ `-t short/long` or `--test=short/long`
Performs a short or long self-test. You can view the test result later using `-l selftest`.

Example

`smartctl -H` or `smartctl --health` indicates whether the hard disk is currently healthy. If `smartctl` doesn't return *PASSED* as a result here, you should *immediately* start performing a full backup!

```
root# smartctl -H /dev/sda
...
SMART overall-health self-assessment test result: PASSED
```

Three self-tests have been carried out for the /dev/sda hard disk so far: one immediately after the installation of the disk (after 40 hours of operation), and the other two after approximately 2,600 hours. No problems have occurred.

```
root# smartctl -t short /dev/sda
root# smartctl -t long /dev/sda
...
root# smartctl -l selftest /dev/sda
Num  Test_Description  Status                   Remaining  LifeTime  LBA
# 1  Extended offline  Completed without error      00%       2592    -
# 2  Short offline     Completed without error      00%       2591    -
# 3  Short offline     Completed without error      00%         40    -
```

> **smbclient** [options]
> **smbclient** directory [password] [options]

smbclient enables interactive access to Windows network directories. In the first syntax variant, you use the program to determine available Windows resources. In the second variant, you enter the desired network directory (e.g., //mars/data if you want to use the data directory on the mars computer). Once the connection has been established, you'll be taken to a shell. As with the ftp command, you can then view directories via ls, change directories using cd, transfer files to the local computer via get (*download*), and save files on the external computer using put (*upload*). You can get an overview of the most important commands via help.

▶ -L *computer name*
  Returns the list of network resources on the specified Windows computer or Samba server.

▶ -m *protocol*
  Specifies the maximum protocol version. -m SMB2 enforces the SMB2 protocol, even if SMB3 would be possible. This option is only necessary in rare cases to avoid compatibility problems.

▶ -N
  Omits the automatic password prompt. This option is only useful if you know that no password is required to access a particular network resource.

▶ -U *username*
  Specifies the username (only necessary if this differs from the current login name).

▶ -W *workgroup name*
  Specifies the name of the Windows workgroup (only necessary if it's not contained in /etc/samba/smb.conf).

Example

First, smbclient establishes a connection to the NAS hard disk, and then ls returns the list of directories and files stored there:

```
user$ smbclient -U name -W wgname //mynas/myshare
Password: xxxxxx
Domain=[wgname] OS=[Unix] Server=[Samba 3.5.2]
smb: > ls
  data               D       0  Apr  5 18:17:11 2025
  file.xy            AR     226  Sep 14 00:00:00 2024   ...
```

In the second example, smbclient establishes a connection to the LibreELEC media center on a Raspberry Pi. It then lists the network directories available there, which are accessible without a password by default. -U none simply specifies an invalid username here.

```
user$ smbclient -L libreelec -U none -N -m SMB2
Sharename       Type      Comment
---------       ----      -------
Music           Disc
Screenshots     Disk
Videos          Disk
...
```

smbpasswd [options] [name]

smbpasswd changes the password of the active or specified Samba user.

▶ -a

Creates a new Samba account. There must be a Linux account with the same name.

▶ -n

Removes the password for the specified account. The network directories are now accessible without a password if the Samba server is configured accordingly (null passwords = yes).

▶ -x

Deletes the specified Samba account.

smbstatus [options]

smbstatus returns the version number of the Samba server running on the computer as well as a list of all currently active connections.

snap [options] command

A few years ago, Ubuntu introduced a new system for installing software packages. It works in parallel to the still dominant Debian package system (see apt and dpkg), and its objective is similar to that of the Flatpak method supported by Red Hat (see flatpak). Like Flatpak, Snap is open to all distributions, but it hasn't yet found any significant distribution outside of Ubuntu.

Snap packages are usually installed via the Gnome program called *Software*. The snap command provides the option of administrating Snap packages in the terminal.

▶ find *search term*
Searches for the specified term in the package descriptions in the standard Snap package source *https://snapcraft.io*.

▶ install [--classic] *name*
Installs the specified package (requires root permissions or sudo). The --classic option needs to be specified only for some packages that require unrestricted access to the file system and are therefore executed with less strict security rules. You can use the --edge, --beta, or --candidate options to install packages from the specified test channels. (Without these options, the release channel is used.)

▶ list
Lists previously installed Snap packages.

▶ refresh
Displays all installed packages.

▶ remove *name*
Uninstalls the specified package.

Example

The following command installs the Atom editor:

```
user$ sudo snap install --classic atom
atom 1.51.0 installed from Snapcrafters
```

snapper [options] command

SUSE distributions use a Btrfs file system for the system partition by default. YaST automatically creates a snapshot before and after each administrative operation. These snapshots can be managed either with the YaST module *Snapper* or with the snapper command briefly presented here. You can prefix the snapper command to be executed with one or more options:

- ► `--iso`
  Formats times in ISO format.

- ► `--utc`
  Displays UTC times (*Universal Time Coordinated*).

The actions that will be executed are specified by subcommands:

- ► `cleanup number|timeline|empty-pre-post`
  Deletes old snapshots, using either a maximum number of snapshots, their age, or their content as a criterion (see also the `/etc/snapper/configs/root` configuration file).

- ► `create`
  Creates a new snapshot.

- ► `delete` *n* or *n1-n2*
  Deletes the specified snapshots. The numbers to be used can be found in the `snapper list`.

- ► `list`
  Lists all snapshots.

- ► `rollback` [*n*]
  Makes the specified or the most recent available snapshot the active snapshot (strictly speaking: the Btrfs default subvolume). A subsequent `reboot` command restores the system to the state it was in at the time the snapshot was created.

- ► `undochange` *n1..n2* [`files`]
  Undoes the changes made between snapshot *n1* and snapshot *n2*—either for all affected files or only for the files specified as parameters.

---

sort [`options`] `file`

`sort` sorts the specified file and displays the result on the screen. The sort order is defined by the language settings and specifically by the `LC_COLLATE` environment variable.

- ► `-c`
  Checks whether the file is sorted or not.

- ► `-f`
  Treats lowercase and uppercase letters as equivalent.

- ► `-k` *n1,n2*
  Takes into account only the characters between the *n1* and *n2* columns. The column numbering starts with 1. Usually, columns are separated by spaces or tab characters, but see `-tz`. If *n2* isn't specified, all characters from *n1* to the end of the line will be considered.

▶ -m

Merges two or more presorted files into one large sorted file. This is quicker than putting the files together first and then sorting them.

▶ -n

Sorts numbers numerically (i.e., 10 after 9 and not after 1).

▶ -o *result file*
Writes the result to the specified file. The file may match the file to be sorted.

▶ -r

Sorts in reverse order.

▶ -t*z*

Specifies the separator between two columns. The default setting is *white space*, that is, any combination of space and tab characters.

▶ -u

Eliminates duplicates from the result (e.g., sort | uniq).

Example

The following command sorts the table of contents passed by ls according to usernames and group names. Files with identical usernames and group names are sorted by size.

user$ **ls -l | sort -k 3**

source file

source runs the specified file in a bash script as if the commands it contains were in the place of the source command. Once the file has been executed, the running shell program is continued in the subsequent line. No new shell gets started to run the specified file. All variables, including the parameters list, therefore also apply to the specified file. If exit is executed in this file, there's *no* return to the program with the source command, but the program execution ends immediately.

The short form ._*file* exists for source, where ‿ represents a space.

speaker-test [options]

The speaker-test command from the alsa-utils package helps you track down problems with the audio system. To do this, it generates test tones that are output via the audio system.

▶ -c *n*

Uses *n* audio channels.

▶ -f *n*

Plays a sine tone at the specified frequency in Hertz. At the same time, the -t sine option must be used.

▶ -s *n*

Tests only the audio channel with the number *n*.

▶ -t pink|sine|wav

Outputs pink noise, a sine wave, or a speech pattern from a WAV file.

▶ -w *file*

Plays the specified WAV file. The -t wav option must be used at the same time.

Example

The following command outputs the words *front left* or *front right* respectively via the left and right audio channels. The speech pattern is loaded from a WAV file. The output can be terminated again by pressing [Ctrl]+[C].

user$ **speaker-test -t wav -c2**

---

split [options] file [target file]

split splits the specified file into multiple individual files. The output file is separated every *n* bytes or lines (depending on the options). If no target file is specified, the command returns the files xaa, xab, and so on. If a target file is specified, then this file name is used for the result files together with the character combination aa, ab, and so on.

▶ -*n* or -l *n* or --lines=*n*

Splits the source file into individual files with *n* lines each.

▶ -b *n* or --bytes=*n*

Separates the output file every *n* bytes. The size can also be specified in KiB, MiB, GiB, or multiples of 1,000 bytes: KB (1,000 bytes), K (1,024 bytes), MB (10^6 bytes), M ($1,024^2$ bytes), GB (10^9 bytes), or G ($1,024^3$ bytes).

▶ -C *n* or --line-bytes=*n*

Like -b, but the files are split at line boundaries and are therefore usually a few bytes smaller than *n*.

Example

In the following example, split splits the archive file backup.tgz into individual files of 4 GiB each and names them dvd.aa, dvd.ab, and so on:

user$ **split -C 4G backup.tgz dvd.**

cat reassembles the individual files into a complete file:

user$ **cat dvd.* > total.tgz**

```
sqlite3 [options] database file
sqlite3 [options] database file "sqlcommand1;sql2;..."
```

sqlite3 is normally used as a shell for editing SQLite databases. After starting, you can interactively run SQL commands, whereby all commands must be terminated with a semicolon. In addition to the usual SQL commands, sqlite3 knows some specific commands that begin with a dot, such as .help to display a help text or .tables to list all tables in the database. .quit or Ctrl+D terminates sqlite3.

In the second syntax variant, sqlite3 runs the specified SQL commands directly.

Example

The following command determines which music albums are stored in an ownCloud database:

```
user$ sqlite owncloud.db
sqlite> select * from oc_media_albums;
1|Nightclub|1|
2|The Monroe Doctrine|2|
3|The Best Of Herbie Mann|3|
...
sqlite> .quit
```

```
ss [options]
```

ss (*socket statistics*) analyzes the network activity of the local computer. If the command is executed without options, it returns a list of all *nonlistening* TCP sockets with an active connection. ss provides similar functionality to netstat, but is considered the more modern command and in some cases provides more detailed information.

▶ -4 or -6

Takes only IPv4 or IPv6 connections into account.

▶ -a

Shows all sockets, both *listening* and *nonlistening*.

▶ -e

Provides even more details (*extended*).

▶ -n

Displays numerical port numbers, not the names of the services.

▶ -p

Specifies which processes use which sockets.

▶ -r

Replaces port numbers with the names of the services.

▶ -s

Displays only summarized information (*summary*).

▶ -t or -u

Takes only TCP or UDP connections into account.

Example

The following command provides a list of all active TCP sockets for IPv6 connections:

```
user$ ss -6 -t -a
State       Recv-Q Send-Q  Local Address:Port          Peer Address:Port
LISTEN      0      100     :::pop3                      :::*
LISTEN      0      100     :::imap2                     :::*
LISTEN      0      128     :::http                      :::*
LISTEN      0      100     :::smtp                      :::*
LISTEN      0      128     :::https                     :::*
LISTEN      0      100     :::imaps                     :::*
LISTEN      0      100     :::pop3s                     :::*
TIME-WAIT   0      0       ::ffff:5.9.22.28:http   ::ffff:188.192.181.195:50045
TIME-WAIT   0      0       ::ffff:5.9.22.28:http   ::ffff:188.192.181.195:50047
ESTAB       0      46200   ::ffff:5.9.22.28:http   ::ffff:46.10.92.30:22893
FIN-WAIT-1  0      1       ::ffff:5.9.22.28:https  ::ffff:203.134.55.243:51106
...
```

```
ssh [options] computer name
ssh [options] computer name command
```

In the first syntax variant, ssh opens a shell on another computer and thus makes it possible to work interactively on this computer. The second syntax variant runs a single command on the second computer. ssh only works if an SSH server is running on the second computer.

▶ -l *loginname*

Uses the specified login name on the second computer instead of the current username. The desired login name can also be entered in the following format: ssh loginname@computername.

▶ -L *localport:host:hostport*

Forms a tunnel between the local computer (localport port) and the second computer (hostport port).

- ▶ -o *option=value*

  Enables the transfer of additional options in the same syntax as in the /etc/ssh/ssh_config file (see man ssh_config). For example, -o StrictHostKeyChecking=no causes SSH to refrain from checking the host data with entries already saved in .ssh/know_host. This can be helpful for the automated administration of many virtual machines.

- ▶ -t

  Forces the use of a pseudo-terminal for the execution of commands in the format ssh -t hostname "commands". This makes it possible to start interactive programs such as top via SSH. The connection isn't stopped until the external command ends (with top this happens by entering Q).

- ▶ -T

  Prevents the activation of a pseudo-terminal. You can find an application of this option in the following examples.

- ▶ -v

  Provides comprehensive debugging information. If you have login problems, ssh -v can often help you find the cause. The option can be specified up to three times (i.e., -v -v -v); ssh then provides even more debugging information.

- ▶ -X

  Enables the execution of X programs.

Examples

The following command shows how you can log in on an external computer, whereby the username remains unchanged. You can then run commands on the external computer (e.g., your root server).

```
kofler@uranus$ ssh kofler.info
kofler@kofler.info's password: ********
```

When you use ssh to connect to another computer for the first time, a warning often appears based on the following pattern:

```
The authenticity of host 'kofler.info (1.2.3.4)' can't be established.
RSA1 key fingerprint is 1e:0e:15:ad:6f:64:88:60:ec:21:f1:4b:b7:68:f4:32.
Are you sure you want to continue connecting (yes/no)? yes
Warning: Permanently added 'kofler.info,1.2.3.4' to the list of known hosts.
```

The following command shows how I copy the entire /var/www directory tree of my kofler.info web server to the local ~/bak directory. Here, ssh is used to run a single command (tar). This assumes that kofler@kofler.info can read all files from /var/www.

Basically, it's also possible to run the tar command with ssh -1 root with root permissions. However, for security reasons, most SSH servers are configured in such a way that a direct root login is impossible.

The results of tar are transferred securely to the local computer via SSH and processed further by the second tar command:

```
kofler@uranus$ ssh kofler.info tar -cf - /var/www | tar -xC ~/bak/ -f -
kofler@kofler.info's password: ********
```

To run multiple commands in a script via SSH on another server, you can use the HERE-DOC syntax of the bash in combination with ssh option -T. The following command changes the password of someaccount on an external host and then runs a few more commands:

```
#!/bin/bash
pw=strictlysecret
...
ssh -T root@host <<ENDSSH
echo someaccount:$pw | chpasswd
rm -f /etc/file1
cp /root/file2 /userxy/file3
ENDSSH
```

ssh-copy-id [options] user@host

ssh-copy-id inserts the public key of the local computer into the .ssh/authorized_keys file of the host computer. If there's more than one key, the ssh-add -L command decides which key will be transferred.

▶ -i [key file]
  Transfers the specified key file or .ssh/id_rsa.pub.

If the ssh-copy-id script isn't available or the key transfer fails due to the security settings of the external SSH server, the public key file (usually .ssh/id_rsa.pub) must be manually transferred to the server and added to the .ssh/authorized_keys file there.

If Fedora, RHEL, or a derivative is running on the host system, the use of ssh-copy-id can lead to the .ssh directory and the authorized_keys file being created without the SELinux context information. One of the following two commands on the host computer can help:

```
root# /sbin/restorecon -r /root/.ssh
root# /sbin/restorecon -r /home/username/.ssh
```

ssh-keygen [options]

ssh-keygen generates, manages, and converts cryptographic keys that can be used for authentication with SSH. If ssh-keygen is called without additional options, it generates an RSA key pair with a key length of 2,048 bits. The private and public parts of the key are stored in the .ssh/id_rsa and .ssh/id_rsa.pub files.

By default, the access to the key is secured by a password; if you simply press `Enter` instead of entering the password, this security feature will be omitted. If a private key that isn't protected by a password falls into the hands of a third party, this third party can easily log on to all computers on which you've installed the public part of the key!

► -b n

Determines the key length in bits: 3,072 bits are recommended for RSA keys (as of spring 2025). With 4,096 bits, you're completely on the safe side. For ECDSA keys, only the values 256, 384, or 521 (not 512!) are possible. With the Curve25519 method, the key length is fixed.

► -p

Allows you to (re)set the password of a key. If the key already had a password, you must enter the old password before you can set a new password. If the new password is empty, the key can be used without a password in future.

► -R hostname

Removes the entries stored in .ssh/known_hosts for the specified host name.

► -t ecdsa|ed25519|rsa

Defines the key type. The default type depends on the version and configuration of the SSH package. In the past, these were mostly RSA keys (named after the developers; Rivest, Shamir, and Adleman). In current versions, ssh-keygen generates cryptographically more secure *Elliptic Curve Digital Signature Algorithm* (ECDSA) keys. ed25519 is a variant of this (Curve25519). Accordingly, the key files are id_rsa*, id_ecdsa*, or id_ed25519*.

Example

The following command generates an ECDSA key:

```
user$ ssh-keygen -t ecdsa
Generating public/private ecdsa key pair.
Enter file in which to save the key (/root/.ssh/id_ecdsa): <Return>
Enter passphrase (empty for no passphrase): ******** <Return>
Enter same passphrase again:  ******** <Return>
Your identification has been saved in /root/.ssh/id_ecdsa
Your public key has been saved in /root/.ssh/id_ecdsa.pub
```

stat [options] files

stat provides detailed information about the specified files, including the access rights in ls notation and as an octal number, the number of blocks used, the time of the last change and the last read access, and so on.

▶ -c formatstring
Formats the output according to the format string.

▶ -f
Provides information about the file system in which the file is located (instead of information about the file). The following codes, among others, can be used in the character string:

 – %a: Access rights in octal notation
 – %A: Access rights in ls notation
 – %C: SELinux context
 – %F: File type (file, directory, block device, etc.)
 – %g: GID of the group to which the file is assigned
 – %G: Name of the group to which the file is assigned
 – %h: Number of hard links
 – %n: File name
 – %s: File size in bytes
 – %u: UID of the file owner
 – %U: Name of the file owner

Example

Without options, stat provides a clear compilation of various metadata of a file. With -c, the result can be reduced to individual pieces of information, and in the second example, the result is reduced to the octal access rights and the file name.

```
user$ stat print.pdf
  file: print.pdf
  size: 390758          blocks: 768         EA block: 4096     regular file
device   : fd01h/64769d    Inode: 397971       links: 1
access   : (0664/-rw-rw-r--)  Uid: ( 1000/  kofler)   Gid: ( 1000/  kofler)
context  : unconfined_u:object_r:user_home_t:s0
access   : 2013-09-24 11:35:17.316000000 +0200
modified : 2013-08-06 17:53:16.083000000 +0200
changed  : 2013-08-06 17:53:16.083000000 +0200
```

```
birth    : -
user$ stat -c "%a %n" print.pdf
664 print.pdf
```

strace [options] command

strace runs the specified command or program and displays all system functions called by the command. Thus, strace is an important tool for debugging compiled programs. Various options can be used to control which (additional) information you want the program to output.

strings [options] file

The command extracts character strings from a binary file and displays them. This is useful, for example, if you're looking for a text (e.g., an error message or a file name) in a program file.

Examples

The following command determines all strings in the binary file /bin/ls that contain the word error:

```
root# strings /bin/ls | grep error
error
error initializing month strings
write error
```

In the following example, I wanted to find out which keyboard configuration file(s) the systemd program systemd-vconsole-setup accesses on openSUSE Leap:

```
user$ strings /usr/lib/systemd/systemd-vconsole-setup  | grep keyboard
/etc/sysconfig/keyboard
Failed to read /etc/sysconfig/keyboard: %s
```

su [options] [user]

su (*substitute user*) without options switches to root mode when the password is entered. This means that root is the active user until the next exit command.

Optionally, another user can be specified for su instead of the default root user. If su is run by root, you don't even have to enter a password when changing users.

▶ -c command
   Runs only the specified command with root permissions.

▶ -l or --login

When the user changes, the new shell is started as the login shell. This means that all login files are also loaded, which is necessary so that environment variables such as PATH are configured correctly.

---

sudo [options] [var1=value1 var2=value2] command

---

sudo runs a command as if it were being executed by different user (usually this is root). Thanks to sudo, ordinary users can perform administrative tasks or run system-critical commands without knowing the root password.

Before sudo allows the execution of a program, this right must be specified for a specific user and for a specific program in the /etc/sudoers file.

▶ -b

Starts the specified command in the background.

▶ -E

Receives the current content of the environment variable. This option is required to start a program with root permissions in graphics mode on Wayland.

▶ -H

Enters the home directory of the user for whose account the command is being executed in the $HOME environment variable (usually /root).

▶ -i or --login

Starts the shell of the target user (-u) or root. The configuration files of the target user are analyzed, that is, .profile, .bashrc, .zshrc, and so on.

▶ -K

Deletes the saved password. The next time sudo is called, the password must be entered again.

▶ -s

Starts the shell specified in local environment variable $SHELL. This makes it possible to run multiple commands in one shell without having to prefix each command with sudo. The shell can be exited using exit or ⌈Ctrl⌉+⌈D⌉.

▶ -u *user*

Starts the command for the specified user (instead of root).

Variables defined in the sudo command are passed on as environment variables. For this reason, sudo var1=x var2=y cmd is a short notation for the following commands:

```
user$ sudo -s
root# export var1=x
root# export var2=y
root# cmd
root# exit
```

Examples

In the following example, sudo is used to run the apt install command with root permissions:

```
user$ sudo apt install gimp
[sudo] password for user: *******
...
```

The second example allows you to work interactively as user martin with his shell and with his local shell settings:

```
user$ sudo -u martin -i
martin$ pwd
/home/martin
martin$ exit
user$
```

Example with Input/Output Redirection

If you try to run a command with sudo that uses input or output redirection, an error may occur:

```
user$ cd /etc/samba
user$ sudo grep -Ev '^#|^;|^$' smb.conf > smb.conf.condensed
-bash: smb.conf.condensed: No authorization
```

The intention was to eliminate all comments from smb.conf and save the resulting file in smb.conf.condensed. However, the problem is that the output redirection isn't executed with sudo permissions, but runs within bash with the rights of the current user. Of course, this user isn't authorized to change any files in /etc. For this type of constructs to work, you must start a shell with sudo and pass the command to this shell:

```
user$ sudo sh -c 'grep -Ev "^#|^;|^$" smb.conf > smb.conf.condensed'
```

swaks [options]

swaks from the package of the same name stands for *Swiss Army Knife for SMTP*. The command helps you to test or debug mail servers. I'll limit myself to two examples here. The first command tests the server's spam detection and sends an email with a sample spam file:

```
user$ swaks --to hawks@a-company.com --server localhost \
        --data /usr/share/doc/spamassassin/examples/sample-spam.txt
```

The second command tests the plain authentication procedure and adds an additional header line to the mail. The outputs of the failed login have been shortened for reasons of space.

```
user$ swaks --to hawks@a-company.com --from maria@a-company.com \
        --auth LOGIN --auth-user maria@a-company.com \
        --header-X-Test "test email"
Password: **********
=== Connected to efd.a-company.com.
<-  220 efd.a-company.com ESMTP Postfix (Ubuntu) ...
<-  250-STARTTLS ...
-> AUTH LOGIN
<-  334 VXNlcm5hbWU6
-> bWFyaWFAZWluZS1maaXJtYS5kZQ==
<-  334 UGFzc3dvcmQ6
-> YXNk
<** 535 5.7.8 Error: authentication failed: UGFzc3dvcmQ6
```

swapon [options] device
swapoff device

swapon activates the specified device (usually a hard disk partition) or the file specified instead as a swap area. The partition or file must first be formatted as a swap area using mkswap. swapon will be automatically run for all swap areas listed in /etc/fstab when Linux gets booted.

swapon -s provides a list of all active swap devices, including usage data. An example of swapon can be found in the description of mkswap.

swapoff deactivates the specified swap file or hard disk partition.

sync

Runs all buffered write operations on the hard disks. If for some reason an orderly shutdown of Linux isn't possible—if the shutdown, reboot, and halt commands can't be executed and the computer doesn't respond to [Alt]+[Ctrl]+[Del]—then sync should be run just before switching off. But this is only an emergency solution!

sysctl options

sysctl reads the status of kernel parameters or changes their value during operation. Permanent changes must be entered in the /etc/sysctl.conf file.

▶ -a

Provides a list of all available kernel parameters, including the current settings.

▶ -n *parameter*

Returns the current value of the specified kernel parameter.

▶ -p [*filename*]

Processes /etc/sysctl.conf or the specified file and activates all settings saved there.

▶ -w *parameter=value*

Changes the specified parameter. The -w option can also be omitted.

Example

The following two commands first activate the forwarding functions of the kernel and then the masquerading for the eth0 network interface:

```
root# sysctl -w net.ipv4.ip_forward=1
root# iptables -A POSTROUTING -t nat -o eth0 -j MASQUERADE
```

systemctl command

systemctl is used to administrate the init process for distributions that use systemd as the init system. This is the case with almost all current distributions. You can use systemctl to determine a list of all processes managed by systemd, stop or restart individual background services, and more.

Among other things, systemctl distinguishes between ordinary services (services), sockets, devices, mount services, and targets (targets, comparable to Init-V runlevels).

▶ daemon-reload

Loads the configuration files. This command must be run for changes to the systemd configuration files and to /etc/fstab to take effect.

▶ enable/disable *name*

Activates the specified service permanently or deactivates it again. When the command is executed, the corresponding links are created or removed again. With current systemd versions, the additional --now option causes the relevant service to be started or stopped immediately.

▶ get-default

Determines the default target.

- ▶ halt/poweroff/reboot
  Shuts down or restarts the system.

- ▶ is-active *name*
  Tests whether the specified service is active.

- ▶ isolate *name*.target
  Activates the specified target and all services that depend on it. Services that aren't required are stopped. Thus, isolate has a function that's similar to a runlevel change in an Init-V system.

- ▶ list-timers
  Provides an overview of active timers, that is, jobs to be run periodically by systemd. For each job, the time of the last and planned next run is specified.

- ▶ list-units
  Provides a list of all services, sockets, targets, and so on managed by systemd. Multipage outputs are routed through less so that they can be read page by page. If you don't want this, you must specify the --no-pager option. The result can be filtered using --type=... . Permitted unit types are service, socket, target, device, mount, automount, and snapshot.

- ▶ reload/restart *name*
  Prompts the process to reload the configuration or restarts the service completely.

- ▶ set-default *name*.target
  Defines the default target for future boot processes.

- ▶ show *name*
  Provides more detailed data than status. The result is structured line by line and can be processed relatively easily by a script.

- ▶ start/stop *name*
  Starts or stops the specified init process or service.

- ▶ status *name*
  Provides status information on the specified service in a clearly legible form.

systemctl applies by default to services that are executed at system level. To edit services at user level, you must pass the --user option. This is useful on modern desktop systems, for example, where Gnome uses systemd.

Examples

The following lines show typical systemctl commands:

```
root# systemctl start    cups  (start CUPS daemon)
root# systemctl stop     cups  (stop CUPS daemon)
root# systemctl restart cups   (restart CUPS daemon)
```

root# **systemctl reload cups**   (reload configuration of CUPS daemon)
root# **systemctl status cups**   (determine status of CUPS daemon)

root# **systemctl enable cups**   (start CUPS daemon automatically in future)
  Created symlink /etc/systemd/system/sockets.target.wants/cups.socket ->
                /lib/systemd/system/cups.socket.
  Created symlink /etc/systemd/system/multi-user.target.wants/cups.path ->
                /lib/systemd/system/cups.path.

root# **systemctl disable cups**   (no longer start CUPS daemon automatically)
  Removed /etc/systemd/system/sockets.target.wants/cups.socket.
  Removed /etc/systemd/system/multi-user.target.wants/cups.path.

root# **systemctl isolate reboot.target** (reboot computer)

systemd-analyze command

systemd-analyze supports troubleshooting in systemd. You can use various commands to analyze aspects of the systemd configuration or the running systemd units. I'll limit myself to one example here. The following command lists all running units, sorted by the time it took to initialize them.

user$ **systemd-analyze blame**
12.425s plymouth-quit-wait.service
 7.271s dnf-makecache.service
 2.702s NetworkManager-wait-online.service
 2.028s sys-devices-pci0000:00-0000:00:07.0-virtio3-...device
 2.028s dev-virtio ports-org.spice space.webdav.0.device
   ...

tac file

tac displays the specified text file on the screen in reverse order, that is, the last line first. The strange command name is the result of twisting the letters of the cat command, which outputs text files in the correct order.

tail [options] file

tail displays the last 10 lines of a text file on the screen.

▸ -n *lines*
  Outputs the specified number of lines.

▶ -n +*offset*

The output starts with the specified line. tail -n +2 file thus outputs the entire file with the exception of the first line.

▶ -f

Reads the file regularly and outputs all new lines. In this form, tail is particularly suitable for monitoring log files.

Example

The following command displays the last 10 lines of syslog. When new lines are added, they are also displayed so that the entire screen is used after a short time—and not just 10 lines. The execution can be terminated via Ctrl + C.

root# **tail -f /var/log/syslog**

```
tar action [options] files
tar action [options] directories
```

tar combines multiple files or entire directories in an *archive* or extracts their components from this archive. tar was originally designed as a tool for reading and writing data on a streamer. tar therefore accesses the installed streamer (usually /dev/tape or /dev/rmt0) by default. If you want to create an archive in a file, you must specify the -f *file* option.

Because tar also compresses the files to be archived depending on the options specified, its function can be compared with the WinZIP program commonly used on Windows. The typical identifier for archive files is .tar. If the archive file is compressed, the identifiers are usually .tgz, .tar.gz, .tar.bz2, or .tar.xz.

tar is controlled in two stages: First, an action must be specified that tar is supposed to perform, and second, this action can be controlled by one or more options. Even if actions and options formally look the same, there's a significant difference: exactly one action must be specified before all other options. While I'll briefly describe all actions in the following lines, I'll only list the most important options (see man tar).

On many Unix systems, tar is controlled with the same commands and options, but the syntax is different: all commands and options are specified as a block without the usual option dashes, such as tar cvf name.tar path. GNU tar understands both syntax variants; that is, it's up to you whether you use option dashes or not.

Actions

▶ -A, --catenate, or --concatenate
Attaches another archive to an existing archive. This option is only suitable for streamers (not for archive files).

▶ -c or --create
Creates a new archive; that is, any existing archive is overwritten.

▶ -d, --diff, or --compare
Compares the files in the archive with the files in the current directory and detects any existing differences.

▶ --delete
Deletes files from the archive. Only suitable for archive files (not for streamers).

▶ -r or --append
Adds additional files to the archive.

▶ -t or --list
Displays the table of contents of the archive.

▶ -u or --update
Adds new or modified files to the archive. The option can't be used for compressed archives. Caution: The archive will always be larger because existing files aren't over-written! The new files are simply appended to the end of the archive.

▶ -x or --extract
Extracts the specified files from the archive and copies them to the current directory. The files aren't deleted from the archive.

Options

▶ -C *directory*
Extracts the files to the specified directory (instead of the current one).

▶ -f *file*
Uses the specified file as an archive (instead of accessing the streamer). If a simple hyphen is specified instead of a file name (-f -), the data is forwarded to the standard output channel or read from the standard input. This is particularly useful if two tar commands are to be linked by | (typically for transporting entire file trees).

▶ -g timestamp-file
Saves metadata about the files saved with tar in the binary timestamp file and only takes into account the files that have changed since then when tar is called again. This option provides an easy way to create incremental backups (see also the last of the following examples).

▶ -j or --bzip2
Compresses or decompresses the entire archive via bzip2 (see also -z!).

▶ `-J` or `--xz`

Compresses or decompresses the entire archive via `xz`. This assumes that the `xz` command is available. Many distributions require the `xz-utils` package to be installed.

▶ `-L` *n* or `--tape-length` *n*

Specifies the capacity of the streamer in KiB. If the capacity is exceeded by the size of the archive, `tar` prompts you to change the magnetic tape.

▶ `-N` *date*, `--after-date` *date*, or `--newer` *date*

Archives only files that are more recent than the specified date. This facilitates the creation of incremental backups.

▶ `-p` or `--preserve-permissions`

Leaves the access rights unchanged when extracting the files. The option applies by default when `tar` is executed by `root`.

▶ `-T` *file* or `--files-from` *file*

Archives or extracts the file names specified in the file.

▶ `-v` or `--verbose`

Displays all file names on the screen while you're working. If `-v` is used in combination with the `t` command, additional information about the files will be displayed (file size and more). If the option is specified twice, the information will be even more detailed.

▶ `-W` or `--verify`

Checks the correctness of the files just archived after writing. These can't be used for compressed archives.

▶ `-z` or `--gzip`

Compresses or decompresses the entire archive using `gzip`. This option is very useful when it comes to creating `*.tgz` files. However, if data is actually stored on a streamer, the option can be dangerous: A single error on the magnetic tape can render the entire archive unusable! Without compression, files are also destroyed, but the damage is usually significantly less. DAT streamers are capable of compressing the data to be processed themselves. This is faster but isn't as efficient as `gzip`.

Examples

In the following example, `tar` archives all files from the `Documents` directory and from all subdirectories in the compressed `myarchive.tgz` file:

```
user$ tar -czf myarchive.tgz Documents
```

`tar -tzf` returns a table of contents of the archive. The files within the archive are arranged arbitrarily.

```
user$ tar -tzf myarchive.tgz | less...
```

`tar -xzf` unpacks the archive and extracts all contained files:

```
user$ cd different-directory/
user$ tar -xzf myarchive.tgz
```

The following command copies all files from /dir1 to /dir2. The advantage over a regular `cp` command is that symbolic links are copied as such (and not the data referenced by the links). The preceding two commands are particularly suitable for transferring entire file trees from one partition to another.

```
root# (cd /dir1 ; tar cf - .) | (cd /dir2 ; tar xvf -)
```

The last example shows the creation of incremental backups for the `images` directory. During the first backup, any existing timestamp file is deleted to perform a complete backup (e.g., always on Sunday nights). The subsequent backups during the night on the following weekdays, and then only take into account the changes that have been made since then.

```
user$ rm backup.timestamps
user$ tar czvf backup-0.tgz -g backup.timestamps images    (Sun: compl. backup)
user$ tar czvf backup-1.tgz -g backup.timestamps images    (Mon: changes only)
user$ tar czvf backup-2.tgz -g backup.timestamps images    (Tue: changes only)
```

For example, if you want to suppress all `tar` output in a cron script, including error messages, you must add the following input/output redirection to the command:

```
user$ tar czf backup.tgz directory >/dev/null 2>&1
```

tasksel

`tasksel` installs or uninstalls predefined package groups. The command is only available on Debian and Ubuntu distributions.

▶ install *group name*
Installs all packages of the specified package group.

▶ --list-tasks
Determines a list of all defined package groups.

▶ remove *group name*
Removes all packages in the specified package group. Caution: With older `tasksel` versions, the packages are removed regardless of dependencies on other packages. This often results in significantly more packages being uninstalled than intended.

▶ --task-packages *group name*
Lists all packages in the specified package group.

Example

The following two commands install the Apache web server on Debian or Ubuntu with frequently required extensions and additional programs:

```
root# tasksel install web-server     (Debian)
root# tasksel install lamp-server    (Ubuntu)
```

tcpdump [options] [filter expression]

tcpdump from the package of the same name reads the network traffic of an interface, filters it according to criteria, outputs it, or saves it in a file. The command not only handles TCP packets, but also UDP and ICMP packets. tcpdump uses the pcap library internally to read and filter the network packets.

Options

▶ -a

Displays package contents in text format (ASCII).

▶ -c n

Terminates the program after n packets.

▶ -i interface

Considers only packets that flow via the specified interface. You can obtain a list of possible interfaces using tcpdump -D.

▶ -n

Displays IP addresses instead of host names.

▶ -q

Displays less information (quiet).

▶ -r file

Reads the packages from a file previously saved using tcpdump -w.

▶ -w file

Saves the packages in binary format (raw) in the specified file. The file can be read and analyzed again later using tcpdump -r or other programs (e.g., Wireshark).

▶ -x

Displays package contents in hexadecimal format.

Filter Expression

The options can be followed by a filter expression, which can be composed of the following keywords, among others:

- greater *n*
  Considers only packets that are larger than *n* bytes.

- host *ipadr* or host *hostname*
  Takes into account only packets that use the specified IP address or the corresponding host as the source or target.

- less *n*
  Considers only packets that are smaller than *n* bytes.

- net *cidr*
  Considers only packets whose source or destination corresponds to the specified address range in CIDR notation (e.g., 10.0.0.0/24).

- port *n* or portrange *n1-n2*
  Considers only packets that use the specified port numbers as source or target.

- proto ether|fddi|tr|wlan|ip|ip6|arp|rarp|decnet|tcp|udp
  Takes into account only packets of the specified protocol. The expression can be prefixed with ip or ip6 if only IPv4 or IPv6 packets are to be analyzed (e.g., ip6 proto udp).

The host, net, and port keywords can optionally be prefixed with dst or src if the specification refers only to the packet destination or source.

You can link multiple filter conditions via and or dust. You must bracket complex expressions using \(and \). Alternatively, you can use simple parentheses, but then you have to put the entire filter expression in apostrophes (e.g., '(port 1 or port 2)'). Other filter options are described in man pcap-filter.

Examples

The following command outputs information about all HTTP packets flowing through the wlan0 interface:

```
root# tcpdump -i wlan0 port 80
tcpdump: verbose output suppressed, use -v or -vv for full protocol decode
listening on wlan0, link-type EN10MB (Ethernet)
10:34:33.681218 IP imac.57402 > bpf.tcpdump.org.http:
   Flags [S], seq 755525464, win 65535, options [mss 1460,nop,wscale 5,nop,nop,
   TS val 595975353 ecr 0,sackOK,eol], length 0 10:34:33.681793 IP imac.57403 >
bpf.tcpdump.org.http:
   Flags [S], seq 2954861158, win 65535, ...
```

The second command records the next 100 HTTP packets flowing from or to the IP address 192.139.46.66 in the dump.pcap file:

```
root# tcpdump -i wlan0 -n -c 100 -w dump.pcap port 80 and host 192.139.46.66
```

The third example determines packets that flow through the bridge br0 and that contain the specified MAC address (as sender or target):

```
root# tcpdump -i br0 -en | grep -i '00:50:56:00:D0:56'
```

tee file

tee duplicates the standard input, displays one copy, and saves the other in a file. In real life, this is useful if the output of a command is to be observed on the screen but saved in a file at the same time. A simple redirection to a file using > would result in nothing being visible on the screen.

Example

The following command displays the table of contents of the current directory on the screen and saves it in the contents file at the same time:

```
user$ ls -l | tee contents
```

telnet [options] host [port]

telnet allows interactive communication with a network service via the Telnet protocol. The command is primarily suitable for testing basic network functions. Because the communication is unencrypted, telnet is too insecure to run commands on external servers. For that kind of task, you should use ssh!

▶ -4 or -6
  Accepts only IPv4 or IPv6 addresses.

▶ -l user
  Uses the specified username (instead of the current username).

Example

The following example shows the communication with a mail server running on the local computer:

```
user$ telnet localhost 25
Trying 127.0.0.1...
Connected to localhost.
Escape character is '^]'.
220 kofler.info ESMTP Postfix (Ubuntu)
 helo kofler.info
250 kofler.info
```

```
mail from:<user@kofler.info>
250 2.1.0 Ok
...
```

Mail servers are often configured in such a way that only the first commands are transmitted unencrypted. The server and client then negotiate the key encryption data (STARTTLS method). At this point, telnet has to opt out; it's therefore only suitable for testing whether the mail server is active on port 25. swaks is more suitable for testing authentication procedures.

test expression

test is used in bash scripts to formulate conditions and is mostly used in *if* queries and loops. Depending on whether the condition is fulfilled, it returns the truth value 0 (true) or 1 (false). Instead of test, you can also use short notation [*expression*]. It's important that you enter spaces before and after the expression!

If you use test or short notation [*expression*] as a condition in a branch or loop, you must end the condition with a semicolon, for example, if [" $1" = "abc"]; then ...

if queries can sometimes be replaced by the following formulation: ["$1" = "abc"] && cmd. In this case, no semicolon is required. The command won't be run until the preceding condition has been fulfilled.

Character Strings

[cs]	True if the string isn't empty
[-n cs]	True if the string isn't empty (like [cs])
[-z cs]	True if the string is empty (0 characters)
[cs1 = cs2]	True if the strings match
[cs1 != cs2]	True if the character strings differ from each other

The character strings or variables should be placed in quotation marks (e.g., ["$1" = "abc"] or ["$a" = "$b"]). Otherwise, errors may occur with character strings containing multiple words.

Numbers

[n1 -eq n2]	True if the numbers are equal (*equal*)
[n1 -ne n2]	True if the numbers aren't equal (*not equal*)

[n1 -gt n2]	True if *n1* is greater than *n2* (*greater than*)
[n1 -ge n2]	True if *n1* is greater than or equal to *n2* (*greater equal*)
[n1 -lt n2]	True if *n1* is less than *n2* (*less than*)
[n1 -le n2]	true if *n1* is less than or equal to *n2* (*less equal*)

Files (Excerpts)

[-d file]	True if it's a directory (*directory*)
[-e file]	True if the file exists (*exist*)
[-f file]	True if it's a simple file (and not a device, directory, etc.) (*file*)
[-L file]	True if it's a symbolic link
[-r file]	True if the file may be read (*read*)
[-s file]	True if the file is at least 1 byte long (*size*)
[-w file]	True if the file may be changed (*write*)
[-x file]	True if the file may be executed (*execute*)
[file1 -ef file2]	true if both files have the same I-node (*equal file*)
[file1 -nt file2]	True if file 1 is newer than file 2 (*newer than*)

Linked Conditions

[! cond]	True if the condition isn't fulfilled
[cond1 -a cond2]	True if both conditions are fulfilled (*and*)
[cond1 -o cond2]	True if at least one of the conditions is fulfilled (*or*)

bash Variant [[expression]]

While test and the short notation [expression] work in this way in most shells, a variant with two square brackets is also available in bash. Among other things, the following additional comparisons or notations are permitted there:

[[*cs = pattern**]]	True if the string starts with *pattern*
[[*cs == pattern**]]	Same as the preceding entry
[[*cs =~ regex*]]	True if the string matches the regular pattern

[[*bed1* && *bed2*]]	True if both conditions are fulfilled (*and*)
[[*bed1* || *bed2*]]	True if at least one of the conditions is fulfilled (*or*)

Another difference is that variables aren't evaluated in the evaluation in [[ ]] (no consideration of wildcard characters such as * and ?, correct handling of file names with spaces). This is why [[-f $filename]] works even if the variable contains spaces or special characters.

Example

The following mini-script doesn't fulfill any meaningful task, it only demonstrates the test syntax in notation [condition] && command. In the for loop, the i variable is supposed to run through the values from 1 to 10. rest contains the remainder of the integer division by 2. With [$rest -eq 1], the script tests whether the remainder is 1, that is, whether i is an odd number. In this case, the following commands are skipped via continue. echo outputs the contents of i. When i reaches the value 6, the loop gets terminated using break. Thus, the program outputs the numbers 2, 4, and 6.

```
#!/bin/bash
for i in {1..10}; do
    rest=$[$i%2]
    [ $rest -eq 1 ] && continue
    echo $i
    [ $i -eq 6 ] && break
done
```

time command

time runs the specified command and then indicates how much time the execution has taken. The output takes place via the standard error channel and can be changed via the TIMEFORMAT environment variable.

Not only does time replace the stopwatch, it also breaks down the execution time into three components:

▸ real
Displays the actual execution time.

▸ user
Indicates the CPU time used during the execution of the program outside the kernel.

▸ sys
Indicates the CPU time used during the execution of the program within the kernel.

The total of user and sys indicates how much CPU overhead the execution of the command has caused. real is often (much) greater than the total of user and sys. There can be two reasons for this: First, a lot of other processes usually run in parallel to the command started by time, which of course also cost computing time. Second, there may be wait times when executing the command, for example, when reading or writing files or when transferring data in the network.

user and sys each refer to *a* CPU core. If the executed command fully utilizes multiple cores, for example, the total of user and sys can be greater than real.

Example

The following command reads the contents of a logical volume and compresses it using the 7zr command (p7zip package). The process takes 24 minutes in idle mode on a multi-core system. If only one core were available, the backup creation would take even longer (approximately 40 minutes).

```
root# time dd if=/dev/vg830/lv3 bs=4M | 7zr a -si lv3.img.7z
real   24m53.424s
user   38m37.841s
sys     0m30.446s
```

Alternatives

time is a command integrated into the shell. The earlier description refers to the bash variant. If you work in zsh, the output of time looks a little different.

---

timedatectl [options] [command]

The timedatectl command changes the time and the active time zone for distributions with the systemd init system. The changed settings take effect immediately; they are also saved permanently in /etc/localtime.

▶ list-timezones
Provides a list of all known time zones.

▶ set-local-rtc 0|1
Indicates whether the computer's clock contains the local time or GMT time.

▶ settime datetime
Resets the date and time. The hardware clock of the computer is also changed accordingly. The time must be entered in the format 2025-12-31 23:59:59.

▶ set-timezone *name*
Sets the desired time zone.

▶ status

Displays the time, time zone, and various other time data, such as the next change between daylight savings time and winter time.

Example

The following commands first show the current time data and then change the time zone:

```
root# timedatectl status
      Local time: 2025-05-18 09:08:37 CEST
  Universal time: 2025-05-18 07:08:37 UTC
        RTC time: 2025-05-17 18:51:42
       Time zone: Europe/Vienna (CEST, +0200)
  ...
root# timedatectl set-timezone Europe/Paris
```

timeout [options] time span command

timeout runs the specified command and ends it automatically if it runs for longer than the specified time. The time span is specified as a floating-point number with the trailing letter s, m, h, or t in seconds, minutes, hours, or days (e.g., 0.5h for half an hour).

▶ -k

Terminates the program that has been running for too long via a KILL signal. By default, timeout uses a TERM signal, but this can be ignored.

Example

In the following command, scp has 30 seconds to transfer a file to another computer. After that, the process gets canceled.

```
user$ timeout 30s scp localfile myserver:destinationdir/
```

tldr command

tldr stands for *too long, didn't read* and is a subtle criticism of the often enormously long man pages. tldr tries to do better and provides a few instructive examples of popular commands. The command is therefore a great addition to this book. The outputs are highlighted in color in the terminal.

```
user tldr rsync
Transfer files either to or from a remote host (but not between two remote
hosts), by default using SSH.  To specify a remote path, use user@host:path/to/
file_or_directory.  More information: https://download.samba.org/pub/rsync/
```

```
rsync.1.
```
- Transfer a file:
  ```
  rsync path/to/source path/to/destination
  ```

- Use archive mode (recursively copy directories, copy symlinks without
  resolving, and preserve permissions, ownership and modification times):
  ```
  rsync --archive path/to/source path/to/destination
  ```

- Compress the data as it is sent to the destination, display verbose
  and human-readable progress, and keep partially transferred files if
  interrupted:
  ```
  rsync --compress --verbose --human-readable --partial --progress \
    path/to/source path/to/destination
  ```

- (five more examples of use)

top [q]

top displays the list of all running processes every five seconds, whereby the processes are sorted according to their share of the computing time. If the optional q parameter is specified during the call, top constantly updates the list and takes up the entire free computing time. Pressing Q terminates the program.

Alternatives

More convenient and visually appealing alternatives to top are the btop and htop commands, which must be installed separately for most distributions. The commands provide various additional functions, color the output, and allow horizontal and vertical scrolling through the running processes.

If you're not interested in the memory consumption and CPU usage, but in the I/O activity or the energy efficiency of the running processes, you should take a look at the description of iotop or powertop.

touch [options] files

touch changes the time of the last change saved with the file. If the command is used without options, the current time is saved as the change time. If the file doesn't yet exist, a new file that's 0 bytes long will be created.

▶ -r *file*
  Uses the saved modification time of the specified file.

▶ -t *time*

Saves the specified time. You must use the format [[CC]YY]MMDDhhmm[.ss] for the time specification, for example, 202512311730 for 12/31/2025 at 5.30 pm. Alternatively, you can also specify the time as a character string with the -d option. More readable formats are permitted, which you can find out about via info touch.

Example

In this example, touch is used to compare the modification times of all files in a directory with the files in a second directory. This is useful, for example, if you've forgotten the -a option for a larger cp command and the copied files now all have the current date.

```
user$ cd target directory
user$ find . -type f -exec touch -r /source directory/{} {} \;
```

tr [options] zk1 [zk2] [< source > target]

tr replaces all characters of string 1 in the specified source file with the corresponding characters of string 2. The two-character strings should be the same length. Characters that don't appear in the first string remain unchanged. It's not possible to replace a single character with multiple characters (e.g., ö with "o)—for this you must use commands such as recode or sed.

▶ -d

Deletes the characters specified in string 1. String 2 doesn't need to be specified.

Example

In the following example, tr replaces all lowercase letters with uppercase letters and displays the result in the terminal (in the standard output):

```
user$ tr a-zäöü A-ZÄÖÜ < text file
```

traceroute[6] [options] target address

traceroute provides a list of all stations on the path of a TCP/IP packet from the local computer to the target address. The total runtime is specified for each station (by default for three attempts). traceroute only works if UDP port 33434 isn't blocked by a firewall on the way to the target address. In this case, traceroute only returns three stars instead of the stopover details.

▶ -4 or -6

Uses the IPv4 or IPv6 protocol only. By default, traceroute automatically uses the appropriate protocol. If a host name is associated with both an IPv4 and an IPv6 address, `traceroute` prefers IPv4. For `traceroute6`, `-6` automatically applies.

▶ `-m` *n*

Determines the maximum number of intermediate stations (30 by default).

▶ `-n`

Shows only the IP number of the intermediate stations (not the host name).

▶ `-p` *n*

Uses the specified UDP port (33434 by default).

▶ `-q` *n*

Sends *n* packets and measures the response times for each packet (3 by default).

Example

The following command shows the stations between my local computer and my web server. At some stations, the IP packets are routed via different routers, which indicates a redundant network connection.

```
root# traceroute kofler.info
traceroute to kofler.info (213.239.211.2), 30 hops max, 60 byte packets
 1  dsldevice.lan (10.0.0.138)  14.116 ms  13.376 ms  13.112 ms
 2  62.47.95.239 (62.47.95.239)  25.650 ms  27.489 ms  29.245 ms
...
 9  hos-tr3.ex3k16.rz6.hetzner.de (213.239.252.8)  48.315 ms
    hos-tr4.ex3k16.rz6.hetzner.de (213.239.252.136)  49.034 ms
    hos-tr2.ex3k16.rz6.hetzner.de (213.239.229.136)  51.231 ms
10  kofler.info (213.239.211.2)  52.069 ms  54.640 ms  55.757 ms
```

trap [command] signal

The `bash` command `trap` runs the specified command when the shell receives the specified signal. If no command is specified, the program or `bash` ignores the signal in question. `trap -l` returns a list of all possible signals and the identification numbers assigned to them. `trap` is usually used for error protection in shell scripts.

Example

The following mini-script runs in an endless loop that can't be interrupted by pressing `Ctrl`+`C`. However, you can end the script using `kill`, for example.

```
#!/bin/bash
trap 'echo "Ctrl+C will be ignored!"' SIGINT
while true
```

```
do
  sleep 1
done
```

tree [options] [start directory]

tree from the package of the same name displays the hierarchy of the directory tree starting from the current or specified start directory.

▶ -a

By default, outputs files too instead of directories only.

▶ -L *n*

Considers only the first *n* hierarchy levels.

▶ -x

Remains in the current file system and doesn't consider mount directories or links to other file systems.

truncate [options] files

truncate reduces or enlarges files. The file is simply cut off when it's reduced in size. Caution: you'll lose data in this process! When the file is enlarged, a corresponding number of 0 bytes are added to the file. With most file systems, this doesn't change the space required on the data medium because the file will be recognized as a *sparse file*. If you want to physically fill a file with zeros, you must use the dd command. If the name of a file that doesn't yet exist is passed to truncate, the command creates the file in the desired size.

▶ -s *n*

Specifies the desired size in bytes. If the number is followed by the letters K, M, G, or T, truncate multiplies the number by 1,024; 1,048,576; and so on. Alternatively, KB, MB, GB, or TB stand for 1,000; 1,000,000; and so on. Thus, 10K corresponds to 10,240 bytes, and 10KB corresponds to 10,000 bytes.

Example

In the following example, file is first filled with 1 MiB of random data. The file is then enlarged to 2 MiB by using truncate. ls -l actually shows a size of 2 MiB, but dd proves that the file only takes up 1 MiB of space on the hard disk or SSD. The space requirement only increases if data other than zeros is written to the file later.

```
user$ dd if=/dev/urandom of=file bs=1024 count=1024
user$ truncate -s 2M file
user$ ls -l file
```

```
-rw-r--r-- 1 kofler users 2097152 19. Nov 14:30 file
user$ du -h file
1,0M    file
```

tty

tty displays the device name of the active terminal (for the text consoles: /dev/tty1 to tty6, for shell windows under X /dev/pts/*n*).

tune2fs [options] device

You can use tune2fs to change various system parameters of an ext2/ext3/ext4 file system.

▶ -c *n*

Specifies after how many mount operations the partition should be checked for errors when booting. 0 means that a check should never be carried out.

▶ -i *n*

Specifies how often (in terms of days) the partition should be checked for errors during booting. 0 means that a check should never be carried out.

▶ -l

Displays information on the specified file system, but doesn't change anything.

▶ -m *n*

Specifies what percentage of the disk should be reserved for root data (usually 5% for mke2fs).

▶ -U *new-uuid* or -U random

Gives the file system a new UUID.

Example

The following command causes the file system in the /dev/sda1 partition to be checked only once a year or after 200 mount operations:

```
root# tune2fs -i 365 -c 200 /dev/sda1
Set the maximum number of mounts to 200
Set the interval between checks to 31536000 seconds
```

type command

The bash command type determines whether the specified command is a shell command (e.g., cd), a self-defined function, or an alias abbreviation. The *command is hashed* message means that this is a Linux command that has already been executed in this session and whose path name has been memorized by the bash in a hashed directory.

Example

cd is a command integrated into the shell:

```
root# type cd
cd is a shell builtin
```

ubuntu-security-status [options]

In the past, ubuntu-security-status displayed information on the maintenance status of the installed packages. The command has been replaced by pro security-status (see that entry).

ufw [options] command

ufw is an Ubuntu-specific command for firewall configuration. ufw is installed by default on Ubuntu and its derivatives, but it isn't activated. If you want to use ufw for firewall configuration, you must first activate the firewall using ufw enable.

The following subcommands are available for configuring the firewall. They can be combined with the --dry-run option. In this case, no changes are made. The option is useful for testing the syntax of complex commands.

▶ allow|deny [on *interface*] *port/protocol/*"*app name*"
Allows or blocks a port (e.g., 22), a protocol (e.g., ssh), or a program for which there are separate rule files (see ufw app list).

The rule can optionally be formulated only for a specific network interface, such as on eth0. man ufw lists various other options for formulating rules that only apply to incoming or outgoing packets or only for specific protocols or IP versions (e.g., for tcp, udp, ipv6).

▶ app info "*app name*"
Provides detailed information on a ufw rule file.

▶ app list
Provides a list of all programs for which there are ufw rule files in the /etc/uf2/applications.d directory.

▶ default allow|deny|reject [incoming|outgoing|routed]
Defines the default behavior of the firewall. If incoming|outgoing|routed isn't specified, the behavior of the input filter will get changed. By default—if the firewall is activated at all—the deny behavior applies for the incoming filter and the allow behavior for the outgoing filter.

▶ `disable`
Deactivates the firewall. All network traffic can flow unhindered.

▶ `enable`
Activates the firewall.

▶ `reload/reset`
Reloads all firewall rules or performs a complete restart of the firewall.

▶ `status [verbose]`
Displays the status of the firewall.

Example

The following commands first allow the `ssh` service and then activate the `ufw` firewall. An existing SSH connection shouldn't be interrupted by this. However, all other server services will be blocked from now on.

```
root# ufw allow ssh
root# ufw enable
root# ufw status verbose
Status: Active
Logging: on (low)
Default setting:  deny (incoming), allow (outgoing), disabled (sent)
For                     Action      From
--                      ------      ---
22                      ALLOW IN    Anywhere
22 (v6)                 ALLOW IN    Anywhere (v6)
```

`ulimit` option limit

`ulimit` limits the system resources used by the shell and the processes utilized by it.

Sizes are generally given in KiB. `ulimit` is often preset in /etc/profile.

▶ `-c` *memory*
Limits the size of coredumps, that is, the memory image that is automatically saved on the hard disk in the event of a program crash.

▶ `-d` *memory*
Limits the memory for the data segment of processes.

▶ `-f` *file size*
Prevents the creation of files that are larger than the specified limit. Doesn't work with all file systems.

▶ `-s` *memory*
Limits the stack memory.

umask [mask]

The umask bash command controls the access rights used to create new files. umask memorizes an octal numerical value for this purpose, which is subtracted from the default access rights of new files or directories.

Linux actually requires that new files be assigned access bits rw-rw-rw (octal 666). New directories and program files created by a compiler are automatically assigned access bits rwxrwxrwx (777). These basic settings are too permissive for practical work. That's why all Linux shells provide for the umask setting. This is a numerical value that specifies the access bits that are *subtracted* from the standard access bits.

The umask setting is predefined across the entire system and is set in /etc/profile or in /etc/bashrc, depending on the distribution. Different values often apply for root and for all other users. In Ubuntu, the umask value is defined by the PAM system; the default value is stored in /etc/login.defs.

Examples

umask without additional parameters shows the current setting:

```
user$  umask   (Debian, openSUSE, Ubuntu)
0022
user$  umask   (Fedora, RHEL)
0002
```

The second example shows the effects on new files and directories if you change the umask value to 027. New files thus receive the access rights 666 - 026 = 640 = rw-r-----, and new directories have 777 - 027 = 750 = rwxr-x---.

```
user$ umask 27
user$ touch new-file
user$ mkdir new-directory
user$ ls -ld new*
-rw-r----- ... user user ... new-file
drwxr-x--- ... user user ... new-directory
```

umount device
umount directory

umount removes a file system from the Linux directory tree. The file system is specified either by the device name of the data medium or by specifying the directory in which the file system is integrated into the directory tree. The command can only be run by root. It results in an error message if there are still open files on the file system.

► -f

Forces the umount process (useful for NFS directories that are no longer accessible).

unalias abbreviation

unalias deletes the specified abbreviation. If the command is called with the -a option, it deletes all known abbreviations. The handling of abbreviations is explained in the description of alias.

uname [options]

uname displays the name of the operating system (i.e., Linux). Other information can also be displayed by specifying options.

► -a

Displays all available information, namely the operating system, version number, date and time, and processor.

► -m

Provides only the CPU platform (e.g., i686 or x86_64 for 64-bit systems).

► -r

Returns only the kernel version.

Example

This computer is running a 64-bit kernel that was compiled in March 2025:

```
root# uname -a
Linux utmf39 6.7.9-200.fc39.aarch64 #1 SMP PREEMPT_DYNAMIC
Wed Mar  6 20:03:23 UTC 2025 aarch64 GNU/Linux
```

uncompress file

uncompress decompresses a file compressed via compress. The file identifier .Z is automatically removed. uncompress is a link to compress, whereby the -d option is automatically activated.

unexpand text file > result

unexpand replaces multiple spaces in the specified text file with tab characters and writes the result to the standard output. The command is the counterpart to expand, which replaces tab characters with spaces.

▶ -a

Replaces all spaces, not just those at the beginning of a line.

▶ -t *n*

Specifies the number of characters per tab position (8 by default).

uniq [options] file

uniq outputs the lines of a text file to the standard output, eliminating consecutive lines with the same name. For presorted files, uniq eliminates all lines that occur more than once.

▶ -c

Indicates for each line how often it was found.

▶ -d

Outputs the duplicates only (but not lines that only occur once).

▶ -u

Outputs only lines that have *no* duplicates (i.e., are *unique*).

Example

sort sorts the test file, and uniq eliminates duplicate lines and saves the result in test1:

user$ **sort test | uniq > test1**

unset variable

The bash command unset deletes the specified variable.

until *condition*; do
   commands
done

until creates loops in bash scripts. The loop keeps running until the specified condition is met. The loop criterion is the return value of the command that is specified as a condition. Comparisons and tests are performed using the test command or its short form in square brackets.

Example

The example shows the formulation of a simple loop with until:

```
user$ i=1; until [ $i -gt 3 ]; do echo $i; i=$[$i+1]; done
1
2
3
```

unxz [options] files

unxz from the xz-utils package decompresses the files previously compressed using xz. The .xz file extension gets removed.

unzip [options] archive.zip [file1 file2]

The command extracts files from ZIP archives, which are particularly common in the Windows world. If the files to be extracted aren't explicitly listed, unzip unpacks all files in the archive. If you want to create your own ZIP archives on Linux, you must use zip.

Example

The following command extracts the print.pdf file from archive.zip. If there are other files in this archive, they won't be affected.

```
user$ zipinfo archive.zip
Archive:  archive.zip
Zip file size: 5870439 bytes, number of entries: 4
-rw-r--r--  3.0 unx  3594936 bx defN 24-Mar-24 19:06 ebook.pdf
-rw-r--r--  3.0 unx  3245639 bx defN 26-Mar-24 12:54 print.pdf
...
user$ unzip archive.zip print.pdf
Archive:  ../archive.zip
  inflating: print.pdf
```

update-alternatives [options] command

update-alternatives manages the links in the /etc/alternatives directory. These links determine the active version of multiple programs installed in parallel with the same function (e.g., editors or Java environments). The command is usually executed by the (de)installation scripts of the relevant packages, but can of course also be used interactively. On Red Hat or Fedora systems, the command is also available as alternatives.

▶ --auto name

Activates the automatic mode for the command. This means the command with the highest priority value is automatically active.

▶ --config *name*

Shows the available alternatives to the specified command. You then interactively specify which alternative should be active in future. update-alternatives then adjusts the links accordingly and switches to manual mode for the command.

▶ --display *name*

Provides a list of all installed alternatives to the specified command.

▶ --install ... or --remove ...

Installs or removes an alternative for a command. These commands are usually only used in the (de)installation script of a package.

▶ --set *name command path*

Sets the command specified via command path as the default program for name.

Example

The following command defines /usr/bin/jmacs as the default editor:

```
root# update-alternatives --set editor /usr/bin/jmacs
```

This example only works on Debian and Ubuntu systems. On Fedora and RHEL, update-alternatives is also available, but the standard editor can't be changed (whereas the Java version, the mail client, and others can). To configure the default editor, you must add the line export EDITOR=/usr/bin/jmacs to .bashrc or .zshrc, replacing jmacs with your favorite program.

update-grub

This script is only available on Ubuntu and Debian. It runs the /etc/grub.d/* configuration scripts and creates the new GRUB menu file; /boot/grub/grub.cfg. On other distributions, you need to run the grub2-mkconfig -o /boot/grub2/grub.cfg command instead of update-grub.

update-initramfs [options]

update-initramfs is responsible for creating, updating, or deleting initrd files in Debian and Ubuntu. The initrd files contain kernel modules that are loaded by GRUB during system startup. These files are configured using the files in the /etc/initramfs-tools directory. For Fedora, RHEL, and SUSE, you can use the dracut command instead of update-initramfs.

▶ -c

Creates a new initrd file for the kernel version that is specified with -k.

▶ -d

Deletes the initrd file for the kernel version specified with -k.

▶ -k *version name*

Specifies the kernel version to be processed. -k all causes update-initramfs to process the initrd files of all installed kernel versions.

▶ -u

Updates the initrd file of the latest kernel version or the kernel version specified with -k.

Example

The following command creates a new initrd file for kernel version 6.6.7. The kernel file has the file name /boot/vmlinuz-6.6.7-generic, and the name of the resulting initrd file is /boot/initrd.img 6.6.7-generic.

```
root# update-initramfs -c -k 6.6.7-generic
```

updatedb

updatedb creates an index directory for the locate command. The index contains a list of all files contained in the entire file system. The command is usually executed automatically once a day by a cron job. Its execution requires root permissions. Depending on the distribution, the file database is saved in the /var/lib, /var/lib/slocate, or /var/lib/mlocate directory.

uptime [options]

Without further options, uptime indicates how long the computer has been running, how many people are currently logged in, and what the average load was in the last minute, in the last 5 minutes, or in the last 15 minutes (*load average*).

▶ -p

Displays the uptime period in a clearly legible format (*pretty*).

▶ -s

Reveals when the computer was started (*since*).

Example

The server on which the following command was executed has been running for more than three months. Such long uptimes are only reasonable if there were no security-relevant kernel patches during this time or if these patches were integrated into the kernel with a live patch system. On Ubuntu servers, you can determine this via canonical-livepatch.

```
user$ uptime
 15:48:11 up 104 days, 23 min,  1 user,  load average: 0.15, 0.12, 0.05
```

useradd [options] name

useradd sets up a new user. On Debian and Ubuntu, it's better to use adduser.

▶ -b "*basedir*"
Defines the basis for the user directories (/home by default).

▶ -c "*full name*"
Specifies the full name of the new user.

▶ -g *group*
Indicates the main group (primary group) of the user.

▶ -G *groupA*,*groupB*,*groupC*
Determines all *supplementary groups* of the user.

▶ -m
If a home directory doesn't yet exist (/home/name), it will be created. All files from /etc/skel will be copied there. On Fedora and RHEL, -m applies by default due to the default settings in /etc/login.defs and doesn't need to be specified. In such cases, the creation of the home directory can be prevented by -M.

▶ -u *n*
Assigns the specified UID (*user identifier*) to the user if it's still available.

Example

The following commands set up the new user Gerald Gersin with the login name gersin, create a home directory, define an initial password, and then force the user to set a new password immediately on first login and every 100 days thereafter:

```
root# useradd gersin -c "Gerald Gersin" -m
root# passwd gersin
Enter a new password: ********
Enter the new password again:  ********
root# chage -d 0 -M 100 gersin
```

userdel name

userdel deletes the specified user account.

▶ -r

Also deletes the entire home directory and the user's mail inbox.

usermod [options] name

usermod changes various properties of the user account, such as the home directory, the group membership, the default shell, and the UID. Most of the options are identical to those in useradd. Changes to the user account don't take effect until after a new login.

▶ -a -G *group*

Adds the user to the specified group.

▶ -L

Temporarily blocks the account. For this purpose, the ! character is placed before the hash code of the password in /etc/shadow, which is why a login is no longer possible.

▶ -U

Releases an account blocked by -L.

Example

The following command adds the user gersin to the docuteam group:

root# **usermod -a -G docuteam gersin**

If you want to remove the user from the group again later, you must run gpasswd -d gersin docuteam.

vcgencmd command

This command enables you to read key data of the device or its CPU on the Raspberry Pi OS and also change a few parameters.

▶ commands

Lists all vcgencmd commands.

▶ get_config *parametername*|int|str

Returns the status of the specified parameter or all integer or string parameters. The parameters are set in the /boot/config.txt file.

▶ get_throttled

Provides information on potential issues with the power supply (from model 4B).

▶ measure_clock arm|core|hdmi|uart

Provides the CPU frequency, the frequency of the graphics cores and the frequency of various other components.

- ▶ `measure_temp`
  Provides the temperature of the CPU.

- ▶ `measure_volts core|sdram_c|sdram_i|sdram_p`
  Specifies the voltage with which the graphics cores and the memory are supplied.

- ▶ `version`
  Provides the firmware version.

Example

The following commands show the operating status of a Raspberry Pi 5 that is currently idle:

```
pi$ vcgencmd measure_clock arm
frequency(0)=1500019456
pi$ vcgencmd measure_temp
temp=48.4'C
```

vgchange [options] [vgname]

The LVM command `vgchange` changes the attributes of a volume group (VG). The most important application is to activate or deactivate VGs.

- ▶ `-a y|n`
  Activates (y) or deactivates (n) all VGs or the specified VG.

vgcreate [options] vgname pvname1 [pvname2 ...]

`vgcreate` creates a new VG from one or more physical volumes (PVs).

Example

In the following example, `pvcreate` first marks the first partition of the hard disks or SSDs /dev/sdb1 and /dev/sdc1 as physical volumes. `vgcreate` creates the volume group vg1 from this. Then, the `lvcreate` command reserves 100 GiB for the logical volume lv1. `mkfs.ext2` sets up a file system in it, and `mount` integrates the file system to the /mnt/lv1 directory just created in the directory tree.

```
root# pvcreate /dev/sdb1
root# pvcreate /dev/sdc1
root# vgcreate vg1 /dev/sdb1 /dev/sdc1
root# lvcreate -L 100G -n lv1 vg1
root# mkfs.ext4 /dev/mapper/vg1-lv1
root# mkdir /mnt/lv1
root# mount /dev/mapper/vg1-lv1 /mnt/lv1
```

vgdisplay vgname

vgdisplay shows detailed information on the specified VG.

vgextend vgname pvname

vgextend adds a PV to a VG.

Example

The following commands define the /dev/sdc2 partition as a new physical volume and then add it to the existing myvg1 volume group, which has become too small:

```
root# pvcreate /dev/sdc2
  Physical volume "/dev/sdc2" successfully created
root# vgextend  myvg1 /dev/sdc2
  Volume group "myvg1" successfully extended
root# vgdisplay myvg1
...
  VG Size            180.64 GB
  Alloc PE / Size    6402 / 22.50 GB
  Free  PE / Size    41324 / 158.14 GB
...
```

vgmerge vgname1 vgname2

vgmerge adds vgname2 to vgname1. Once the command has been successfully executed, there's only vgname1. It consists of all PVs that previously formed the storage pool for vgname1 and vgname2.

vgreduce [options] vgname [pvname1 pvname2 ...]

vgreduce removes the specified PVs from the storage pool of the VG. This only works if the PVs are unused. The -a option removes all inactive PVs.

vgrename oldvgname newvgname

vgrename gives a VG a new name. Instead of oldvgname, the UUID of the VG can also be specified.

vgscan

vgscan lists all volume groups.

vipw [options]
vigr [options]
visudo [options]

The vipw, vigr, and visudo commands start the vi/vim editor (see the section on keyboard shortcuts) and open the /etc/passwd, /etc/group, or /etc/sudoers file. When saving, a check is carried out to ensure that you're adhering to the syntax of the respective file. As long as this isn't the case, saving isn't possible.

▶ -r *directory*
Uses the specified directory as the root directory and edits the files relative to this directory.

▶ -s
Opens the /etc/shadow file (vipw command) or the /etc/gshadow file (vigr command).

You can use the three commands even if you're not a fan of the vi editor. In this case, you must first set the VISUAL or EDITOR environment variable and enter the path to your favorite editor there.

virsh [[-c connection] command]

You can use the virsh command to administrate virtual KVM and Xen machines. virsh can be used in two ways: either to run a virsh command directly or run it interactively as a shell. In the first variant, you can specify a connection string with the -c option:

```
root# virsh -c qemu:///session list --all ...
```

The most important virsh commands are briefly described here. If you read more details in the man page, then note that virtual machines are called *domains* in the virsh terminology.

▶ attach-device *name device.xml* [--persistent]
Adds an additional hardware component to a virtual machine (e.g., a USB device), which is described in libvirt XML format. The --persistent option causes the device to be permanently connected to the virtual machine and saved in the virtual machine's XML file.

▶ attach-disk *name source target*
Adds a data medium to a virtual machine, where *source* is the device name on the host system, and *target* is the device name in the guest. The command can be

supplemented by numerous options that specify the driver (--driver), the caching method (--cache), and more. The --persistent option permanently connects the data medium to the virtual machine.

▶ **attach-interface** *name type source*
Adds a network interface to a virtual machine. *type* specifies the type of interface, such as network or bridge. *source* specifies the interface name on the host computer (e.g., br0). The details of the interface can be set using additional options (--mac, --model, and more).

▶ autostart [--disable] *name*
Specifies that the virtual machine should be started automatically during the boot process of the host computer. By using the --disable option, you can deactivate the automatic start again.

▶ connect qemu:///session
Establishes a user connection to the libvirtd instance of the current user. This allows you to manage your own virtual machines.

▶ connect qemu:///system
Establishes a connection to the system instance of libvirtd. If you run virsh with root permissions, virsh automatically establishes this connection.

▶ connect qemu+ssh://*user@hostname*/system
Establishes a connection to the libvirtd instance of another computer (*hostname*). The communication takes place via an SSH tunnel.

▶ console *name*
Allows you to operate the specified virtual machine directly in the console. This assumes that a getty process for the serial */dev/ttyS0* interface is running in the virtual machine. To terminate the connection, you must press [Ctrl]+[ ].

▶ define *xml file*
Sets up a new virtual machine whose key data is summarized in the specified XML file. Caution: if a virtual machine with the same name already exists (according to the <name> element in the XML file), its definition will be overwritten!

▶ destroy *name*
Terminates the virtual machine immediately. This is similar to unplugging the power cable on your computer, and it can have the same consequences (i.e., a destroyed file system, etc.)!

▶ detach-device *name device*.*xml*
Removes the hardware component described by an XML file from the virtual machine.

▶ detach-disk *name target*
Removes the data medium from the virtual machine. *target* specifies the device name in the guest.

- detach-interface *name type* --mac=xxx
  Detaches a network interface from the virtual machine.

- domstatus *name* and dominfo *name*
  Provide information on a virtual machine.

- edit *name*
  Loads the XML file describing the virtual machine into an editor, whereby the $EDITOR environment variable is taken into account.

- list [--inactive or --all]
  Lists all running virtual machines. If you only want to list the machines that aren't currently active or all machines, you need to enter the --inactive or --all options.

- managedsave *name*
  managedsave-remove *name*
  Saves the status of the virtual machine (i.e., the contents of the RAM, the CPU registers, etc.) in a file in the /var/lib/libvirt/save/ directory and then stops the execution of the machine. To reactivate the virtual machine, you can simply use start. Then, the status file gets deleted automatically. If you want to discard the saved state and restart the virtual machine completely, you should delete the state file using managedsave-remove.

- net-create *xmlfile*
  net-start *netname*
  net-destroy *netname*
  net-undefine *netname*
  net-list

  Support the management of virtual networks. These are private network areas that can be connected to the host system via NAT, for example, the default network of the libvirt tools.

- pool-define *xmlfile*
  pool-define-as *poolname type* --target *path*
  pool-start *poolname*
  pool-auto-start *xmlfile*
  pool-destroy *poolname*
  pool-delete *poolname*
  pool-list
  pool-info *poolname*

  Support the administration of libvirt storage pools. pool-define creates a new pool whose properties are described in an XML file. pool-define-as also creates a new pool, whereby the key data is passed directly as a parameter. Permitted pool types include dir (a local directory), netfs (a network directory), logical (a volume group), disk (a hard disk), or iscsi (an iSCSI server).

The new pool must then be started via pool-start. If the pool is to be started automatically in the future, you must also run pool-autostart.

The commands for deleting a pool are somewhat confusing: A pool must be deactivated using pool-destroy prior to the deletion. Although the command suggests something bad, the pool is merely stopped and can be restarted later by using pool-start. Only pool-delete deletes the pool. pool-delete assumes that all volumes in the pool have been deleted first, which means that only an empty pool can be deleted.

▶ qemu-monitor-command --hmp *name* '*command*'
Runs the specified QEMU monitor command for the virtual machine specified by *name*. The --hmp option is required because the command must otherwise be specified in JSON format.

▶ restore *file*
Reactivates a virtual machine saved using save. The state file can be deleted afterward.

▶ save *name file*
Saves the state of the virtual machine (i.e., essentially the contents of the RAM) in a file and then stops the execution of the machine. To reactivate the virtual machine, you must use restore.

▶ schedinfo [*options*] *name*
Displays or changes the scheduler parameters of the virtual machine (--set parameter=value). These parameters can be used to control how many hardware resources a virtual machine is allowed to use. Resource control requires the activation of the cgroups functions on the host system.

▶ setmaxmem *n*
setmem *n*

Sets the limit for memory usage (RAM) and the currently allocated memory. The additional --config, --current, and --live options allow you to specify whether the settings are to be saved permanently or only executed currently.

▶ shutdown/reboot *name*
Shuts down or reboots the virtual machine. The virtual machine receives a shutdown signal via ACPI. The ACPI daemon acpid must be installed in the virtual machine for the signal to be processed.

▶ snapshot-create *vmname* [*xml-file* [*--redefine*]]
snapshot-create-as *vmname snapshotname* [*description*]
snapshot-list *vmname*
snapshot-delete *vmname snname*
snapshot-revert *vmname snname* [--running]

Creates a snapshot of a running virtual machine, lists all snapshots, deletes a snapshot, or applies the contents of a snapshot to the image file. The snapshot function can only be used for virtual machines that use QCOW2 image files.

▶ start *name*

Starts the specified virtual machine. If you want to communicate with the machine in graphics mode, you should use either a VNC client (the connection data is determined by the virsh command vncdisplay, see the vncdisplay item in this list) or the virt-viewer program.

▶ suspend/resume *name*

Stops the specified virtual machine temporarily or resumes execution. However, the stopped virtual machine continues to use RAM! Thus, only the virtual CPU is stopped.

▶ ttyconsole *name*

Specifies which device of the host computer is used to access the serial interface of the guest system (e.g., /dev/pts/5).

▶ undefine *name*

Deletes the XML file that describes the virtual machine. The image file with the virtual hard disk is retained. undefine can't be run until all snapshots of the virtual machine have been deleted.

▶ vcpuinfo *name*

Provides information on the (virtual) CPUs available to the virtual machine, as well as information on the CPU time used so far. The command can also be used to check the CPU pinning.

▶ vcpupin *name guest-cpu-nr host-cpu-list*

Links the physical CPU cores of the host system with the virtual CPUs of the guest. vcpupin vm1 0 2 permanently assigns the third CPU core of the host to the first virtual CPU of virtual machine vm1.

▶ vol-create *xmlfile*
  vol-create-as *poolname newvolname size*
  vol-delete *volname*
  vol-list
  vol-info *volname*

Support the administration of data media in storage pools. vol-create creates a new data medium whose properties are described in the specified XML file. vol-create-as creates a new data medium of the desired size, whereby the k, M, G, and T suffixes are permitted for KiB, MiB, GiB, and TiB. When you create an image disk, you can use --format raw/qcow2/qed to specify the desired format (raw by default).

--allocation *size* determines how much of the memory is to be allocated in advance; with raw volumes, the entire memory is always reserved.

If there are multiple storage pools, you must specify which pool you're referring to for all commands with the `--pool poolname` option.

▶ `vncdisplay name`
Returns the IP address (empty for `localhost`) and port number for the VNC display of the virtual machine. You can then start any VNC client to interact with the virtual machine. For security reasons, VNC access only works from `localhost` by default (see the `/etc/libvirt/qemu.conf` file).

`vncdisplay` doesn't return a result if the virtual machine doesn't share its graphics system via VNC at all, but uses the more modern Spice system instead. In this case, you can operate the virtual machine using the `virt-viewer` program. You can directly pass the virtual machine name to this program.

However, should the need arise to determine the Spice port number, it will get difficult. `virsh` lacks a command that determines the spice port of a virtual machine, similar to `vncdisplay`. The following command, which I found in a forum on *https:// ubuntuusers.de/*, provides a solution by extracting the port number from the process list:

```
spice_port=$(ps aux | grep vm_name | grep -oP "(?<=-spice port=).*?(?=,)")
```

Examples

Once you've started the `libvirt` shell via the `virsh` command, you can run commands to manage all virtual machines on the local computer:

```
root# virsh
virsh# list --all
 Id Name          Status
-------------------------------
 13 fedora        running
  - ubuntu        shut down
  - windows       shut down
virsh# start windows
Domain windows started
virsh# vncdisplay windows
:1
virsh# exit
```

You can also connect to the `libvirtd` daemon on another machine via SSH. If a `root` login with password via SSH isn't possible on the KVM host for security reasons, you must set up your public SSH key on the KVM host before the first connection is established.

```
virsh# connect qemu+ssh://user@hostname/system
user@hostname's password: *******
```

virt-clone [options]

virt-clone copies a shutdown virtual machine that is managed by libvirt tools. This creates a new XML definition file and a copy of the image file. The other hardware components remain largely unchanged.

▶ --auto-clone

Creates a new virtual machine. The VM name is composed of the previous name plus -clone. -clone is also inserted into the names of the image files, so that rhel9.img becomes rhel9-clone.img, for example. With --auto-clone, the explicit specification of --name and --file can be omitted.

▶ --connect *hypervisor*

Establishes a connection to the specified virtualization system.

▶ -f *file* or --file=*file*

Specifies the desired name of the new image file. If the original virtual machine has multiple virtual hard disks, you must use this option multiple times. Instead of an image file, you can also specify the name of a device file, for example, if you use logical volumes of the host system as data storage.

▶ --mac *nn:nn:nn:nn:nn:nn*

Specifies the desired new MAC address of the network adapter. If this information is omitted, virt-clone automatically uses a unique random MAC address.

▶ --name *reliving*

Specifies the name of the new virtual machine.

▶ --original *name*

Specifies the name of the original virtual machine.

▶ --preserve-data

Prevents the image file from being copied. This means that the new virtual machine uses the same image file as the previous virtual machine. It's therefore not permitted to run the old and the new virtual machine at the same time! This option is useful if you want to test a system with a new virtual hardware or network configuration.

Example

The following command copies Ubuntu server installation userver5. The new virtual machine is named userver6, and the new image file is saved in the /var/lib/libvirt/images/userver6.img file. Make sure to create the new image file in a libvirt storage pool. Otherwise, the SELinux rules on RHEL/Fedora will prevent the virtual machine from running.

```
root# virt-clone --original userver5 --name userver6 \
           --file /var/lib/libvirt/images/userver6.img
```

`virt-install` [options]

`virt-install` is a Python script that helps you to set up new virtual machines.

▶ `--arch` *architecture*
Specifies the desired CPU architecture, for example, `i386`, `i686`, or `x86_64`. By default, `virt-install` uses the same architecture as the one used on the host system.

▶ `--cdrom` *file*
Specifies the file name of the ISO file or the device name of the CD/DVD drive from which the installation data is loaded.

▶ `--connect` *hypervisor*
Establishes a connection to the specified virtualization system. If `virt-install` is started on a KVM host with `root` permissions, the script automatically establishes a connection to `qemu:///system`.

▶ `--disk` *file*
Specifies the file name of the image file in which the virtual machine is to be saved. If you want to use multiple virtual storage devices, you must specify the option multiple times.

▶ `--disk` *opt1=value1,opt2=value2,...*
Allows you to specify multiple image files from different `libvirt` storage pools. Permitted options include `path` to specify an image file or a block device, `pool` to specify a previously created `libvirt` storage pool, `vol` to specify an existing image file in a storage pool (`vol=poolname/imagename`), `size` to specify the desired size of new image files (in GiB), `bus` to select the bus system (`ide`, `scsi`, or `virtio`) and `cache` to select the caching (`writethrough`, `writeback`, or `none`).

▶ `--graphics` *type,opt1=value1,opt2=value2 ...*
Specifies how the graphics system of the virtual machine is to be implemented. The possible types are `vnc` or `spice`. VNC is used as standard. Various connection parameters can be specified with the other options, such as `port`, `listen`, and `password`.

▶ `--import`
Causes `virt-install` to not perform a new installation, but to create the new virtual machine on the basis of an existing image file.

▶ `--name` *type, opt1*
Specifies the name of the virtual machine.

▶ `--network` *type,opt1=value1,opt2=value2,...*
Specifies the key data of a network device of the virtual machine. The option can be specified multiple times if required. `type` specifies how the network connection to the host computer should be established. Permissible values are `bridge=name` for a previously set up network bridge, `network=name` for a virtual network defined using

virsh, or simply user, if the virtual machine is to communicate with the host system via NAT.

The other options determine the parameters of the network device: model specifies which adapter is to be emulated, for example, e1000, rtl8139, or virtio. mac defines the MAC address of the device. If this parameter isn't specified, virt-install generates a random MAC address itself. MAC addresses for KVM must start with 52:54:00!

▶ --nodisk

Enables the installation without an image file (e.g., if the installation is to take place in a network storage device).

▶ --noreboot

Prevents the automatic restart after the installation has been completed.

▶ --os-type *name*

Specifies the type of operating system to be installed, such as linux, unix, windows, or other. The information is used to optimize various hardware parameters.

▶ --os-variant *name*

Specifies which operating system is to be installed, for example, fedora28, virtio26, rhel7, or win10. A complete list of all known operating system names is provided by virt-install. If you use this option, you don't need to specify --os-type.

▶ --ram *n*

Indicates the size of the RAM in MiB.

▶ -serial pty

Equips the virtual machine with a serial interface. When the virtual machine is started, the serial interface is connected to a pseudo-TTY device on the host computer. In virsh, you can determine the device name using ttyconsole *vmname*.

▶ --soundhw=ac97/es1370/sb16

Equips the virtual machine with a sound card.

▶ --vcpus=*n*

Specifies the desired number of CPU cores (one by default).

▶ --video=cirrus/vga/vmvga

Specifies which graphics adapter is to be emulated (cirrus by default).

Example

Before you run virt-install, you must create an image file for the virtual hard disk:

```
root# virsh
virsh# vol-create-as default disk.qcow2 10G --format qcow2
virsh# exit
```

The `virt-install` command enables you to set up a new virtual machine:

```
root# virt-install --name myvmname --ram 2048 --cdrom install.iso \
      --os-variant rhel8  --disk vol=default/disk.qcow2 --graphics vnc \
      --noreboot
```

virt-top [options]

virt-top shows how much CPU capacity and memory space the running virtual machines require. While virt-top is running, you can sort the display via ⓪, ①, ②, and ③ according to virtual machines, CPU cores, network interfaces or disk devices. Ⓠ terminates the command.

▶ --connect hypervisor
Establishes the connection to the specified virtualization system. By default, virt-top communicates with qemu:///system if it has been started with root permissions.

▶ -o sort
Specifies the desired sort order, such as cpu, mem, or time, to sort the list according to CPU performance, memory requirements or total computing time.

Example

```
root# virt-top
virt-top 14:36:17 - x86_64 8/8CPU 1600MHz 15961MB
5 domains, 3 active, 3 running, 0 sleeping, 0 paused, 2 inactive D:0 0:0 X:0
CPU: 0.1%  Mem: 6144 MB (6144 MB by guests)

   ID S RDRQ WRRQ RXBY TXBY %CPU %MEM    TIME    NAME
   3 R    0   22    0    0  1.3 12.0  35:20:50 kofler.info
   4 R    0    0    0    0  0.1 12.0 274:52.67 pi-book.info
   1 R    0    0    0    0  0.0 12.0  39:11:52 ubuntu-book.info
   -                                           (michael-kofler.com)
```

virt-viewer [options] name/id/uuid

virt-viewer is a VNC and Spice client that displays the graphics system of a virtual machine. The name, ID, or UUID of the virtual machine is normally simply passed to the command.

▶ -c hypervisor
Establishes a connection to the specified virtualization system. This option is only required if it's an external KVM host.

wait [process number]

The bash command wait waits for the end of the specified background process. If no process number is specified, the command waits for the end of all running background processes started by the shell.

Example

The use of wait is particularly useful in shell scripts if multiple background processes are to be started first and then have to wait for the end of all these processes:

```
#!/bin/bash
command1 &
command2 &
command3 &
command4 &
wait
# only continue here when command 1 to command 4 are finished
...
```

wakeonlan [options] mac

wakeonlan from the package of the same name (available for Debian and Ubuntu) sends a special network packet for a MAC address and attempts to "wake up" the identified device from sleep mode. This only works if the device supports the *Wake-on-LAN* standard.

▶ -i *host*
Sends the Wake-on-LAN packet only to the specified host or address and not to the entire network.

watch command

watch runs the specified command periodically and displays the output.

▶ -d or --differences
Marks the differences compared to the preceding result.

▶ -e or --errexit
Terminates watch if the called command returns an error.

▶ -n *err exit*
Specifies after how many seconds the command should be run again (by default, it's every two seconds).

Example

The following command displays all changes in the list of running mysqld processes:

user$ **watch -d 'ps aux | grep mysqld'**

> **wc** files

wc counts the number of lines, words, and characters in the specified files. If multiple files are entered using wildcards, wc also calculates the total of the three entries. In addition, wc is also well suited for combination with other programs.

Example

find returns one line for each regular file. wc counts the lines and thus returns the total number of all files in the directory tree.

user$ **find / -type f -print | wc**

> **wget** [options] url1 url2 . . .

wget downloads files from HTTP, HTTPS, and FTP servers. wget can resume interrupted downloads, can recursively follow links, and is suitable for automating downloads.

The files to be transferred are specified as URLs (*Uniform Resource Locators*), for example, as http://myserver.com/file.iso. The wget command is controlled by countless options; only the most important of these are summarized here:

▶ -b

Performs the download as a background process and writes all status messages to the wget-log file.

▶ -B *url* or --base=*url*

Prefixes all relative links within the file you've specified with -I file with the basic *url* address.

▶ -c

Resumes an interrupted download. Caution: if the file has changed on the server in the meantime, the downloaded file is incorrect and contains some data from the old file and some data from the new file!

▶ --force-html

Interprets the file specified with -i as an HTML file and loads all files referenced by links in the HTML file.

▶ --ftp-user=*user* --ftp-password=*pw*

Specifies the username and password for FTP downloads. This information can also be transferred in the URL (ftp://user:password@server/file.txt).

▶ -i *file* or --input-file=*file*
Reads the files (URLs) to be transferred from the specified text file. If you pass the - character instead of a file, wget expects the URLs from the standard input. The -i option only accepts local files as parameters, not URLs. (-i http://server/name.html therefore doesn't work!)

▶ --limit-rate=*n*
Limits the download quantity per second. The letters k and m denote kilobytes and megabytes respectively (i.e., --limit-rate=0.25m).

▶ --no-check-certificate
Doesn't check HTTPS certificates. This makes sense if you know that the relevant server uses self-signed certificates.

▶ -q
Dispenses with the output of status messages (*quiet*).

▶ --retry-connrefused
Makes further attempts to download the file even after the *connection refused* error. This option is only useful for unreliable download servers that occasionally disappear from the network and reappear a short time later.

▶ --spider
Tests whether all links in the file specified by -i *htmlfile* --force-html are still valid. However, the files indicated by links aren't downloaded.

▶ -t *n* or --tries=*n*
Makes *n* attempts to reconnect if the connection is lost (the default setting is 20). With -t 0, wget tries its luck until the download is successful or you cancel the command.

▶ -w *n* or --wait=*n*
Specifies how many seconds wget should wait prior to downloading the next file. This option prevents the download server from being overloaded by countless, almost simultaneous, download requests.

Recursive Downloads

The following options control recursive downloads:

▶ -E or --html-extension
Adds the .html extension to the file names of all downloaded files of the *application/ xhtml+xml* or *text/html* type whose names don't end with .html or .htm.

▶ -H or --span-hosts
Also follows links to other websites.

- ▶ -k or --convert-links
  Changes the links in the downloaded HTML files so that they reference the local files. This makes it possible to view the pages offline later.

- ▶ -l *n* or --level *n*
  Limits the recursion level (5 by default). -l inf deactivates the limitation.

- ▶ -L or --relative
  Tracks only relative links (but no absolute links to the start website).

- ▶ -r
  Activates recursive downloads. For HTTP downloads, wget tracks all HTTP links on the start page, downloads the files specified in this way, tracks their links, and more. wget only takes into account relative links and links to pages on the start website. For FTP downloads, wget reads all subdirectories.

- ▶ -p
  Downloads all files required to view the source file (including files that exceed the maximum recursion level).

Examples

In its basic form, wget simply downloads the specified file:

```
user$ wget ftp://myftpserver.com/name.abc
```

If the download gets interrupted for any reason, it can be resumed without any trouble by using -c:

```
user$ wget -c ftp://myftpserver.com/name.abc
```

The following recursive download command (-r option) helps to read a website offline later or to archive its current status. The recursion depth is limited to four levels by -l 4.

```
user$ wget -r -l 4 -p -E -k http://website.com
```

whatis file

whatis outputs a short description of the specified command or keyword. whatis descriptions only exist for topics for which man texts are installed. If whatis doesn't work, the underlying databases that can be created via mandb or makewhatis are probably missing.

whereis file

whereis searches all common paths for binary files, man files, and source code for the specified file name. whereis is therefore less thorough than find, but much faster. The man page for whereis lists the directories that are searched.

which command

which searches all paths specified in PATH for the command. which returns the full name of the command that would be executed if the command were called without path information. This is particularly useful if multiple versions of a command exist in different directories.

In bash, type can also be used instead of which. The type command helps you find out whether command is actually a built-in shell command, an alias, or a self-defined function.

Example

which determines where the ls command is stored in the file system:

```
user$ which ls
/usr/bin/ls
```

while condition; **do**
   commands
done

while creates loops in bash scripts. The loop runs until the specified condition is no longer fulfilled for the first time. The loop criterion is the return value of the command that is specified as a condition. Comparisons and tests are performed using the test command or its short form in square brackets.

Examples

The following loop outputs the numbers 0 to 5:

```
#!/bin/bash
i=0
while [ $i -le 5 ]; do
    echo $i
    i=$[$i+1]
done
```

To process a text file line by line, you must formulate your bash script as follows:

```
#!/bin/bash
while read line; do
  echo $line
  ...
done < textfile.txt
```

who [options]

who displays a list of all system users currently logged in. Even if you're the only person working on your computer, you can log in with different names on different text consoles. If you want to know who was last logged in on this computer, you must run the last command.

▶ -a

Provides detailed information on each user, including the console used and the login time. For SSH connections, who also shows the computer on which the login occurred.

▶ -m

Specifies the username of the currently active console. The who am i command has the same meaning.

Example

In addition to the TTY processes for the six consoles, there are two active shell sessions of the user kofler on the test computer:

```
user$ who -a
            system boot   2025-04-09 08:57
            run-level 2   2025-04-09 08:57
LOGIN       tty5          2025-04-09 08:57          971 id=5
LOGIN       tty2          2025-04-09 08:57          976 id=2
LOGIN       tty3          2025-04-09 08:57          977 id=3
LOGIN       tty1          2025-04-09 08:57         1802 id=1
kofler    + pts/1         2025-04-24 11:45 02:02    351 (xxx.telekom.at)
kofler    + pts/2         2025-04-24 13:47   .      2354 (xxx.telekom.at)
```

whois [options] host name

whois performs DNS queries for the specified host name and displays the result.

Example

```
user$ whois kofler.info
Domain Name: KOFLER.INFO
...
Domain Status: ok https://icann.org/epp#ok
Registrant Name: Kofler Michael
...
```

wl-copy [options] text
wl-paste

wl-copy from the wl-clipboard package copies the text passed as a parameter to the clipboard of a Wayland system. If no text is transferred, the command expects the data from the standard input channel.

wl-paste outputs the contents of the clipboard via the standard output channel.

▶ --type mime/type
  Defines the MIME data type for wl-copy.

wlr-randr [options]

wlr-randr without additional options lists all permissible graphic resolutions for a Wayland system. With --mode, you can change the current resolution for a monitor output. wlr-randr is the Wayland counterpart to the X command (see xrandr).

Example

The following command changes the resolution of the monitor connected to the second HDMI output to 2,560 × 1,440 pixels:

```
user$ wlr-randr --output HDMI-A-2 --mode 2560x1440
```

wol [options] mac

wol from the package of the same name (available for Fedora and RHEL) sends a special network packet for a MAC address and tries to "wake up" the device from sleep mode. This only works if the device supports the *Wake-on-LAN* standard.

▶ -h *host* or -i *host*
  Sends the Wake-on-LAN packet only to the specified host or address and not to the entire network.

`wpa_passphrase` ssid

`wpa_passphrase` helps to set up a configuration file for the Wi-Fi authentication service; `wpa_supplicant`.

Example

`wpa_supplicant` expects the name of a Wi-Fi network (i.e., *service set identifier*) as a parameter. You must then enter the password for the Wi-Fi. The command then outputs multiple lines in the syntax of `wpa_supplicant`. You now only need to add these lines to the end of /etc/wpa_supplicant/wpa_supplicant.conf:

```
user$ wpa_passphrase my-wifi-name
# reading passphrase from stdin
strictlySecret
network={
        ssid="my-wifi-name"
        #psk="strictlySecret"
psk=9a184914197f550e1c6b350cc49b09a5fab532a8ec991f997ee77fd0a5e78d96
}
```

`write` username

`write` allows you to send a message to another user. After running the command, all characters entered prior to Ctrl+D are transferred to the terminal of the specified user.

`wsl` [options] [command]

`wsl` isn't a Linux command, but a Windows command. It has been included in this book because it helps you manage the *Windows Subsystem for Linux* (WSL), that is, to start and remove Linux installations on Windows, and more. The command is only available if you activate the **Windows subsystem for Linux** option in the **Activate Windows Features** program.

Without further options, `wsl` starts the default Linux installation or runs the command passed to `wsl`.

► `-d` or `--distribution` *name*
Starts the specified distribution. Its further use takes place interactively in a shell.

► `-l` or `--list`
Lists all Linux installations. With the additional `--verbose` option, the command also reveals which Linux instances are currently running and whether they are WSL1- or WSL2-compatible. (WSL2 uses a real Linux kernel to run the respective distribution,

which has speed advantages. However, there are restrictions with WSL2 compared to WSL1 regarding the network connection.)

▶ -s or --set-default *name*
Specifies which Linux installation is the default Linux system.

▶ --set-default-version *name* 1|2
Determines whether new Linux installations are run with WSL1 or WSL2 by default.

▶ --set-version *name* 1|2
Specifies how the specified Linux distribution should be run (WSL1 or WSL2).

▶ --shutdown
Shuts down all running Linux distributions.

▶ -t or --terminate *name*
Stops the execution of the specified Linux distribution.

▶ -u or --user *name*
Runs the command or login for the specified user.

▶ --unregister *name*
Deletes the specified Linux installation.

wsl doesn't provide the option of installing a new Linux distribution. To do this, you need to consult the Microsoft Store, where various Linux distributions are available for free download. The first time you start a WSL distribution, you must set up a user and password. This user has sudo privileges.

Examples

There are four Linux installations on the test computer, with Ubuntu as the default distribution:

```
> wsl --list --verbose
NAME                    STATE        VERSION
* Ubuntu                Stopped      2
kali-linux              Stopped      1
docker-desktop-data     Stopped      2
docker-desktop          Stopped      2
```

wsl runs the id command twice, once in the standard account and once with root permissions. No password is required for this.

```
> wsl id
uid=1000(kofler) gid=1000(kofler) groups=1000(kofler),4(adm),...

> wsl -u root id
uid=0(root) gid=0(root) groups=0(root)
```

The last example starts Kali Linux for interactive use:

```
> wsl -d kali-linux
$ cat /etc/os-release
PRETTY_NAME="Kali GNU/Linux Rolling"
NAME="Kali GNU/Linux"
...
$ ip a
8: eth0: <BROADCAST,MULTICAST,UP> mtu 1500 group default qlen 1
   link/ether 04:d4:c4:92:65:f5
   inet 192.168.178.52/24 brd 192.168.178.255 scope global dynamic
   ...
16: eth1: <BROADCAST,MULTICAST,UP> mtu 1500 group default qlen 1
   link/ether 00:15:5d:7e:85:98
   inet 172.31.96.1/20 brd 172.31.111.255 scope global dynamic
   ...
```

xargs command

xargs forwards the data coming from the standard input to the command specified as a parameter. xargs is usually used in combination with a command preceded by a pipe, that is, in the format command1 | xargs command2. In this way, the results of the first command can be processed with the second command. If the results of the first command are so extensive that they can't be passed in a command line, command2 is called in several steps.

The difference between command1 | command2 is that xargs passes the standard input in the form of parameters to command2. xargs must be used if command2 only processes the transferred parameters. Ordinary pipes, on the other hand, are suitable if command2 processes the data from the standard input.

▶ --null

Expects 0 bytes to separate file names (but not spaces and tab characters). This option is suitable for processing find results if find was run with the -print0 option.

Example

The following command searches the current directory and all subdirectories for backup files ending with the ~ character and deletes them. This also works for file names that contain spaces.

```
user$ find -name '*.~' -print0 | xargs --null rm
```

xdg-open file

xdg-open opens a suitable program in the graphics system for editing the specified file. The command analyzes the MIME data stored in the XDG system; that is, it opens an audio player for an MP3 file, but a graphics program for a PNG file. Although XDG stands for *X Desktop Group*, this command also works if the graphics system uses Wayland.

On Debian and Ubuntu systems, xdg-open can be run using the short name open, as is the case on macOS.

xdpyinfo [options]

xdpyinfo provides comprehensive information on the running X server in graphics mode. You can use grep to extract the details relevant to you.

Example

The following commands determine the release number of the X server, the resolution of the screen and its pixel density (*dots per inch* [DPI]):

```
user$ xdpyinfo | grep release
vendor release number:    11702000
root# xdpyinfo | grep -C 1 dimensions
screen #0:
  dimensions:    1600x1200 pixels (411x311 millimeters)
  resolution:    99x98 dots per inch
```

xfs_admin [options] device/imagefile

xfs_admin shows or changes parameters of an XFS file system.

▶ -c 0|1

Deactivates or activates the *lazy counter* function of the file system, which means that the superblocks of the file system are updated less frequently, noticeably speeding up certain file operations.

▶ -f

Indicates that the file system is located in an image file and not on a data medium device.

▶ -l

Outputs the label of the file system.

▶ -L name

Sets the name (label) of the file system again.

▶ -u

Shows the UUID of the file system.

▶ -U *uuid*

Assigns a new UUID to the file system.

xfs_growfs [options] mount-directory

xfs_growfs enlarges an XFS file system. This assumes that the underlying device (e.g., a partition or a logical volume) has previously been enlarged.

The enlargement will take place during ongoing operation. For this purpose, the file system must be integrated into the directory tree. There are no plans to reduce the size of XFS file systems.

xfs_growfs has various options that enable you to control which areas of the file system will be enlarged and by how much. As a rule, however, it's not necessary to specify these options. xfs_growfs recognizes how large the underlying data medium is and decides for itself how large the various areas of the XFS file system should be.

xfs_info device

xfs_info outputs a summary of the key data of an XFS file system. The command is equivalent to xfs_growfs -n.

xfs_repair [options] device/imagefile

xfs_repair attempts to restore a damaged XFS file system to a consistent state.

▶ -d

Repairs an active read-only file system. Usually, xfs_repair can only repair file systems that aren't currently in use. This option is useful if you've started your Linux computer in single-user or emergency mode and need to repair the system partition. After the repair, you must restart the system. The man page describes the option as *dangerous*. It's safer to start the computer with a live or emergency system and perform the repair from there.

▶ -f

Indicates that the file system is located in an image file and not on a data medium device.

▶ -n

Doesn't make any changes, but shows which repairs are carried out if the command is run without this option.

xhost +/-hostname

With xhost +hostname, the graphics server accepts X connections from the specified host name. Similarly, X blocks connections from this host after xhost -hostname.

For security reasons, the X graphics system usually only allows local connections. to work from an external computer, an exception must be explicitly defined for this host via xhost +name. Even more liberal, xhost + completely deactivates access control, making it possible to log in from any host.

Example

In real life, running xhost is often necessary, especially if you first want to establish a connection to the graphics server as a normal user via VNC and then run a graphical program with root permissions using sudo or su. This isn't allowed by default. The solution is xhost +localhost (because the connection has already been established thanks to VNC and it's now only a matter of running programs under a different account).

The following example assumes that you first establish a VNC connection, then run xhost in a terminal window, and finally start a program in graphics mode via sudo—in this case, it's the gedit editor:

```
user$ xhost +localhost
user$ sudo gedit          (start editor with root permissions)
```

xinput [command]

xinput supports the configuration of input devices (e.g., keyboard, mouse, and touchpad) for the X window system.

▶ list
  Lists all recognized input devices.

▶ list-props *deviceid*
  Lists the parameters of a device. You should determine the device ID beforehand by using xinput list.

▶ set-button-map *deviceid mapbutton1 mapbutton2* ...
  Changes the assignment of the buttons of a mouse.

xkill [options]

xkill terminates a stuck or half-crashed X program. The relevant window must be clicked with the mouse after starting the command.

`xrandr` [options]

xrandr changes the screen resolution, frame rate, and other settings during operation. As the name implies, xrandr only works on Xorg. An alternative for Wayland is wlr-randr.

▶ --addmode *output name*
Adds a new graphics mode, possibly previously defined with --newmode, for the specified display output.

▶ --dpi *n*
Indicates how many pixels per inch are displayed.

▶ --left-of output *name*, --right-of *name*, --below *name*, or --above *name*
Activates the signal output specified by --output and specifies how the screens are positioned relative to each other. This only works if a sufficiently large virtual resolution has been set in xorg.conf so that both screens can be covered.

▶ --newmode *name freq x1 x2 x3 x4 y1 y2 y3 y4*
Defines a new mode (a new resolution). The syntax corresponds to ModeLine lines in xorg.conf. You can use the gtf command to determine the correct number combination.

▶ --off
Switches off the signal output specified with --output.

▶ --orientation *x*
Changes the image position. Permitted settings are normal, inverted, left, and right.

▶ --output *name*
Indicates that the other options relate to a specific signal output. This is useful if multiple monitors are connected. You can determine the names of the active signal outputs via xrandr -q.

▶ --primary
Makes the screen selected via --output the primary screen on which Gnome or KDE displays the panel, dock, and more.

▶ -q
Indicates which setting options are available. The result is heavily dependent on the graphics driver used.

▶ --rate *n*
Changes the frame rate for the current resolution. *n* is the desired frame rate in Hertz.

▶ --size *n*
Changes the resolution. *n* is a number from the result list of xrandr -q or the desired resolution in the format *widexheight*.

Example

The following command sets the resolution to 1280 × 1024 pixels:

```
user$ xrandr --size 1280x1024
```

The next command activates both the DVI and the VGA signal output. (The names of the outputs vary depending on the graphics card and driver.) The same image is displayed on both monitors. The --auto option causes each monitor to be operated in the optimum resolution and frame rate for it.

```
user$ xrandr --output DVI-I-0 --auto --output VGA-0 --auto
```

The following three commands define a new graphics mode with a resolution of 1,280 × 720 pixels and activate it for the HDMI1 output:

```
user$  xrandr --newmode 1280x720 74.18 1280 1390 1430 1650 720 725 730 750
user$  xrandr --addmode HDMI1 1280x720
user$  xrandr --size 1280x720
```

xset [command]

xset changes various settings of the X server.

▸ +dpms or -dpms
Activates or deactivates the energy-saving functions (*display power management signaling*).

▸ dpms *n1* [*n2* [*n3*]]
Specifies after how many seconds the DPMS modes *Standby*, *Suspend*, and *Off* of the monitor should be activated.

▸ q
Provides a list of the current settings.

▸ s *n*
Activates the screen saver after *n* seconds.

xz [options] files

xz from the xz-utils package compresses the specified files. xz is a relatively new compression command that provides even smaller files than bzip2. The compressed files are given the extension .xz by default.

▸ -0 to -9
Specifies how well xz should compress. The best results (i.e., the smallest files) are obtained with -9, but xz then requires the most memory. The default is -6. In this

case, xz requires a maximum of 100 MiB when compressing and a maximum of 10 MiB when decompressing.

▶ -d or --decompress
Decompresses the specified file (corresponds to unxz).

In spring 2024, it became known that a backdoor had been installed in some versions of xz, which fortunately weren't yet widespread. Although this wouldn't have affected the xz command directly, it would have affected the underlying xz library, which is used by SSH, among others.

yay [options]

yay is an alternative to pacman. The command manages both normal Arch Linux packages and AUR packages. These are additional packages that are only provided in the form of build scripts. The same options apply for yay as for pacman, so see that entry as well.

yum [options] command

yum installs, updates and removes RPM packages on RHEL. Behind the scenes, however, the dnf command first established in Fedora has been used since version 8. For this reason, all commands and options are described in dnf in this book.

z directory1 [directory2]

z is a modern alternative to cd. To install it, you need to download the z.sh script from *https://github.com/rupa/z*, save it in any directory, and then add the following line to .bashrc or .zshrc:

```
# at the end of .bashrc or .zshrc
. /path/to/z.sh
```

After a new login, you can use z or cd to change the active directory. z remembers when and how often you use which directories. This database is then used by z to complete incomplete commands. Thus, once you've entered the /home/<name>/Downloads/iso directory, z iso will simply work.

If there are multiple iso directories (e.g., also /usr/share/iso and /var/lib/libvirt/images/iso), z prioritizes the most frequently or most recently visited directory. The current order of the directories reveals z if you run the command without any further parameters.

To specify the desired directory more precisely, you can pass two character strings to z. z searches for the first character string in the front part of the path and for the second character string in the back part:

```
user$ z home iso    # corresponds to cd /home/<user>/Downloads/iso
user$ z var iso     # corresponds to cd /var/lib/libvirt/images/iso
user$ z usr iso     # corresponds to cd /usr/share/iso
```

zcat file.gz
zless file.gz
zmore file.gz

The three commands work like cat, less, and more. The only difference is that compressed files can be read directly via gzip, that is, without prior decompression using gunzip.

Instead of zless, less can be used directly with many distributions. This command is usually configured in such a way that it can be used without distinction for compressed and uncompressed files.

zenity [options]

zenity displays simple GTK dialogs for entering text, selecting a tag in the calendar, or selecting a file. The command returns the result for further processing. With zenity, you can give your bash scripts a more modern look—provided that the script is run in a graphical desktop system.

The following options merely illustrate the concept of the command. With more than 50 options, the command provides considerably more input and control options.

▶ --calendar
Displays a calendar and allows you to select a tag.

▶ --entry
Enables text input.

▶ --file-selection
Allows you to select a file.

▶ --password
Allows you to enter a password.

▶ --text *zenith*
Specifies which text is to be displayed within the dialog.

▶ --title *text*
Gives the dialog a title.

Example

The following lines of a script first prompt the user to enter and then display the password:

```
pw=$(zenity --pasword --title 'Your password please ...')
echo "That was your password: $pw"
```

zip [options] archive.zip file1 file2 ...

The command creates the ZIP archive archive.zip and inserts all the specified files into it. The archive is particularly suitable for exchanging data with Windows users. To extract the files again on Linux, you must use unzip.

► -r

Recursively archives the content of directories.

Example

The following command inserts all HTML files passed as parameters into myarchive.zip:

```
user$ zip myarchive.zip *.html
```

zipinfo [options] file.zip

The command returns the table of contents of a ZIP archive.

zramctl [options] command

In modern distributions (e.g., Fedora), part of the RAM is used as swap space. This is useful because data blocks stored there are compressed. Assume that a virtual machine has 4 GiB of RAM. The ZRAM memory may hold a maximum of 2 GiB of data. If an average compression by a factor of 2 is achieved, the full swap space only uses 1 GiB of RAM. The system can therefore use a total of 5 GiB of memory. Of course, swapping and compressing involves a loss of efficiency, but in practice, the system works amazingly well.

Without any additional parameters, zramctl provides information about the use of the system. In this case, ZRAM can hold 2.8 GiB of data. The swap space currently only contains 752 MiB of data. This data can be compressed to 222 MiB. Including various metadata, the swap space takes up 238 MiB.

```
user$ zramctl
NAME        ALGORITHM DISKSIZE   DATA COMPR  TOTAL STREAMS MOUNTPOINT
/dev/zram0 lzo-rle      2,8G 752,3M  222M 238,3M        6 [SWAP]
```

- -a or --algorithm lzo|lzo-rle|lz4|lz4hc|842|zstd
  Uses the desired compression algorithm.

- -f or --find
  Uses the first free ZRAM device.

- -s or --size *n*
  Creates a ZRAM device of the specified size (e.g., -s 1GiB).

Example

The following commands show how to manually set up a (further) ZRAM area for 1 GiB of data:

```
root# zramctl --find --size 1GiB
/dev/zram1
root# mkswap /dev/zram1
root# swapon /dev/zram1
```

With Fedora, systemd takes care of activating the ZRAM system.

zypper [options] command

The SUSE-specific zypper command helps with RPM Package Manager. It installs and updates packages, sets up package sources, and so on. zypper requires the libzypp library, which is an integral part of all current SUSE distributions. The following commands can be abbreviated as long as the meaning is clear, for example, zypper in *name* instead of zypper install *name*.

- addrepo [options] *uri name*
  Sets up a new package source and gives it an alias. The *Uniform Resource Identifier* (URI) describes the package source, for example, in the following form:

  http://download.opensuse.org/distribution/leap/15.0/repo/oss/

  By default, package sources are activated (*enabled* column in zypper repos) and marked for refreshing (*refresh* column). However, the package source isn't loaded into the cache for the first time. For this to happen, you must run zypper refresh.

  For package sources whose content doesn't change, you can avoid regular updates by using the -n (*no refresh*) option.

  When you set up a package source, you have the option of importing the key with which the packages are signed. If you're convinced of the authenticity of the package source, you should agree to the relevant queries.

zypper addrepo doesn't provide the option of specifying the name of a package source. zypper repos therefore displays the alias again in the name column. If you want to set the name and alias separately, you must change the file describing the package source in the /etc/zypp/repos.d/ directory directly.

▶ clean [--all]
Deletes downloaded packages from the cache. The --all option also deletes the metadata cache, that is, the description of the packages available in the repositories. This data must be downloaded again before the next package gets installed.

▶ dup
Performs a distribution update, for example, from openSUSE 15.0 to version 15.1. You must first perform a regular update and then convert the package repositories to the new version. A restart is required after the update.

▶ info *name*
Provides information on the specified package, patch, product, and so on.

▶ install [-y] *name*
Installs the specified package. Instead of a package, the name of a product, a patch, or a language may also be specified. To install all packages of a predefined pattern, you must run zypper install -t pattern *name*. With the additional -y option, zypper dispenses with queries.

▶ lifecycle
Provides a tabular overview of how long the installed packages will continue to be supplied with updates and for which packages this is no longer the case.

▶ list-updates [-t package]
Shows all available patches, with -t package listing both patches and updates.

▶ patches
Lists all available patches.

▶ patch-check
Determines the number of all available patches.

▶ patch-info [*name*]
Provides information on all patches or the specified patch.

▶ refresh
Tests whether the tables of contents of the package sources have changed and reloads them if necessary—even for package sources that have been set up with the -n (*no refresh*) option. Using the -f (*force*) option, you can reaccess the table of contents even if zypper believes that nothing has changed.

▶ remove [-y] *name*
Removes the specified package. The additional --clean-deps option also uninstalls dependent packages that are no longer required. With -y, zypper waives queries.

▶ removerepo *name*
Removes the package source specified by the URI or alias.

▶ renamerepo *old new*
Assigns a new alias to the package source.

▶ repos
Lists all installed package sources. The command displays the name and alias of each package source, but not the address (URI). For package sources set up using YaST, the address is used as an alias, which is why the alias column often looks like an address column. However, this doesn't apply to package sources that are set up using zypper addrepo.

▶ search [options] *expression*
Searches for packages with the specified search expression. By default, only the package names are searched. With -d, zypper also searches the package description. -i restricts the search to packages that have already been installed, and -u restricts searches to packages that haven't yet been installed. You can use -t pattern to determine a list of predefined package groups.

▶ update [-t package] [-y]
Updates all packages for which patches are available. With the -t package option, zypper also takes updates, that is, newer program versions, into account.

Here's a brief explanation: *Updates* are complete packages that are available in a newer version than the one installed. *Patches*, on the other hand, are supplementary or update packages (delta RPMs) that only contain the changes and are therefore much smaller.

Example

The first command updates all package sources, the second installs the nano editor, and the third determines which updates are available. The output is heavily abridged due to space limitations.

```
root# zypper refresh
All repositories have been refreshed.
```

```
root# zypper install nano
Reading installed packages...
>The following NEW package is going to be installed:  nano
Overall download size: 335.0 K. After the operation, additional 1.2 M
will be used.
Continue? [YES/no]: yes
```

```
root# zypper list-updates
...
Repository              Name       Current Version  Available Version  Arch
Haupt-Repository (OSS)  7zip       23.01-4.2        23.01-5.1          aarch64
Haupt-Repository (OSS)  argyllcms  3.1.0-2.1        3.1.0-3.1          aarch64
Haupt-Repository (OSS)  avahi      0.8-31.3         0.8-32.1           aarch64
...
```

#&%! (Special bash Characters)

Both when entering commands and in shell programming, you can use a vast number of special characters for various actions. Table 1 summarizes the meaning of the most important special characters.

Character	Meaning
;	Separates multiple commands
:	Shell command that does nothing
.	Starts the shell program without a separate subshell (. file, corresponds to source file)
#	Introduces a comment
#!/bin/sh	Identifies the desired shell for the shell program
&	Runs the command in the background (com &)
&&	Conditionally executes the command (com1 && com2)
&>	Redirects standard output and errors (corresponds to >&)
|	Creates pipes (com1 | com2)
||	Conditionally executes the command (com1 || com2)
*	Wildcard character for file names (any number of characters)
?	Wildcard character for file names (any character)
[abc]	Wildcard character for file names (one character from abc)
[expression]	Short notation for test expression
[[expr]]	Extended test syntax, bash-specific
(...)	Executes commands in the same shell ((com1; com2))
{...}	Groups commands
{,,}	Assembles strings (a{1,2,3} → a1 a2 a3)
{a..b}	Assembles strings (b{4..6} → b4 b5 b6)
~	Abbreviation for the home directory

Table 1 Special bash Characters

Character	Meaning
>	Redirects output to a file (com > file)
>>	Redirects output to append to existing file
>&	Redirects standard output and errors (corresponds to &>)
2>	Redirects standard error output
<	Redirects input from a file (com < file)
<< end	Redirects input from active file to end
$	Marks variables (echo $var)
$!	PID of the last started background process
$$	PID of the current shell
$0	File name of the currently executed shell script
$1 to $9	The first nine parameters passed to the command
$#	The number of parameters passed to the shell program
$* or $@	The totality of all transferred parameters
$?	Return value of the last command (0 = OK or error number)
$(...)	Command substitution (echo $(ls))
$((...))	Arithmetic analysis (echo $((2+3)))
${...}	Various special functions for editing strings
"..."	Prevents analysis of most special characters
'...'	Prevents analysis of all special characters
`...`	Command substitution (echo `ls`, corresponds to echo $(ls))
\character	Cancels the effect of the special character

Table 1 Special bash Characters (Cont.)

Configuration Files

This section of the *Linux Command Reference* summarizes the syntax of the most important Linux configuration files. The files are sorted by name, not by the path of the directory in which they are located.

Most of the files presented here apply to all distributions. I'll explicitly point out distribution-specific peculiarities. For more details on configuration files, see man 5 name. The number 5 indicates that you want to read the documentation of a configuration file, not that of a command that may have the same name.

/etc/adduser.conf

The Debian/Ubuntu–specific file /etc/adduser.conf contains default settings for setting up new users and groups using the commands adduser and addgroup. The adduser.conf file contains line-by-line settings in the form parameter=value.

▶ ADD_EXTRA_GROUPS=0|1
Specifies whether new users should automatically be assigned to the secondary groups listed in EXTRA_GROUPS.

▶ DHOME
Specifies the base directory for the home directories (usually /home).

▶ DIRMODE
Specifies the access bits with which new home directories are to be set up (755 by default).

▶ DSHELL
Specifies the default shell.

▶ EXTRA_GROUPS="group1,group2,..."
Contains a list of secondary groups to which new users are assigned by default (ADD_EXTRA_GROUPS=1) or via the adduser option, --add_extra_groups.

▶ FIRST_GID and LAST_GID
Determines the range of GIDs for common groups.

▶ FIRST_SYSTEM_GID and LAST_SYSTEM_GID
Determines the range of GIDs for the groups of system accounts.

▶ FIRST_SYSTEM_UID and LAST_SYSTEM_UID
Determines the range of UIDs for system accounts.

▶ FIRST_UID and LAST_UID

Determines the range of UIDs for ordinary users.

▶ GROUPHOMES=yes|no

Specifies whether the home directories of new users should be set up at the location /home/groupname/username.

▶ LETTERHOMES=yes|no

Determines whether home directories should be placed in subdirectories with the initial letter, such as/home/k/kofler. This can be useful for computers with a large number of accounts.

▶ NAME_REGEX=yes|no

Contains a regular expression (see grep) to which new usernames must correspond.

▶ SKEL

Specifies the directory from which default files are to be copied to a new user directory (usually that's /etc/skel).

▶ USERGROUPS=yes|no

Specifies whether a separate group should be set up for each user.

Example

The following lines show excerpts of the default settings in adduser.conf that are valid on Ubuntu:

```
DSHELL=/bin/bash
DHOME=/home
GROUPHOMES=no
LETTERHOMES=no
SKEL=/etc/skel
FIRST_SYSTEM_UID=100
LAST_SYSTEM_UID=999
FIRST_SYSTEM_GID=100
LAST_SYSTEM_GID=999
FIRST_UID=1000
LAST_UID=29999
FIRST_GID=1000
LAST_GID=29999
USERGROUPS=yes
USERS_GID=100
DIR_MODE=0755
...
```

/etc/aliases

The /etc/aliases file lists forwarding addresses for local email accounts. The email server is responsible for analyzing the file; that is, the file is only relevant if a mail transfer agent (MTA) is running on the computer, such as sendmail or postfix.

As a rule, aliases are primarily intended for local forwarding, for example, to forward emails from webmaster@hostname to adminxy@hostname. As far as the syntax is concerned, it's also permissible to forward local mails to external hosts; in practice, however, this often fails due to spam protection measures of the external hosts.

Many mail servers don't analyze /etc/aliases directly, but instead consider a database file that is generated from the aliases file. For this reason, changes in the aliases file often don't take effect until after the newaliases command gets executed.

/etc/aliases contains the redirection rules line by line:

```
name: alias1, alias2, alias3 ...
```

The name must be a local mail account (without @hostname!). There are multiple options for specifying the alias, whereby it depends on the mail server which variants are supported:

▶ *name*
  The email gets redirected to another local mail account.

▶ *adresse@host*
  The email gets redirected to an external email address.

▶ */path/file*
  The email gets added at the end of the specified file.

▶ *|command*
  The email gets transferred to the specified command.

▶ :include:*/path/file*
  The alias list is read from the specified file.

If multiple aliases are specified, the email gets redirected to all addresses.

Example

In the following example, emails addressed to root, postmaster, and webmaster are forwarded to the local users michael and ben. Emails sent to support are saved in the support mailbox and forwarded to michael and peter.parks@abc-company.com.

```
root:       michael
postmaster: michael
webmaster:  michael, ben
support:    support, michael, peter.parks@abc-company.com
```

/etc/bashrc

/etc/bashrc contains default settings for bash. The file is executed as a bash script. All language constructs supported by bash are therefore permitted. Usually, the prompt environment variable PS1 is set in bashrc; in some distributions, the umask value is also used.

The settings in /etc/bashrc are supplemented by those in /etc/profile. /etc/profile applies to all shells, whereas bashrc only contains the settings that apply specifically to bash. User-specific settings in .bashrc are also taken into account.

/boot/firmware/config.txt

config.txt is a Raspberry Pi–specific file for configuring the hardware. It must be located together with other boot files in the first partition of the SD card. On the Raspberry Pi OS, the file is accessible in the /boot/firmware directory (for older versions of the Raspberry Pi OS, the file is accessible directly in /boot). The file is analyzed directly by the System-on-a-Chip (SoC) of the Raspberry Pi during the boot process before the kernel gets started.

▶ arm_64bit=1
  Loads the 64-bit kernel at system startup (models 3B+, 4B, 400, and 5).

▶ arm_freq=*n*
  Specifies the desired maximum CPU frequency in MHz. The frequency is automatically reduced when the CPU has little to do. If you increase the frequency compared to the default frequency of the CPU (*overclocking*), you should also increase the voltage slightly (over_voltage_delta parameter).

▶ camera_auto_detect=1
  display_auto_detect=1

  Activates DSI cameras and displays.

▶ cmdline=*string*
  Passes the character string to the kernel. The parameter is analyzed instead of the /boot/cmdline.txt file.

▶ gpu_freq=*n*
  Indicates the clock frequency of the graphics unit in MHz.

▶ decode_MPG2=0x12345678
  decode_WVC1=0x12345678

  Specifies codes to enable the MPG-2 and VC-1 decoder. The keys can be purchased at *www.raspberrypi.com* and must match the serial number of the CPU. The serial number is stored in the /proc/cpuinfo file.

The decode parameters are only relevant for models up to 3B+, but no longer for 4B, 400, and 5. The newer models have an H264 hardware decoder. Other video formats are decoded by software. License keys aren't required at all.

▶ dtoverlay=*xxx*
Activates the device tree overlay for a hardware component and loads the corresponding driver modules.

▶ dtparam=audio=on
Activates the audio system and loads driver snd_bcm28xx.

▶ dtparam=i2c_arm=on
Activates the I^{2}C bus.

▶ dtparam=spi=on
Activates the SPI bus.

▶ over_voltage_delta=*n*
Increases the voltage for the CPU (in micro V, 50000 corresponds to 0.05V).

▶ sdram_freq=*n*
Indicates the clock frequency for the main memory (RAM) in MHz.

▶ usb_max_current_enable=1
Activates the USB boot process even with a weak power supply (i.e., with a weak power supply unit). This option is only available from model 5 onward and can result in an unstable operation.

Changes to config.txt won't take effect until the next restart. You can read many config.txt parameters during operation using vcgencmd get_config. You can read more config.txt details here:

https://elinux.org/RPiconfig

/etc/crontab

The /etc/crontab file is analyzed by the background process crond. crontab contains line-by-line information about when which commands are to be executed for which user account. The cron system makes it possible to perform tasks automatically at predefined times, for example, to start a mirroring process of the file system to an external backup server at 2:30 a.m. on the first Monday in January, April, July, and October.

Each crontab entry consists of seven columns in one line. The columns are separated from each other by spaces or tabs. The last line *must* end with a line break character, or it will be ignored. Comments are introduced with the # character at the beginning of the line. It's not permitted to enter a comment following a crontab entry.

▶ **First column (minute)**
Specifies in which minute (0–59) the program is supposed to be executed.

- ▶ **Second column (hour)**
  Specifies the desired hour (0--23).

- ▶ **Third column (day)**
  Specifies the day of the month (1–31).

- ▶ **Fourth column (month)**
  Specifies the month (1--12).

- ▶ **Fifth column (weekday)**
  Specifies the day of the week (0--7, 0, and 7 both represent Sunday).

- ▶ **Sixth column (user/account)**
  Specifies for which user the command is executed (often that's **root**)

- ▶ **Seventh column (command)**
  Contains the command to be executed. The command may contain spaces; the entire text up to the end of the line is interpreted as a command.

If a * is entered in the first five columns instead of a number, this field will be ignored. For example, 15 * * * * means that the command should always be executed 15 minutes after the full hour, in every hour, on every day, in every month, regardless of the day of the week. Likewise, 29 0 * * 6 means that the command is executed every Saturday at 0:29 am.

For the time fields, the notation */n is also allowed. This means that the command will be executed every nth minute/hour, and so on. Therefore, */15 * * * * would mean that the command is executed every quarter of an hour (n:00, n:15, n:30, and n:45).

In the time fields, multiple entries separated by commas and from-to ranges are also permitted. 1,13 in the second column means that the command is to be executed at 1 am and 1 pm. In the second column, 2,4,6,8-18,20,23 causes the command to be executed at 2 o'clock, 4 o'clock, 6 o'clock, 8 o'clock, 11 o'clock, and every hour between 6 am and 6 pm.

Instead of the five time columns, the @ abbreviations summarized in Table 1 may also be used. Another additional rule states that a minus sign at the beginning of the first column prevents syslog from logging the command execution. However, this is only allowed if the sixth column contains root.

Shortcut	Code	Meaning
@reboot	–	Run after each reboot
@yearly	0 0 1 1 *	Run once a year
@annualy	0 0 1 1 *	Like @yearly
@monthly	0 0 1 * *	Run once a month

Table 1 crontab Interval Abbreviations Replace the First Five Columns

Shortcut	Code	Meaning
@weekly	0 0 * * 0	Run once a week
@daily	0 0 * * *	Run once a day
@hourly	0 * * * *	Run once per hour

Table 1 crontab Interval Abbreviations Replace the First Five Columns (Cont.)

SHELL, PATH, and MAIL

Prior to the lines just described for the periodic start of programs, /etc/crontab usually contains three lines for setting environment variables. On CentOS, the default settings look as follows:

```
SHELL=/bin/bash
PATH=/sbin:/bin:/usr/sbin:/usr/bin
MAILTO=root
```

This means that the command string is executed by the bash, that the PATH directories are taken into account when searching for commands, and that emails with output or error messages are sent to root.

Example

The following three lines in /etc/crontab cause a maintenance script to be called 15 minutes after every full hour, a backup tool to be started every Sunday at 00:30 am, and a backup copy of a MySQL database to be created every day at 3:15 am:

```
15  *  *  *  *  root   /usr/bin/maintenance-script
30  0  *  *  0  root   /usr/bin/backup-tool
15  3  *  *  *  mysql  /usr/bin/mysql-backup
```

Other cron Files

In addition to the system-wide /etc/crontab file, you have various other options for defining cron jobs, depending on the distribution:

▶ **/etc/cron.d**
In addition to /etc/crontab, the cron daemon crond also analyzed all files in the /etc/cron.d directory in most distributions. The same syntax rules apply to these files as for /etc/crontab.

▶ **User-specific crontab files**
Depending on the distribution, user-specific crontab files are stored in the /var/spool/cron/ or /var/spool/cron/tabs directory. The name of the file indicates which user the file applies to. The sixth column is omitted in these files. The crontab command is provided for changing user-specific cron entries.

▶ **cron.hourly, cron.daily, cron.weekly, und cron.monthly**
In most distributions, the default configuration in /etc/crontab contains some entries that cause all script files in /etc/cron.hourly/* to be executed with root rights once an hour and the script files in /etc/cron.daily/* and so on once a day. These files are ordinary script files to which no crontab syntax rules apply. But remember to set the *execute* bit (chmod a+x file)!

The file names of custom scripts in cron.daily, cron.weekly, and cron.monthly may only consist of numbers, letters, hyphens, and underscores. As soon as the file name contains even one dot, the script will be ignored! The scripts from /etc/cron.daily, -.weekly, and -.monthly aren't executed if the Anacron program is installed.

/etc/deluser.conf

The Debian/Ubuntu–specific file /etc/deluser.conf contains default settings for deleting users and groups using the deluser and delgroup commands. The deluser.conf file consists of line-by-line entries in the form parameter=value.

▶ BACKUP=0|1
Specifies whether backups are to be made of the files that are deleted by the REMOVE parameter.

▶ BACKUP_TO
Specifies the location of the backup directory.

▶ EXCLUDE_FSTYPES="(fs1,fs2,...)"
Specifies which file systems should be ignored when searching for the user's files to create a backup.

▶ ONLY_IS_EMPTY=1
Causes groups to be deleted only if they have no members.

▶ REMOVE_HOME=0|1
Specifies whether the account's home directory should also be deleted.

▶ REMOVE_ALL_FILES=0|1
Specifies whether user files outside the home directory should be deleted, such as spooling files and the email inbox.

Example

The following lines show the default settings in deluser.conf that are valid on Ubuntu:

```
REMOVE_HOME = 0
REMOVE_ALL_FILES = 0
BACKUP = 0
BACKUP_TO = "."
ONLY_IF_EMPTY = 0
EXCLUDE_FSTYPES = "(proc|sysfs|usbfs|devpts|tmpfs|afs)"
```

/etc/dnf/dnf.conf

The basic settings of the package management system DNF, which is commonly used on Fedora and Red Hat, can be found in /etc/dnf/dnf.conf. The /etc/yum.repos.d/*.repos files also contain detailed settings for the individual package sources.

Basic Settings

► clean_requirements_on_remove = true/false

Specifies whether packages that were installed due to dependencies should be auto-matically removed again as soon as all packages that reference the package are unin-stalled. Even if this option is true, dnf never removes packages that were explicitly installed using dnf install without being asked.

► color=never

Causes the dnf command in the terminal not to use any colors. If you want to use colors, use color=always.

► exactarch=1

Causes dnf to only consider updates where the architecture matches the already installed package. Therefore, i386 packages can't be replaced by newer x86_64 packages.

► gpgcheck=1

Causes dnf to ensure the authenticity of the packages with a key. gpgcheck can also be set individually for each package source in deviation from the setting in dnf.conf. plugins decides whether dnf takes plug-ins into account.

► installonly_limit=*n*

Controls how many old versions of kernel packages are installed at the same time. The setting affects all packages for which updates are installed in the form of new packages. This procedure ensures that older versions of the affected package are always available as a fallback.

► keepcache=0

Causes downloaded packages to not be archived after installation. In general, this is a useful setting because the space required for the packages becomes quite large over time and there's normally no reason to install the packages a second time. However, it can happen that dnf detects a problem during the installation and aborts the installation process. In that case, the downloaded packages will also get deleted. If you can fix the problem and then repeat the update, all packages must be down-loaded again. You can avoid this situation by using keepcache=1. To explicitly delete the downloaded packages in /var/cache/dnf, you must run dnf clean packages.

In current Fedora versions, the file contains only a few settings:

```
# File /etc/dnf/dnf.conf on Fedora
[main]
gpgcheck=1
installonly_limit=3
clean_requirements_on_remove=True
best=False
skip_if_unavailable=True
```

Package Sources

The package sources are configured using *.conf files in the /etc/yum.repos.d directory. The directory wasn't renamed to /etc/dnf.repos.d when switching from yum to dnf.

▶ [*repovariant*]
  Specifies which part of the package source the following settings relate to.

▶ baseurl=*url* or mirrorlist=*url*
  Specifies where the package files are located. With mirrorlist, the url refers to a list of mirror servers. dnf automatically decides on one of the mirrors. dnf replaces the $releasever, $arch, and $basearch variables in the url with the version number of the Linux distribution and its architecture.

▶ enabled=0/1
  Indicates whether the package source is active.

▶ name=*reponame*
  Specifies the name of the package source.

▶ metadata_expires
  Controls how long the metadata downloaded from a package source is valid. dnf saves the metadata in a cache and refrains from downloading it again if the metadata isn't yet outdated. This saves time and download volume, but can lead to dnf ignoring recent changes in the package source. If necessary, you can use dnf clean metadata to force the deletion of the local metadata. This forces dnf to reimport the metadata of all package sources the next time.

The following lines show an excerpt from the file /etc/yum.repos.d/#fedora.repo. This file contains the definition of the package source with the basic packages of the Fedora distribution.

```
[fedora]
name=Fedora $releasever - $basearch
metalink=https://mirrors.fedoraproject.org/metalink? repo=fedora-$releasever&
arch=$basearch
enabled=1
countme=1
```

```
metadata_expire=7d
repo_gpgcheck=0
type=rpm
gpgcheck=1
gpgkey=file:///etc/pki/rpm-gpg/RPM-GPG-KEY-fedora-$releasever-$basearch
skip_if_unavailable=False

[fedora-source]
name=Fedora $releasever - Source
...
```

/etc/fstab

/etc/fstab contains line-by-line entries for all file systems and swap partitions that are to be integrated into the directory tree or activated when the computer is started. Each entry consists of six columns separated by spaces or tabs. Comments are introduced using the # character.

▶ **First column** (Device)
Contains the device name of the data medium. Instead of the device name, you can also specify the name or ID of the file system using LABEL=xxx, UUID=xxx, or the ID of a GPT partition by using PARTUUID=xxx.

▶ **Second column** (Path)
Specifies the directory in which the data medium is integrated into the file tree. The directories specified in the second column must already exist. For swap partitions, you should enter the keyword none here.

▶ **Third column** (File system type)
Specifies the file system, such as ext4, btrfs, or vfat. An overview of the most important Linux file systems can be found in the description of the mount command. Due to the auto entry, Linux attempts to recognize the file system itself. For swap partitions that don't contain an internal file system, you must enter the swap keyword.

▶ **Fourth column** (Options)
Contains the desired mount options. A reference of the options for all important file systems can be found in this book in the mount command. If you don't require any options, you should enter the defaults keyword instead. Multiple options are only separated by commas, not by spaces!

▶ **Fifth column** (Dump)
Contains information for the Unix program dump. This column is ignored on Linux. It's common to enter 1 for the system partition and 0 for all other partitions or disks.

437

▶ **Sixth column** (fsck)
Specifies whether the file systems are to be checked at system startup and in which order. Often 1 is entered for the system partition and 0 for all other partitions. This means that only the system partition is checked for errors at computer startup and, if necessary, repaired by fsck. If you want further partitions to be checked automatically, you must enter the number 2 for these partitions. This means that the check should take place after the system partition has been checked. If entries in the fifth and sixth columns in **/etc/fstab** are missing, 0 is assumed.

Example

The following example shows the fstab file of a server with LVM and RAID. The boot partition is located on RAID device /dev/md/0. The system and data directories use logical volumes as a storage location. The file systems for /, /var, and /backup are checked on restart. If an error occurs in the root file system, the file system is mounted in read-only mode.

```
/dev/md/0         /boot    ext2   defaults            0  0
/dev/vg0/root     /        ext4   errors=remount-ro   0  1
/dev/vg0/var      /var     ext4   defaults            0  2
/dev/vg0/backup   /backup  ext4   defaults            0  2
/dev/vg0/swap     swap     swap   defaults            0  0
```

/etc/group

/etc/group contains a list of all group names with the corresponding group identification numbers (GIDs) and a list of all users who belong to the group. To administrate the groups, you usually use the commands groupadd, groupmod, and groupdel.

The line-by-line entries in the group file consist of four columns, each separated by colons:

▶ **First column** (Group name)
Indicates the name of the group.

▶ **Second column** (unused)
Contained the hash code of the group password in older Linux versions. Current Linux distributions save the x character here. If group passwords are defined, they are saved in /etc/gshadow.

▶ **Third column** (GID)
Contains the group identification number. In most distributions, GIDs smaller than 1000 are reserved for system accounts.

▶ **Fourth column** (Accounts)
Contains a comma-separated list of all users who are members of this group. However, only secondary memberships are taken into account, but the primary group of each user is stored in the fourth column of /etc/passwd.

Example

The following lines show excerpts from /etc/group on a computer that serves as a web server and development server. Each system account and each regular user has its own group. The members of the admin group can use sudo to perform administration work. The members of the devel group have special access rights to the version management system.

```
root:x:0:
bin:x:1:
admin:x:109:kofler,huber
...
kofler:x:1000:
huber:x:1001:
miller:x:1002:
devel:x:1023:kofler,huber,miller,gruber,smith
...
```

/etc/default/grub

/etc/default/grub contains default settings for the Linux boot loader GRUB 2. These settings are only taken into account when the GRUB configuration file /boot/grub/grub.cfg is regenerated—either automatically during a kernel update or manually using the command update-grub (Debian, Ubuntu) or grub-mkconfig (Fedora, openSUSE).

The following list describes the most important keywords:

▶ GRUB_CMDLINE_LINUX and GRUB_CMDLINE_LINUX_DEFAULT
Taken into account by 10_linux and specify which options are to be passed to the kernel. The GRUB_CMDLINE_LINUX options apply to every boot; the GRUB_CMDLINE_LINUX_DEFAULT options are added additionally for standard boot, but not for recovery mode.

▶ GRUB_DEFAULT=0
Specifies which GRUB menu entry should be selected by default. The "saved" setting means that the last selected menu item is activated. However, this only works if the GRUB files are located in an ordinary partition! If, on the other hand, LVM or RAID is involved, GRUB can't save any environment variables after the menu selection.

Another option is to assign the menuentry string of the desired menu item to GRUB_DEFAULT. However, you must be careful to keep the spelling exact.

439

► `GRUB_DISABLE_OS_PROBER=false`

Activates the search for other operating systems when recreating `boot.cfg`. (There's nothing like a double negative.)

Since around 2022, GRUB has been configured in such a way that only entries for the active distribution appear in the GRUB menu, but not entries for starting other distributions or Windows. These entries used to be necessary (BIOS) but are now largely obsolete (EFI). In case something goes wrong with the EFI configuration, the GRUB menu entries are a good backup/redundancy.

► `GRUB_DISABLE_LINUX_UUID=true`

Causes GRUB to pass the root directory to the kernel as the device name. Without this option, GRUB passes the UUID of the file system. This setting applies only to the start of the active distribution (`10_linux` script), not to other distributions.

► `GRUB_DISABLE_RECOVERY=true`

Prevents `update-grub` or `grub2-mkconfig` from adding menu entries for starting Linux in recovery mode to `grub.cfg`.

► `GRUB_DISTRIBUTOR`

Is analyzed by the `/etc/grub.d/10_linux` script and specifies the name of the current distribution.

► `GRUB_ENABLE_BLSCFG=true`

Causes the menu items from the `/boot/loaders/entries` directory to be taken into account when booting in addition to `grub.cfg`. This makes the configuration files clearer and corresponds to the procedure of systemd-boot (an alternative boot procedure).

► `GRUB_GFXMODE`

Specifies the graphics mode in which GRUB should run (by default in a resolution of 640 × 480 pixels).

► `GRUB_HIDDEN_TIMEOUT`

Is important if GRUB only recognizes a single Linux distribution on your computer during installation. In this case, `GRUB_HIDDEN_TIMEOUT` specifies how long the user has to press ⬆ to display the GRUB menu. The screen remains black during this wait time. If multiple operating systems are installed, GRUB ignores the `GRUB_HIDDEN_TIMEOUT` #setting and displays the menu.

► `GRUB_HIDDEN_TIMEOUT_QUIET=true`

Prevents a countdown counter from being displayed during the `GRUB_HIDDEN_TIMEOUT` wait time. If you want the GRUB menu to always be displayed, set the lines `GRUB_HIDDEN_TIMEOUT=...` and `GRUB_HIDDEN_TIMEOUT_QUIET=...` with the comment character #.

▶ GRUB_TIMEOUT=n

Specifies how many seconds GRUB waits for a menu item to be selected. If this time elapses without user input, GRUB starts the selected operating system. The time set here only comes into effect when the GRUB menu appears at all.

▶ GRUB_TERMINAL=console

Causes GRUB to run in text mode. However, no Unicode characters can be displayed there.

Example

On Ubuntu, the configuration file contains the following settings:

```
# File /etc/default/grub
GRUB_TIMEOUT=5
GRUB_DISTRIBUTOR="$(sed 's, release .*$,,g' /etc/system-release)"
GRUB_DEFAULT=saved
GRUB_DISABLE_SUBMENU=true
GRUB_TERMINAL_OUTPUT="console"
GRUB_CMDLINE_LINUX=
GRUB_DISABLE_RECOVERY="true"
GRUB_ENABLE_BLSCFG=true
```

/boot/grub/grub.cfg

Depending on your distribution, either /boot/grub/grub.cfg or /boot/grub2/grub.cfg will be the configuration file for the current GRUB version 2. This file is generated automatically, and direct changes aren't recommended. To influence the behavior of GRUB, you can change various default settings in /etc/default/grub or insert statements in /etc/grub.d/*custom, which are then adopted directly in grub.cfg. To recreate grub.cfg afterward, you can run one of the following commands:

```
root# update-grub                              (Debian and Ubuntu)
root# grub2-mkconfig -o /boot/grub2/grub.cfg   (Fedora, openSUSE)
```

Modern GRUB versions only save the basic settings in grub.cfg. The actual menu entries for starting Linux are stored in separate files (/boot/loader/entries/*.conf; see also GRUB_ENABLE_BLSCFG=true in /etc/default/grub).

/etc/gshadow

The /etc/gshadow file contains the hash codes of the group passwords and other group administration data. For security reasons, only root may read and modify this file.

The line-by-line entries in the gshadow file consist of four columns, each separated by colons:

- **First column** (Group name)
  Indicates the name of the group.

- **Second column** (hash code)
  Contains the hash code of the group password or an asterisk (no password). The use of group passwords is unusual for security reasons. If you do decide to do this, you need to set the group password via the gpasswd command. Anyone who knows the group password can then switch to this group using newgrp, even if they aren't a member of the group.

- **Third column** (Administrators)
  Contains a comma-separated list of group administrators (account names). Group administrators can change the group password and add or remove other members from the group.

- **Fourth column** (Members)
  Contains a comma-separated list of group members who are allowed to activate the group *without* entering a password.

On many Linux systems, the gshadow file is unused and reads as follows:

```
root:*::
daemon:*::
bin:*::
sys:*:: ...
```

/etc/host.conf

/etc/host.conf controls the resolver library, which specifies how unknown host names are resolved during network operations.

- multi on|off
  Specifies whether the resolver library returns all entries for each host (on) or only the first matching entry (off).

- order a,b,c...
  Specifies in which order which procedures are to be used for host name determination. bind means that the resolver library falls back on the nameserver set in /etc/resolv.conf. hosts allows the evaluation of /etc/hosts. nis causes the obsolete Network Information Service to be used.

Example

The standard configuration for host.conf looks like the following listing for most distributions. The resolver library first analyzes the /etc/hosts file and then contacts the nameserver set in `resolv.conf`. The `multi` line allows multiple IP addresses to be assigned to a host name specified in /etc/hosts.

```
order hosts, bind
multi on
```

/etc/hostname

The /etc/hostname file contains the full name of the computer, which consists of the host name and the domain name, for almost all common Linux distributions. /etc/hostname on the Ubuntu server installation for my web server therefore contains the following text:

```
kofler.info
```

/etc/hostname is evaluated when the computer is started. Changes can be made using the `hostname` command. With systemd-compatible distributions, the host name can also be set permanently using the `hostnamectl` command.

/etc/hosts

/etc/hosts assigns IP addresses to their host names. The file normally only contains entries for `localhost` and, in the case of statically configured servers, for the host name. In the past, the file was used to list other computers within the local network. Today, however, even in small networks, this function is performed by a router or a local nameserver, such as Bind or dnsmasq.

The /etc/hosts file contains line-by-line entries based on the following pattern:

```
ip-addr   hostname1 hostname2 hostname3 ...
```

The minimal variant looks like this:

```
127.0.0.1   localhost
```

In most Linux distributions, `localhost` is also defined for IPv6. The following lines show the default settings on Fedora and RHEL:

```
127.0.0.1   localhost localhost.localdomain localhost4 localhost4.localdomain4
::1         localhost localhost.localdomain localhost6 localhost6.localdomain6
```

In the case of a static network configuration, for example, on a root server, hosts can also contain an entry with the IP address and the host name of the computer:

```
211.212.213.214  abc-company.com abc-company
```

```
/etc/network/interfaces
```

The /etc/network/interfaces file (ENI for short) contains the network configuration for Debian distributions. With Ubuntu, the NetworkManager and Netplan take care of the network configuration. In current versions of the Raspberry Pi OS, the NetworkManager has sole control over the network.

Interaction with the NetworkManager

For desktop installations, the file is normally almost empty and only contains the settings for the loopback interface:

```
# loopback interface
auto lo
  iface lo inet loopback
```

The NetworkManager is responsible for all other interfaces.

IPv4 Configuration

The situation is completely different for server installations or on computers with a static network configuration. In this case, the auto name line initiates a configuration block for each network interface that will be activated when the computer is started. The subsequent iface *name options* line describes the basic configuration of the interface. The other parameters then follow depending on the configuration type.

If the interface obtains the IP data from a DHCP server, the configuration is very brief:

```
# IPv4 network interface with DHCP
auto eth0
iface eth0 inet dhcp
```

In a static configuration, the network parameters are specified line by line using multiple keywords, the meaning of which is self-explanatory. The dns-nameservers keyword is only used on Ubuntu to configure the nameservers. The resolvconf package is responsible for the evaluation. On Debian, however, you must set the nameserver yourself in /etc/resolv.conf.

```
# IPv4 network interface, static configuration
auto eth0
iface eth0 inet static
  address         211.212.213.37
  netmask         255.255.255.224
  gateway         211.212.213.1
  dns-nameservers 211.222.233.244 212.232.234.245
```

You can activate changes to a single interface using ifup xxx, replacing xxx with the interface name.

IPv6 Configuration

If you also want to use IPv6, you can simply define the relevant interface a second time in /etc/network/interfaces using the inet6 keyword. The auto keyword specifies that the IPv6 configuration takes into account the *router advertisement* of the gateway or IPv6 router.

```
# IPv4 configuration via DHCP, IPv6 configuration via router advertisement
auto eth0
iface eth0 inet dhcp
iface eth0 inet6 auto
```

If the IPv6 gateway uses a DHCPv6 server, the correct method is dhcp. If, in addition, the router address is to be configured via *router advertisement*, the additional accept_ra 1 option is required. This is the case, for example, if you use dnsmasq with the enable-ra option as the DHCP server.

```
# IPv4 and IPv6 configuration via DHCP
auto eth0
iface eth0 inet dhcp
iface eth0 inet6 dhcp
 accept_ra 1
```

For a static configuration, interfaces must look as follows:

```
# IPv4 and IPv6 configuration static
auto eth0
iface eth0 inet static
  ... (IPv4 configuration as before)
iface eth0 inet6 static
  address 2a01:4f8:161:107::2
  netmask 64
  gateway fe80::1
```

Wi-Fi and WPA Configuration

NetworkManager normally takes care of Wi-Fi interfaces and WPA authentication on notebooks. However, it's also possible to make the configuration directly in /etc/interfaces. This is particularly common on embedded devices or minicomputers such as the Raspberry Pi.

Wi-Fi interfaces are basically configured in the same way as Ethernet interfaces. You control the interaction with the wpa_supplicant program for authentication using wpa parameters. In the simplest case, the configuration can look as follows:

```
# File /etc/network/interfaces
...
auto wlan0
iface wlan0 inet dhcp
  wpa-ssid "wlan-name"
  wpa-psk "top secret"
```

You use `wpa-ssid` to specify the name of your Wi-Fi network. (SSID stands for *Service Set Identifier*.) `wpa-psk` specifies the Wi-Fi password, whereby it's assumed that the encryption is performed using the *Wi-Fi Protected Access* (WPA) method with preshared keys. This is the rule in the home sector.

If more parameters need to be set, it's common practice to save the authentication data in an external file. The `interfaces` parameters then has the following structure:

```
# /etc/network/interfaces (manual WiFi configuration with WPA)
auto wlan0
iface wlan0 inet dhcp
  wpa-conf /etc/wpa_supplicant/wpa_supplicant.conf
```

Here, `wpa-conf` specifies the location of a WPA configuration file, which must correspond to the syntax of /etc/wpa_supplicant/wpa_supplicant.conf.

If there's a graphical user interface for Wi-Fi configuration to manage the connection data to *multiple* Wi-Fi networks, as on the Raspberry Pi OS, you must use the `wpa-roam` keyword:

```
# /etc/network/interfaces (WiFi default configuration on Raspberry Pi OS)
allow-hotplug wlan0
iface wlan0 inet manual
  wpa-roam /etc/wpa_supplicant/wpa_supplicant.conf
iface default inet dhcp
```

The `allow-hotplug` line allows other programs to control the Wi-Fi interface. The line that starts with `wpa-roam` specifies the file in which the WPA keys of various Wi-Fi networks are stored. As soon as a suitable network is within range, a connection is automatically established, and the IP data is obtained via DHCP. The `iface default inet dhcp` statement states that every network defined in `wpa_supplicant.conf` should use DHCP for IP configuration by default.

Network Bridges and Special Settings

To configure virtual bridges or to fulfill special configuration requests, there are a number of other keywords that are documented on the `man` page for `interfaces`. In particular, you can use the `pre-up`, `up`, and `post-up` keywords as well as `pre-down`, `down`, and `post-down` keywords to specify commands that are to be executed immediately before, during, or after setting up or stopping a network interface. The following lines show a network bridge that is connected to physical interface eth0 and redirects network packets to four virtual machines:

```
# IPv4 network bridge
auto  br0
iface br0 inet static
  # static configuration
  address     210.211.212.18
  broadcast   210.211.212.31
  netmask     255.255.255.224
  gateway     210.211.212.1
  pointopoint 210.211.212.1

  # Bridge
  bridge_ports    eth0
  bridge_stp      off
  bridge_fd       0
  bridge_maxwait  0

  # additional IPs for KVM
  up route add -host 210.211.212.26 dev br0
  up route add -host 210.211.212.27 dev br0
  up route add -host 210.211.212.28 dev br0
  up route add -host 210.211.212.29 dev br0

  # static routes for IPv6
  up ip -6 addr add 2a01:1234:567:890::2/64     dev br0
  up ip route   add default via fe80::1         dev br0
  down ip route del default via fe80::1         dev br0
  down ip -6    del add 2a01:1234:567:890::2/64 dev br0
```

/etc/systemd/journald.conf

The *journal* log service belonging to systemd is controlled by the following configuration files:

```
/etc/systemd/journald.conf
/etc/systemd/journald.conf.d/*.conf /run/systemd/journald.conf.d/*.conf [/usr]/
lib/systemd/journald.conf.d/*.conf
```

In many distributions, only /etc/systemd/journald.conf exists, and this file also only contains commented-out example lines. In this case, the default settings apply.

To change the configuration, you can use the following keywords within the [Journal] configuration group:

- ForwardToSyslog

  Specifies whether messages logged by the journal should also be forwarded to a traditional syslog service.

- MaxFileSec

  Specifies the latest time after which a new logging file should be started. The default setting is one month. This setting is only relevant if the logging files grow slower than the SystemMaxUse, SystemMaxFileSize, and SystemKeepFree limits specify.

- MaxLevelStore

  Indicates the priority level up to which messages are saved in the journal. The default setting is debug.

- SystemKeepFree

  Specifies what percentage of the file system must remain free. The default setting is 15%.

- SystemMaxFileSize

  Specifies the maximum size of a logging file. The default setting is one eighth of SystemMaxUse. This results in automatic *rotating*, creating a maximum of seven older files in addition to the current logging file.

- SystemMaxUse

  Specifies the maximum percentage of the file system that the logging files can occupy. Before this limit is exceeded, old logging files will be deleted. The default setting is 10%.

You can look up countless other options via man journald.conf.

/etc/locale.conf

The /etc/locale.conf file contains the language and character set settings for distributions that use a current systemd version as the init system. The localectl command helps you change the settings. The following list describes the most important localization parameters that can be set:

- LANG=xxx

  Determines the default value for all LC variables that haven't been set.

- LC_ALL=xxx

  Overwrites all individual LC settings.

- LC_COLLATE=xxx

  Determines the sorting order.

- ▶ LC_CTYPE=xxx

  Determines the character set.

- ▶ LC_MESSAGES=xxx

  Determines the display of messages, error messages, and so on.

- ▶ LC_MONETARY=xxx

  Determines the representation of monetary amounts.

- ▶ LC_NUMERIC=xxx

  Determines the representation of numbers.

- ▶ LC_PAPER=xxx

  Determines the paper size.

- ▶ LC_TIME=xxx

  Determines the display of date and time.

A list of all possible settings is provided by locale -a. The notation *spr_land.zs* is usually used, where *spr* denotes the language by two letters, and *land* denotes the country by two letters. *zs* specifies the character set. In German-speaking countries, for example, you would use de_DE.utf8.

Depending on the distribution, these or other localization settings are saved in other locations:

Debian, Ubuntu	/etc/default/locale
SUSE	/etc:/sysconfig/language

Example

In the following setting in locale.conf, German is the default language, and UTF-8 is the default character set:

LANG="de_DE.utf8"

/etc/login.defs

/etc/login.defs contains various settings that are taken into account when creating new users and groups. The settings apply to the commands from the shadow-utils package, for example, for adduser and useradd as well as for addgroup and groupadd. The login.defs parameters, on the other hand, have no influence on commands that use the pluggable authentication modules (PAM); this includes in particular passwd!

The settings in login.defs are made line by line. In each line, first a keyword is entered, then one or more blank or tab characters, and finally the desired setting.

- ► CREATE_HOME
  Specifies whether a home directory should also be created automatically when new users are created (yes/no).

- ► ENCRYPT_METHOD
  Specifies which algorithm is to be used to store the hash codes in /etc/shadow. Permitted settings are DES, MD5, SHA256, and SHA512 (most secure).

- ► ENV_PATH and ENV_SUPATH
  Contains a default setting for the PATH environment variable for normal users and for root or su. For most distributions, additional settings for the PATH variable are made in /etc/profile or in /etc/bashrc.

- ► GID_MIN and GID_MAX
  Specifies the minimum and maximum group ID for common groups.

- ► LOG_OK_LOGINS
  Records successful logins with syslog (yes/no).

- ► MAIL_DIR
  Specifies the location of the directory for storing local emails.

- ► PASS_MAX_DAYS, PASS_MIN_DAYS, and PASS_WARN_AGE
  Specifies how long passwords are valid and the earliest period after which they may be changed (see chage).

- ► PASS_MIN_LEN
  Defines the minimum password length.

- ► SYS_GID_MIN and SYS_GID_MAX
  Specifies the minimum and maximum GID for system accounts.

- ► SYS_UID_MIN and SYS_UID_MAX
  Specifies the minimum and maximum UID for system accounts.

- ► SYSLOG_SU_ENAB and SYSLOG_SG_ENAB
  Records su and newgrp commands with syslog (yes/no).

- ► UID_MIN and UID_MAX
  Specifies the minimum and maximum user ID for normal users.

- ► USERGROUPS_ENAB
  Controls whether a separate group should be created for each user (yes/no).

Example

The following lines show the default settings in login.defs on Fedora. You should note, however, that the settings for PASS_MIN_LEN and UMASK aren't relevant for the operation of Fedora! The minimum requirements for passwords are specified by PAM module pwquality, while umask is preset to 022 in /etc/profile.

```
MAIL_DIR        /var/spool/mail
PASS_MAX_DAYS           99999
PASS_MIN_DAYS               0
PASS_MIN_LEN               5
PASS_WARN_AGE             7
UID_MIN                 1000
UID_MAX                60000
SYS_UID_MIN              201
SYS_UID_MAX             999
GID_MIN                 1000
GID_MAX                60000
SYS_GID_MIN             201
SYS_GID_MAX             999
CREATE_HOME             yes
UMASK                   077
USERGROUPS_ENAB         yes
ENCRYPT_METHOD        SHA512
```

/etc/mailname

On Debian and Ubuntu systems, /etc/mailname contains the host name to be used by the mail server. The setting is analyzed by postfix (myorigin = /etc/mailname) and the mutt program, among others. This setting is particularly important for sending email. It completes the mail address of local mails. So, when huber sends an email, huber@mailhostname is used as the sender, whereby mailhostname is read from /etc/mailname.

In other distributions, the host name is set directly in the configuration files of the mail server; with postfix, this is in /etc/postfix/main.cf with myorigin = *mailhostname*.

/etc/mdadm/mdadm.conf

The /etc/mdadm/mdadm.conf file summarizes the configuration of the computer's software RAID system. If the file is missing or lost, it can be partially recreated from the metadata stored in RAID partitions using the following command:

```
root# mdadm --examine --scan > /etc/mdadm/mdadm.conf
```

The following list shows the most important keywords in mdadm.conf:

► ARRAY device1 metadata=... UUID=... name=... devices=...
Specifies which components make up a RAID array.

► CREATE owner=... group=... mode=... auto=...
Specifies which default settings should apply when activating a RAID device. The auto parameter corresponds to the --auto option of the mdadm command.

- DEVICE device1 device2 | partitions | containers
  Specifies which hard disks or partitions can contain RAID components. If the DEVICE specification is missing, DEVICE partitions containers applies. The mdadm command scans all SSD and hard disk partitions as well as all logical volumes.

- MAILADDR name
  Specifies the local email address to which messages, warnings, and error messages are to be sent if mdadm is running in monitor mode.

Example

The following lines show the configuration file of a Linux server with two RAID-1 devices:

```
CREATE owner=root group=disk mode=0660 auto=yes
MAILADDR root
ARRAY /dev/md/0 metadata=1.2 UUID=0860...f0a2 name=rescue:0 ARRAY /dev/md/1
metadata=1.2 UUID=2546...e6e2 name=rescue:1
```

/etc/modules

The Debian/Ubuntu-specific file /etc/modules contains the names of kernel modules that are to be loaded at computer startup, line by line. Usually, Linux itself recognizes which kernel modules need to be loaded. Only if this doesn't work do you need to add the relevant module names to the /etc/modules file.

Alternative Procedures

Kernel modules that are already required during the boot process must be installed in the initrd file. The dracut (Fedora, CentOS, RHEL) and update-initramfs (Debian, Ubuntu) commands help with this.

Another option is to load the kernel module at the end of the init process using the /etc/rc.local script. The Raspberry Pi OS uses device trees to manage the kernel modules that match the hardware. The configuration is performed using /boot/firmware/config.txt. The device tree files with the *.dtb identifier are located in the /boot/firmware and /boot/firmware/overlays directories.

/etc/netplan/netplan.yaml

Netplan (*https://netplan.io*) is used in Ubuntu. The framework configures and controls other network backends such as NetworkManager and systemd component networkd. The configuration files for Netplan are located in /etc/netplan, /lib/netplan, and /run/netplan, and they have the extension *.yaml. For configuration changes to take effect, you must run netplan apply. The configuration files are parsed in alphabetical order.

On desktop systems, /etc/netplan only contains a minimal configuration file by default, which delegates all further work to NetworkManager:

```
# File /etc/netplan/01-network-manager-all.yaml (desktop systems)
network:
  version: 2
  renderer: NetworkManager
```

For server installations in local networks that use DHCP, the configuration reads as shown in the following lines. networkd is responsible for the actual implementation of the network connection. (The line renderer: networkd can be omitted because networkd is the default backend for Netplan.)

```
# File /etc/netplan/01-netcfg.yaml (server/DHCP)
# Configure interface enp0s3 via DHCP
network:
  version: 2
  renderer: networkd
  ethernets:
   enp0s3:
    dhcp4: true
```

You can use this template as a guide for static network configuration including name-server specification:

```
# File /etc/netplan/01-netcfg.yaml (server/static)
# Configure interface ens3 static
network:
  version: 2
  renderer: networkd
  ethernets:
   ens3:
   addresses: [ 138.201.20.182/26 ]
   gateway4:    138.201.20.176
   nameservers:
      search: [ ubuntu-book.info ]
      addresses:
         - "213.133.100.100"
         - "213.133.98.98"
```

Of course, IPv6 addresses are also permitted in the configuration files. In some cases, you must specify IPv4- and IPv6-specific keywords, such as gateway4 versus gateway6 or dhcp4 versus dhcp6. You can find many more examples including routing, bridging, bonding, VLANs, and so on, at the following site:

https://netplan.io/examples

```
/etc/systemd/network/networkd.network
```

The network backend networkd, which is part of systemd, is currently installed in many distributions, but is rarely active. The main exception is Ubuntu Server, where networkd is controlled by Netplan.

networkd parses configuration files in the /etc/systemd/network, /lib/systemd/network, and /run/systemd/network directories. The files must have the extension *.network.

The following lines show an example of how you can set up an Ethernet interface with a static IP address in a virtual machine:

```
# File /etc/systemd/network/01-mystatic.network
[Match]
Name=ens3

[Network]
Address=138.201.20.182/26
Gateway=138.201.20.176
DNS=213.133.100.100
DNS=213.133.98.98
Domains=ubuntu-book.info
```

The second example also shows a static configuration, this time for a root server at Hetzner, including routing and IPv6 setup:

```
[Match]
MACAddress=90:1b:0e:8e:48:86

[Network]
Address=2a01:4f8:171:2baf::2/64
Gateway=fe80::1
Address=138.201.20.176/26
Gateway=138.201.20.129

[Route]
Destination=138.201.20.128/26
Gateway=138.201.20.129
```

If a DHCP server is available, a possible configuration looks as follows:

```
# File /etc/systemd/network/01-dhcpclient.network
[Match]
Name=enp0s3

[Network]
DHCP=ipv4
```

```
[DHCP]
UseMTU=true
RouteMetric=100
```

You can find further details on the syntax of the networkd configuration in man systemd.network. The Arch Linux wiki also contains very useful information:

https://wiki.archlinux.org/index.php/systemd-networkd

/etc/NetworkManager/system-connections/<interface>.nmconnection

Almost all Linux distributions rely on the NetworkManager for Wi-Fi configuration. For other network distribution, distribution-specific procedures are often used in parallel. Exceptions include the current Fedora and RHEL versions (from RHEL 9), which rely entirely on the NetworkManager, even for static configurations on servers. The ifcfg files in /etc/sysconfig/network-scripts that used to be common in RHEL are obsolete.

With NetworkManager, the configuration of the network interfaces is saved in the /etc/NetworkManager/system-connections directory in the keyfile syntax. The syntax of these files is documented here:

www.linux.org/docs/man5/nm-settings-keyfile.html

https://people.freedesktop.org/~lkundrak/nm-docs/nm-settings.html

If possible, you should configure the network using graphical tools. In Gnome, that's in the **network** module of the settings. nmcli and nmtui work well in the console or terminal. (nmtui stands for *NetworkManager Text User Interface* and is a menu-guided and dialog-guided configuration program for text mode.)

It's of course also possible to change the configuration files directly. For NetworkManager to take your configuration files into account, they must be owned by root and must be readable only by root. Make sure to execute chmod 600 filename and, if necessary, also chown root:root filename!

When experimenting with your own configuration files, dynamically generated files in the /var/run/NetworkManager directory often interfere. Perform a reboot! NetworkManager creates dynamic configuration files only for those interfaces for which there's no explicit configuration in /etc/NetworkManager.

For changes to the configuration to take effect, you must ask NetworkManager to reread existing configuration files. Then, you explicitly tell it your wish to actually apply the changes for a specific interface—here, enp1s0:

```
root# nmcli connection reload      (reload configuration)
root# nmcli dev reapply enp1s0     (apply new configuration)
root# nmcli connection show        (interface overview)
```

DHCP Configuration

An Ethernet interface that receives its network configuration via DHCP doesn't actually need a configuration file. NetworkManager automatically takes care of the configuration. However, you can still write the configuration in a file like in the following sample:

```
# File /etc/NetworkManager/system-connections/enp1s0.nmconnection
[connection]
id=myInterfaceName
uuid=27afa607-ee36-43f0-b8c3-9d245cdc4bb3
type=802-3-ethernet
autoconnect=true
interface-name=enp1s0
[ipv4]
method=auto
[802-3-ethernet]
mac-address=52:54:00:64:9b:85
```

method=auto is responsible for DHCP to be used. You can create the mapping to the interface either via interface-name or via mac-address. You can determine the required hexadecimal address using ip addr. autoconnect=true causes the network interface to be activated automatically when the computer is booted.

id is a freely selectable name for the connection. This name must be used in some nmcli commands to select the interface. It's often convenient to simply use the interface name (here, enp1s0) as the ID.

You're also free to decide on the file name for the configuration file. It's advisable to follow the <interface name>.nmconnection method that applies on Fedora and RHEL. NetworkManager takes into account all files that aren't clearly identifiable as backup files.

Static Configuration

For a static IPv4 configuration, you need to change a few lines in the [ipv4] section:

```
# File /etc/NetworkManager/system-connections/enp1s0.nmconnection
...
[ipv4]
address1=192.168.122.244/24,192.168.122.1
dns=192.168.122.1
method=manual
```

method=manual activates the static configuration. address1=... assigns an IP address, the mask, and (if required) the corresponding gateway address to the interface. dns sets the nameserver you want. If required, you can also specify multiple nameserver addresses separated by commas.

IPv6 Configuration

The relevant lines for a static IPv6 configuration looks as follows:

```
# File /etc/NetworkManager/system-connections/enp1s0.nmconnection
...
[ipv6]
addr-gen-mode=default
address1=2abc:1234:1234::10/64,2abc:1234:1234::1
dns=2001:4860:4860::8888,2001:4860:4860::8844
method=manual
```

In the [ipv6] section, address1 again assigns the IP address including mask as well as the corresponding gateway address. The public Google nameservers are used as nameservers in this example.

Firewall Zones

You can include the desired zone for the firewalld program in the [connection] section of the configuration files:

```
# File /etc/NetworkManager/system-connections/enp1s0.nmconnection
[connection]
...
zone=trusted
```

Without this specification, the firewall system automatically uses the default public zone. You can determine the currently valid zones for all interfaces using firewall-cmd --get-active-zones.

/etc/nsswitch.conf

The /etc/nsswitch.conf file controls how host names, usernames, and group names are resolved via the name switch functions in the GNU C library. Changes to this file are rarely necessary, unless LDAP or another network service is used in the local network for user administration and authentication.

The file contains entries line by line, with the first column indicating the name of a database. The term "database" refers to a group of information, such as the list of all users, including login name, full name, home directory, default shell, and so on. Permitted keywords for the first column include aliases (mail aliases), group, hosts, networks, passwd, rpc, and shadow.

The database name is followed by a colon and then a list of services separated by spaces. The most important services are as follows:

- `files` reads the data from local configuration files, for example, /etc/passwd or /etc/group.

- `compat` has a similar meaning to `files`, but allows the interpretation of additional information in /etc/passwd.

- db reads the data from database files.

- nis analyzes the outdated network information service.

- dns queries the nameserver (for host names).

- mdns4_minimal uses Zeroconf to resolve host names.

- ldap communicates with an LDAP server.

- wins uses a WINS server.

In addition, commands can be specified in the form [STATUS=ACTION] bzw. [!STATUS= ACTION], which are then to be executed when a certain event occurs or doesn't occur. Permissible STATUS values are success, notfound, unavail, and tryagain. Possible actions are return (return result immediately, don't fall back on other options) and continue (use the next lookup service).

Example

The following lines show an excerpt from the default configuration on Fedora. Only the /etc/passwd, /etc/shadow, and /etc/group files are taken into account for the user accounts. Host names are read from /etc/hosts, requested from the nameserver and determined via Zeroconf (Avahi).

```
passwd:     files
shadow:     files
group:      files
hosts:      files dns mdns4_minimal
...
```

/etc/os-release

For current distributions that use systemd as the init system, /etc/os-release contains information about the installed distribution. The following lines show the syntax of the file:

```
NAME="AlmaLinux"
VERSION="9.3 (Shamrock Pampas Cat)"
ID="almalinux"
ID_LIKE="rhel centos fedora"
VERSION_ID="9.3"
PLATFORM_ID="platform:el9"
PRETTY_NAME="AlmaLinux 9.3 (Shamrock Pampas Cat)"
...
```

Strictly speaking, the file isn't a configuration file, but an information file. It therefore makes no sense to change the file. However, it may be updated from time to time in the course of updates.

/etc/passwd

The /etc/passwd file contains data on all local accounts on the computer, including the login name, the full name, the default shell, and the home directory. However, for security reasons, the hash code of the password isn't included; it's stored in the separate /etc/shadow file. If possible, you should use the commands provided for managing the accounts, such as useradd, usermod, or userdel.

The passwd file consists of seven columns, each separated by colons:

▶ **First column** (Account name)
Contains the login name. It must not contain any spaces and should be free of special characters. It's customary to use only lowercase letters.

▶ **Second column** (Password)
In the past, this contained the hash code of the password. All current Linux distributions now store the x character in this column, the hash codes can be found in /etc/shadow.

▶ **Third column** (UID)
Contains the internal user ID, which is used to manage access rights and processes. In most distributions, UIDs smaller than 1000 are reserved for system accounts, and UIDs from 1000 are intended for regular users.

▶ **Fourth column** (Primary group)
Indicates the number of the user's primary group. The user can be assigned to other groups; secondary groups are saved in /etc/group.

▶ **Fifth column** (Comment)
Intended for a comment string, which can consist of several parts. In practice, only the full username is usually saved here. If this wasn't specified when the user was created, the fifth column is the same as the first.

▸ **Sixth column** (Home directory)
Contains the path to the home directory.

▸ **Seventh column** (Shell)
Specifies the shell that is automatically active after login; often this is /bin/bash for normal users. For system accounts for which no login is provided, /sbin/nologin or /bin/false is often used instead of the shell. The default shell can be changed via the chsh command.

Example

The following lines show some system accounts and a user account on a Fedora system:

```
root:x:0:0:root:/root:/bin/bash bin:x:1:1:bin:/bin:/sbin/nologin dae-
mon:x:2:2:daemon:/sbin:/sbin/nologin adm:x:3:4:adm:/var/adm:/sbin/nologin
...
kofler:x:1000:1000:Michael Kofler:/home/kofler:/bin/bash
```

/etc/profile

/etc/profile and /etc/profile.d/*.sh contain system-wide settings for environment variables. Variables such as PATH, MAIL, and HOSTNAME are preset there. In addition, many distributions use umask to set which access bits are to be set for newly created files.

/etc/profile is a shell script file. Syntactically, therefore, everything is permitted that the common denominator of the most important shells allows. On Ubuntu, in addition to /etc/profile, the /etc/environment file is provided for the system-wide setting of environment variables. However, this file usually only contains the settings for the PATH variables.

The profile settings are supplemented by user-specific settings in .profile and by shell-specific settings. For example, bash loads the files /etc/bashrc, /etc/bash.bashrc, .bashrc, and .alias. Depending on the distribution, there may be additional configuration files. For example, PATH in Debian and Ubuntu is set by /etc/environment. For the environment variables valid in the X graphics system, the files in the /etc/X11/Xsession.d/ directory are also processed if this exists.

/etc/rc.local
/etc/rc.d/rc.local
/etc/init.d/boot.local

Many distributions run the script file rc.local for compatibility with old Init-V systems. The file has different locations depending on the distribution:

Debian, Raspberry Pi OS, Ubuntu	`/etc/rc.local`
CentOS, Fedora, RHEL	`/etc/rc.d/rc.local`
openSUSE/SUSE	`/etc/init.d/boot.local`

The file must be executable (chmod +x)! In addition, with some distributions, you must run `daemon-reload` after setting up the `systemctl` file so that the file is taken into account from the next reboot.

`rc.local` is a relic of the old Init-V system, which was replaced by systemd many years ago. Depending on the distribution, the file is empty or doesn't exist by default, but it will be taken into account if you create it correctly. However, it's conceivable that this will change in the future. Then, you must set up a "real" systemd unit instead of `rc.local` (see `systemd.service`).

Example

The following script saves the time of the last boot process in `/var/log/boottime`:

```
#!/bin/bash
# File /etc/rc.local or /etc/rc.d/rc.local or /etc/init.d/boot.local
date > /var/log/boottime
```

To ensure that the script in `/etc` or `/etc/rc.d` or `/etc/init.d` is taken into account by systemd, you must run the following commands:

```
root# chmod +x /etc/rc.local      (change file name depending on distribution!)
root# systemctl reload-daemon
root# reboot
```

/etc/resolv.conf

`/etc/resolv.conf` controls how the IP addresses for unknown network names (host names) are determined. "Unknown" means that the names aren't defined in `hosts`. The file usually consists of only two or three lines with the following keywords:

▶ nameserver *ipaddr*
Specifies the IP address of a nameserver. The keyword can be used multiple times to specify alternative nameservers. A maximum of three IPv4 and three IPv6 nameservers are permitted.

▶ domain *mydomain*
Specifies the local domain name. This enables the shortened specification of host names, that is, `name` instead of `name.mydomain`.

▶ search *domain*

Specifies a domain name that is taken into account for search queries. A total of up to six domain names can be specified, each with its own search entry. During name resolution, the specified domain names are tested in sequence; that is, for ping name, it's first name, then name.domain1, then name.domain2, and so on. The default setting for search is the local domain name.

domain and search are mutually exclusive. If resolv.conf contains both keywords, the latter applies.

/etc/resolv.conf is often generated dynamically, especially if the IP configuration is carried out via DHCP.

An increasing number of distributions (Fedora, Ubuntu) are setting up a local nameserver with systemd-resolv. /etc/resolv.conf then references this server (often 127.0.0.53). If you want to know which external nameserver the local service is accessing, you must run resolvectl or systemd-resolv --status.

Example

The following lines show a manual nameserver configuration for IPv4 and IPv6:

```
# /etc/resolv.conf
domain mylan                       # Domain name of the LAN
nameserver 211.212.213.1           # IPv4 DNS
nameserver 211.212.214.1           # Replacement DNS (if the first one fails)
nameserver 2001:4860:4860::8888    # IPv6-DNS
```

/etc/rsyslog.conf

Many Linux distributions use the syslog-compatible program rsyslogd to log messages from the kernel and various network services. This program must be configured using the /etc/rsyslog.conf file and the supplementary /etc/rsyslog.d/*.conf files. These configuration files contain rules that consist of two parts:

▶ **Selector**
The first part of each rule specifies what is supposed to be logged.

▶ **Action**
The second part controls what should be done with the message.

Rules can be spread across multiple lines using the \ character. It's possible that multiple rules apply to one message. In this case, the message is logged or passed on several times. For changes to the syslog configuration to take effect, the syslog service must be restarted via service rsyslog restart!

Selectors

Each selector consists of two parts separated by a dot: *service.prioritylevel*. It's allowed to specify multiple selectors separated by a semicolon. Furthermore, multiple services can be separated by commas in *one* selector. All Linux programs that use syslog must assign a service and priority to their messages.

Syslog recognizes the following services (facilities): auth, authpriv, cron, daemon, ftp, kern, lpr, mail, news, syslog, user, uucp, and local0 to local7. The * character comprises all services.

Syslog also knows the following priority levels (in increasing importance): debug, info, notice, warning = warn, err = error, crit, alert, and emerg = panic. The warn, error, and panic keywords are considered obsolete—you should use warning, err, and emerg instead. The * character comprises all priority levels. The none keyword applies to messages that aren't assigned a priority.

The specification of a priority level includes all higher (more important) priority levels. Thus, the mail.err selector also includes crit, alert, and emerg messages from the mail system. If you explicitly want only messages of a certain priority, you must prefix the = character (i.e., mail.=err).

Actions

The name of a logging file is normally specified as the action. Logging files are usually synchronized after each output. If the file name is preceded by a minus sign, syslog won't synchronize. This is much more efficient, but messages that haven't yet been physically saved will be lost in the event of a crash.

Syslog can also forward messages to *First In First Out* (FIFO) files or pipes. In this case, you need to prefix the file name with the | character. The file /dev/xconsole, which appears in the following listing, is a special FIFO file for passing messages to the X graphics system.

The * character means that the message will be sent to all users logged into consoles or via SSH. As this is very disruptive, it's only used by default for critical messages.

Journal Configuration

Current distributions use the *Journal* program belonging to the systemd project instead of or in addition to rsyslogd. It's configured using /etc/systemd/journald.conf and a few other files.

Example

The following lines show the standard Ubuntu syslog configuration in a slightly shortened and more clearly formatted form:

```
# File /etc/rsyslog.d/50-default.conf bei Ubuntu
# Selector                  Action
auth,authpriv.*             /var/log/auth.log
*.*;auth,authpriv.none      -/var/log/syslog
kern.*                      -/var/log/kern.log

mail.*                      -/var/log/mail.log
mail.err                    /var/log/mail.err
*.emerg                     :omusrmsg:*

daemon.*;mail.*; news.err; *.=debug;*.=info; *.=notice;*.=warn  |/dev/xconsole
```

In plain language, the preceding configuration means:

▶ /var/log/auth.log contains authentication messages of all priority levels. These include failed and successful login attempts (also via SSH), PAM messages, sudo commands, and so on. As the only logging file, auth is synchronized immediately with each message.

▶ /var/log/syslog contains *all* messages logged via syslog, including authentication messages that aren't assigned a priority. The all-encompassing approach is both an advantage and a disadvantage. On one hand, you can extract all conceivable information from a single file; on the other hand, it's of course particularly difficult to find relevant entries in this hodgepodge.

▶ /var/log/kernel.log contains all kernel messages.

▶ The messages from the mail system (e.g., postfix) are distributed across several files. In mail.log *all* messages are stored; in mail.err only error messages.

▶ Critical system messages, for example, about an imminent shutdown or kernel errors, are forwarded to all users via :omusrmsg:*, specifically to all terminal windows and consoles. omusrmsg is a rsyslog module for sending messages to users.

▶ Various warnings and error messages are also forwarded to the X system. To track these messages on X, you must start the xconsole program. It looks like a small terminal window, but doesn't allow any input.

/etc/services

The /etc/services file contains a list of all common network services and the IP ports and protocol types assigned to them. Changes to this file are only necessary in exceptional cases. The syntax of the file is clearly shown in the following sample lines:

```
# service-name  port/protocol  [aliases]   [# comment]
tcpmux          1/tcp                       # TCP port service multiplexer
tcpmux          1/udp                       # TCP port service multiplexer
```

```
rje          5/tcp                          # Remote Job Entry
rje          5/udp                          # Remote Job Entry
echo         7/tcp
echo         7/udp
...
ftp          21/tcp
ftp          21/udp          fsp fspd
ssh          22/tcp                         # The Secure Shell (SSH) Protocol
ssh          22/udp                         # The Secure Shell (SSH) Protocol
telnet       23/tcp
telnet       23/udp
...
```

/etc/shadow

/etc/shadow adds the password hash to the user information from passwd as well as data indicating how long the account is valid and when the password should be renewed. To change this data, you usually use the passwd and chage commands. The shadow file consists of nine columns separated by colons:

▶ **First column** (Account name)
Matches the first column of /etc/passwd.

▶ **Second column** (Password hash)
Allows the password to be checked. However, it's impossible to reconstruct the password from the hash code. In current Linux distributions, the hash code is generated via sha. As a random initialization value is used when generating each hash code (the *salt*), two users who use the same password will still have different hash codes. This also makes dictionary attacks impossible in which pregenerated hash codes are compared with the content of /etc/shadow.

For accounts that don't require a login, the second column simply contains an asterisk. For locked accounts, the hash code is preceded by two exclamation points. The code becomes invalid. To reactivate the account, the exclamation points are removed again.

▶ **Third column** (Last change)
Indicates when the password was last changed. This is calculated in days from January 1, 1970. The value 0 means that the user must change the password immediately at the next login.

▶ **Fourth column** (Minimum age)
Indicates the earliest number of days after which a password may be changed. 0 allows a change at any time.

- **Fifth column** (Maximum age)
  Specifies the maximum number of days after which a password must be changed. 0 allows unlimited use of the password.

- **Sixth column** (Warning time)
  Specifies how many days before the password expires the user is notified. 0 deactivates these warnings.

- **Seventh column** (lockout time)
  Specifies how many days after the password expires the account will be locked. It can then only be reactivated by the administrator using passwd -u.

- **Eighth column** (Expiry time)
  Specifies when the account expires. The date is given in days from January 1, 1970. If the account is to be valid indefinitely, this column remains empty. The value 0 isn't used!

- **Ninth column**
  Reserved for future extensions.

Example

The following lines show an excerpt from /etc/shadow, whereby the long hash codes have been shortened:

```
root:$6$.cc11McOB::0:99999:7:::
bin:*:18095:0:99999:7:::
daemon:*:18095:0:99999:7:::
adm:*:18095:0:99999:7:::

...
kofler:$y$j9T$1...:19793:0:99999:7:::
```

To find out when the password was last changed, you must run date:

```
user$ date -d "1970-01-01 + 19793 days"
```

```
Mon Mar 11 00:00:00 CET 2025
```

The other chage entries for the users root and kofler indicate that the password can be changed at any time (0) and is valid almost indefinitely (99999 = approximately 273 years). The user is warned seven days before the password expires when logging in.

/etc/apt/sources.list

The APT package sources are defined in /etc/apt/sources.list and /etc/apt/sources.list.d/*. The syntax of each line looks as follows:

```
package type uri distribution [component1] [component2] [component3] ...
```

The package type is deb for ordinary Debian packages or deb-src for source packages. The second column specifies the root directory of the package source. In addition to HTTP and FTP directories, APT also supports ordinary directories as well as RSH or SSH servers.

The third column indicates the distribution. All other columns indicate the components of the distribution that can be taken into account. The component names depend on the distribution and the package source! For example, Ubuntu distinguishes between *main*, *restricted*, *universe*, and *multiverse* packages, while Debian differentiates between *main*, *contrib*, *non-free*, and similar components.

The package sources mentioned first are preferred: thus, if a particular package is available for download from multiple sources, APT downloads it from the first source.

Example

The following listing shows some Ubuntu package sources. For space reasons, each entry was spread across two lines.

```
# File /etc/apt/sources.list
deb http://de.archive.ubuntu.com/ubuntu/ mantic           \
                                 main restricted universe multiverse
deb http://de.archive.ubuntu.com/ubuntu/ mantic-updates  \
                                 main restricted universe multiverse
deb http://security.ubuntu.com/ubuntu    mantic-security \
                                 main restricted universe multiverse
```

/etc/sudoers

The /etc/sudoers file and the additional files in /etc/sudoers.d define which users are allowed to run which commands with which rights via sudo. The file also contains various general settings that control the basic behavior of sudo. The settings vary greatly depending on the distribution.

The sudoers file should only be changed using the visudo command. It calls the editor defined by the environment variables VISUAL or EDITOR (by default, this is vi), performs a syntax check before saving, and thus ensures that you don't exclude yourself from further administration work due to a sudoers file that contains errors. This is especially important for distributions such as Ubuntu, which don't provide a root login.

Basic Settings

► Defaults always_set_home

Changes the HOME environment variable when the user changes, so that it refers to the new user's home directory. Usually, this happens automatically. This option is only relevant if !env_reset applies or env_keep contains the PATH variable.

- Defaults env_keep="*var1 var2 var3*"

  Specifies which environment variables are to be retained when the user changes and are therefore excluded from env_reset.

- Defaults env_reset

  Causes all environment variables to be reset when the user changes. This setting applies by default. To deactivate it, you must enter Defaults !env_reset.

- Defaults mail_badpass

  Causes a warning email to be sent to the administrator after an incorrect login attempt.

- Defaults secure_path="*pfad1:pfad2:pfad3*"

  Defines the content of the PATH environment variable for sudo commands.

- Defaults targetpw

  Indicates that the password for the account in which the command is to be run must always be entered, that is, usually the root password. Without this setting, sudo expects the password of the current user.

- Defaults timestamp_timeout=*n*

  Specifies the time after which sudo asks for the password again. The default setting is five minutes.

Rights

Three-column entries in /etc/sudoers define which users are allowed to run which programs from which computer. The following line indicates that the user katherine is allowed to run the /sbin/fdisk command on the computer named uranus. The ALL keyword indicates that katherine can run the command under any account, that is, as root, as news, as lp and so on.

```
katherine uranus=(ALL) /sbin/fdisk
```

If the first column of sudoers is prefixed with the % character, the entry applies to all members of the specified group. The following line means that all users in the wheel group can run all commands from any computer as any user:

```
%wheel   ALL=(ALL) ALL
```

It's possible to allow a specific user to run sudo without specifying a password. However, the NOPASSWD keyword is only valid if there are no other sudoers lines that require a password from the same user. This also applies to group entries, such as %admin.

```
kofler ALL=(ALL) NOPASSWD: ALL
```

The following rule, which reads as follows, is extremely liberal: anyone can do anything. This rule only makes sense in combination with the Defaults targetpw option, which always requires the password of the user on whose behalf a command is to be run to be entered.

```
ALL ALL=(ALL) ALL
```

Example

The following lines show the default configuration on Ubuntu. In this distribution, root is set up without a valid password. A root login is therefore impossible! In addition, su or ssh -l root don't work. The only way to run admin commands is thus provided by sudo. All members of the admin and sudo groups have this permission, whereby the admin group is only included for reasons of compatibility with older Ubuntu versions. In current Ubuntu versions and in most other distributions, sudo is the usual group for users with sudo rights.

```
# Default configuration in /etc/sudoers on Ubuntu
Defaults    env_reset
Defaults    mail_badpass
Defaults    use_pty
Defaults    secure_path=\
                "/usr/local/sbin:/usr/local/bin:/usr/sbin:/usr/bin:/sbin:/bin"
root        ALL=(ALL:ALL) ALL
%admin      ALL=(ALL) ALL
%sudo       ALL=(ALL:ALL) ALL
@includedir /etc/sudoers.d
```

/etc/sysctl.conf

/etc/sysctl.conf contains default settings for kernel parameters. The file is analyzed at system startup, and all the parameters listed there are set accordingly. For subsequent changes, you can use sysctl -p to ensure that all settings are loaded from sysctl.conf and set.

The kernel parameters to be set are specified line by line in the form name.name.name= value. Comment lines start with ; or with #.

Example

The following setting in sysctl.conf causes forwarding to be activated for IPv4 and IPv6 when the computer is started. This is necessary if the computer is supposed to work as a router.

```
net.ipv4.ip_forward=1 net.ipv6.conf.all.forwarding=1
```

```
systemd.service
```

The init system systemd is controlled by a whole range of configuration files, which are located in the following three directories:

```
/etc/systemd/system
/run/systemd/system
[/usr]/lib/systemd/system
```

If systemd is also used to manage user-specific services, there are also *.service files in /home/name/.config/systemd/user and in some other directories.

The *.service files described in this section define how a service—for example an SSH server—is started, monitored, and, if necessary, stopped again by systemd. *.service files consist of several sections that are introduced by lines with the content [Unit], [Service], [Install].

[Unit] Keywords

▶ Before/After=*name*
  Names other services or units that are to be started before or after. Before and After provide systemd with information on the sequence in which the services should be started.

▶ Description=*description*
  Describes the service using a character string.

▶ OnFailure=*name*
  Names one or more services that are to be run if the start of the current service fails.

▶ Wants/Requires=*name*
  Names other services that should or must be started beforehand. These keywords define dependencies. systemd attempts to start the relevant services beforehand. With Wants, a possible failure is silently ignored, while with Requires it means that the current service isn't started and an error is triggered instead.

[Service] Keywords

▶ Environment=*var1*='*value1*' *var2*='*value2*' ...
  Defines several environment variables that are passed on to the process to be started.

▶ EnvironmentFile=*filename*
  Loads the file and takes into account the variable assignments contained in it line by line.

▶ ExecReload=*/path/program options*
Runs the specified command to reload the configuration of the service during operation.

▶ ExecStart=*/path/program options*
Runs the specified command to start the service. In combination with Type=oneshot, multiple ExecStart lines are permitted, which are then executed in sequence.

▶ ExecStop=*/path/program options*
Runs the specified command to end the service. If you don't use this keyword, the service will be terminated by a KILL signal. If required, the KillMode and KillSignal keywords, which are listed in the systemd.kill man page, describe which signal is to be sent and how. ExecStop is only possible in combination with Type=oneshot.

▶ RemainAfterExit
Often used in combination with Type=oneshot. systemd remembers the currently activated state.

Without this option, systemd believes that the action has ended with the end of the last StartExec command after systemctl start name and immediately sets the status back to stop. Explicitly executing systemctl stop would then have no effect.

▶ Type=simple|forking|oneshot|dbus|notify|idle
Specifies the type of service or how the program should be started.

If the type specification is missing, simple applies. systemd then assumes that the command specified with ExecStart starts a background service. Not until this command ends does systemd consider the service to be regularly terminated.

Type=notify works in a similar way, but systemd expects an explicit notification that the start process is complete. The background service must call the sd_notify function or an equivalent function.

Type=forking also has similarities with simple, but here, the ExecStart command is responsible for starting a background process that is detached from the start command. As soon as the ExecStart command is completed, systemd assumes that the service is running in its own background process.

With the oneshot type, systemd assumes that the command to be executed is completed so quickly that it can be waited for before other actions are performed. The status of the service is reset to stop at the end, unless the service file contains the keyword RemainAfterExit.

Details on the other type variants are provided by systemd.service.

▶ User=*username* and Group=*groupname*
Determines under which account the command is executed. By default, system services run with root rights.

[Install] Keywords

▶ WantedBy= or RequiredBy=*target*

Specifies for which target the service is desirable or required, for instance, for multi-user.target or for reboot.target. In practice, WantedBy is usually used. This means that the target is reached even if the start of individual services fails. RequiredBy, on the other hand, results in an error if there are problems. Then, the relevant target can't be activated.

Example

The following lines show the httpd.service file in the default configuration of Fedora. It's responsible for starting the Apache web server. A prerequisite for the start is the successful activation of the network and any network file systems. The web server is supposed to be started automatically as part of the multiuser target.

```
# File /lib/systemd/system/httpd.service (Fedora)
[Unit]
Description=The Apache HTTP Server
Wants=httpd-init.service
After=network.target remote-fs.target nss-lookup.target httpd-init.service
Documentation=man:httpd.service(8)

[Service]
Type=notify
Environment=LANG=C

ExecStart=/usr/sbin/httpd $OPTIONS -DFOREGROUND
ExecReload=/usr/sbin/httpd $OPTIONS -k graceful
# Send SIGWINCH for graceful stop
KillSignal=SIGWINCH
KillMode=mixed
PrivateTmp=true
OOMPolicy=continue

[Install]
WantedBy=multi-user.target
```

Setting Up Custom systemd Services

To set up your own systemd service, you need to create a *.service file in the /etc/systemd/system directory. The best way to do this is to use a service file from another, comparable service. Suitable files can be found in the /lib/systemd/system or /usr/lib/systemd/system directories, depending on the distribution.

For efficiency reasons, systemd keeps the contents of important configuration files in a cache. After configuration changes, you must explicitly reload the configuration files using systemctl:

```
root# systemctl daemon-reload
```

To start the service immediately and in future every time the multiuser target is reached, the following command is required:

```
root# systemctl enable --now myservice
```

In case of issues, it's best to use journalctl -u myservice for debugging.

Other systemd Configuration Files

In addition to the *.service files, there are also files for systemd configuration with the identifiers *.mount, *.path, *.socket, *.target, and *.wants, which aren't discussed any further in this book. Their syntax is documented in the man pages service.mount, service.path, and so on. All of these files together are considered unit files. The general syntax of those units is explained in man systemd.unit.

If you want to set up jobs that are run periodically by systemd, you must familiarize yourself with the format of the *.timer files (see systemd.timer).

systemd.timer

systemd can take care of the regular execution of processes in a way similar to cron. The configuration is performed using a name.timer file that must have a service file with the same name (i.e., name.service). For custom-defined jobs, both files are usually located in the /etc/systemd/system directory.

The structure of a *.timer file is similar to that of a *.service file (see systemd.service). It usually consists of three sections: the familiar [Unit] and [Install] sections, as well as a [Timer] section, whose settings control when and how often the job is executed.

[Timer] Keywords

▶ AccuracySec=*time span*

Specifies the accuracy within which the jobs are to be run. By default, a time window of one minute is provided for this, within which the start time is determined randomly. Shorter time spans are possible, but should only be used if this is really necessary.

▶ OnBootSec=*time span*

Runs the job for the first time after the specified period of time after the boot process. If the time is given without a unit, seconds are meant. The syntax of the time specifications is documented in man systemd.time. Permissible time spans include, for example, 2h 15min or 2weeks or 4months.

▶ OnCalendar=*time*

Specifies the time at which the job is to be run. For example, 12.30 means daily at 12.30 pm. The syntax for absolute time specifications is somewhat complex. The details are documented in detail in the **Calendar Events** section of man systemd.time. For example, OnCalendar=Sun 2025-*-* 17:15 means that a job is executed every Sunday at 5:15 pm in 2025.

▶ OnUnitActiveSec=*time span*

Restarts the job as soon as the specified period of time has passed since the job was last started.

▶ OnUnitInactiveSec=*time span*

Restarts the job as soon as the specified period of time has passed since the job was last completed.

Example

The following dnf-automatic.timer file ensures that the service described in the dnf-automatic.service file is run once a day. Both files are part of the Fedora package dnf-automatic, which automatically performs updates with dnf.

```
# File /usr/lib/systemd/system/dnf-automatic.timer
# (Fedora, if the dnf-automatic package is installed automatically)
[Unit]
Description=dnf-automatic timer
ConditionPathExists=!/run/ostree-booted

[Timer]
OnCalendar=*-*-* 6:00
RandomizedDelaySec=60m
Persistent=true

[Install]
WantedBy=basic.target
```

If you want to set up your own timer, you must create both a *.service and a *.timer file in /etc/systemd/system. Then, ask systemd to reload the configuration and activate the timer:

```
root# systemctl daemon-reload

root# systemctl enable --now mytimer.timer
```

/etc/vconsole.conf

The /etc/vconsole.conf file contains the keyboard and font settings for working in text consoles for distributions that use a current systemd version as the init system. The localectl command helps you change the settings. This command has the advantage over a direct change via vconsole.conf in that it also saves the desired keyboard layout in /etc/X11/xorg.conf/00-keyboard.conf so that the setting also applies to the graphics system.

vconsole.conf usually consists of just two lines with the following parameters:

▶ FONT=*name*
   Specifies the font to be used in text consoles. Suitable font files can be found in the /lib/kbd/consolefonts directory in most distributions.

▶ KEYMAP=*name*
   Specifies the keyboard layout. You can determine the settings available for selection via localectl list-keymaps.

Example

In a German Fedora installation, vconsole.conf contains the following settings:

```
KEYMAP="de-nodeadkeys"
FONT="eurlatgr"
```

/etc/wpa_supplicant/wpa_supplicant.conf

The background service wpa_supplicant is responsible for authentication in wireless networks (Wi-Fi). In desktop distributions, the program is used by NetworkManager as a backend; the configuration is performed via settings dialogs of KDE, Gnome, and so on.

The situation is different if you want to set up a Wi-Fi connection on a Raspberry Pi or another device (embedded device) without a graphical user interface. Then, you must enter the Wi-Fi authentication data in the /etc/wpa_supplicant/wpa_supplicant.conf file. The following lines represent an example of this:

```
ctrl_interface=DIR=/var/run/wpa_supplicant GROUP=netdev
update_config=1
country=DE

network={
ssid="wlan-sol3"
  psk="strengGeheim"
  key_mgmt=WPA-PSK
}
```

The first three lines contain the default settings on the Raspberry Pi OS. The correct country setting is crucial. It ensures that the Wi-Fi adapter complies with the regulations of the respective country. This is followed by any number of network groups, one for each wireless network.

When writing wpa_supplicant.conf, make sure that you *don't* enter *any* spaces before and after the equal signs! You can read a lot of details about the permitted keywords as well as further examples with man wpa_supplicant.conf and in the following documentation file and the following website:

/usr/share/doc/wpasupplicant/README.modes.gz

https://wiki.ubuntuusers.de/WLAN/wpa_supplicant

In simple cases, you can also create the required configuration lines using wpa_passphrase or perform the configuration directly in /etc/network/interfaces.

Keyboard Shortcuts

The last section of the *Linux Command Reference* deals with the keyboard shortcuts of the most important editors and other commands that are usually operated via the keyboard. These include, for example, bash, man, info, less, and mutt.

Almost all programs offer the option of defining custom keyboard shortcuts. This section refers to the default configuration, which is the standard configuration for most Linux distributions.

bash

Table 1 summarizes which keyboard shortcuts you can use within the Bourne Again Shell (bash) when entering commands. The keyboard shortcuts actually originate from the readline library. You can change the configuration of this library in /etc/inputrc or .inputrc.

Keyboard Shortcut	Function
Ctrl + A	Moves the cursor to the beginning of the line (like Pos1)
Ctrl + C	Cancels the current command
Ctrl + E	Moves the cursor to the end of the line (like End)
Ctrl + K	Deletes the rest of the line from the cursor position
Ctrl + Y	Reinserts the most recently deleted text
Ctrl + Z	Interrupts the current command (can be continued via fg or bg)
Tab	Completes file and command names
↑ / ↓	Scrolls through the commands executed so far

Table 1 Keyboard Shortcuts for Entering Commands in bash

emacs

emacs is one of the most feature-rich and complex editors available on Linux. There are hundreds of keyboard shortcuts and commands, of which only the most important are presented here.

In general, there are three ways to enter emacs commands: the menu, using keyboard shortcuts (mostly a combination with [Ctrl] or [Alt]), or typing the entire command name. The third variant is initiated with [Alt]+[X] or [Esc], for example, [Alt]+[X] delete-char [Enter]. The input of commands and other parameters is facilitated by two mechanisms:

▶ During input, you can add [Tab] to the name of an emacs command. You can also complete file names in the same way.

▶ Commands previously specified using [Alt]+[X] can be accessed with [Alt]+[P] (*previous*) and [Alt]+[N] (*next*) after the new command has been introduced with [Alt]+[X].

In the emacs documentation, keyboard shortcuts are presented somewhat differently: DEL refers to [Backspace]! C stands for Control ([Ctrl]) and M for [Meta]. There is no direct equivalent of the [Meta] key on a standard PC keyboard. M-x can be emulated on a PC keyboard in two ways: by [Esc] and [X] (in succession) or by [Alt]+[X].

Table 2 summarizes the basic functions. You can use the cursor keys and various keyboard shortcuts to move the cursor (see Table 3).

Keyboard Shortcut	Function
[Ctrl]+[X], [Ctrl]+[F]	Loads a new file
[Ctrl]+[X], [Ctrl]+[S]	Saves the current file
[Ctrl]+[X], [Ctrl]+[W]	Saves the file under a new name
[Ctrl]+[G]	Cancels the input of a command
[Ctrl]+[X], [U]	Undoes the last change (*undo*)
[Ctrl]+[X], [Ctrl]+[C]	Exits emacs (with prompt to save)

Table 2 Basic Emacs Commands

Keyboard Shortcut	Function
[Alt]+[F]/[Alt]+[B]	Moves the cursor one word forward or backward
[Ctrl]+[A]/[Ctrl]+[E]	Places the cursor at the beginning or end of the line
[Alt]+[<]/[Alt]+[⌂]+[>]	Moves the cursor to the beginning or end of the text
[Alt]+[G] *n* [Enter]	Moves the cursor to line *n*

Table 3 Cursor Movement

Table 4 specifies how you can select, delete, and reinsert text, and Table 5 summarizes how to search and replace.

Keyboard Shortcut	Function
Ctrl + [ ]	Sets an (invisible) marker point
Ctrl + W	Deletes the text between the marker point and the current cursor position
Ctrl + Y	Reinserts the most recently deleted text
Ctrl + X , Ctrl + X	Swaps cursor position and marker point
Alt + D	Deletes the next word or the end of the word from the cursor
Alt + Backspace	Deletes the preceding word or the beginning of the word up to the cursor
Ctrl + K	Deletes the rest of the line from the cursor position
Alt + O , Ctrl + K	Deletes the beginning of the line before the cursor position
Alt + M	Deletes the paragraph that follows
Alt + Z , x	Deletes all characters until the next occurrence of x, which is also deleted
Ctrl + Y	Inserts the last deleted text at the cursor position

Table 4 Selecting, Deleting, and Reinserting Text

Keyboard Shortcut	Function
Ctrl + S	Incremental search forward
Ctrl + R	Incremental search backward
Alt + P	Selects a previously used search text (*previous*)
Alt + N	Selects a search text to be used later (*next*)
Ctrl + G	Cancels the search
Ctrl + X , Ctrl + X	Swaps the marker point (start of the search) and the current cursor position
Ctrl + Alt + S	Incremental pattern search forward
Ctrl + Alt + R	Incremental pattern search backwards
Alt + %	Search and replace without pattern
Alt + X query-replace-r Enter	Search and replace with pattern

Table 5 Search and Replace

gnome-terminal

If you run shell commands on Gnome, you'll most likely use the gnome-terminal program. So that you can use the usual keyboard shortcuts in the bash, you should first deactivate the **Menu shortcut letters** option to control the menus using [Alt] shortcuts with **Edit • Settings**. You can then still use the [F10] key for menu control if required—unless you also deactivate the processing of this key in the configuration dialog just mentioned. Some gnome-terminal–specific keyboard shortcuts remain available in any case; they are summarized in Table 6.

Keyboard Shortcut	Function
[⇧]+[Ctrl]+[C]	Copies the selected text to the clipboard
[⇧]+[Ctrl]+[F]	Searches for a text in the terminal outputs
[⇧]+[Ctrl]+[G]	Repeats the backward search
[⇧]+[Ctrl]+[H]	Repeats the forward search
[⇧]+[Ctrl]+[N]	Opens a new terminal window
[⇧]+[Ctrl]+[Q]	Closes the window
[⇧]+[Ctrl]+[T]	Opens a new terminal tab
[⇧]+[Ctrl]+[V]	Pastes the contents of the clipboard
[⇧]+[Ctrl]+[W]	Closes the tab
[⇧]+[Ctrl]+[+]	Enlarges the font
[⇧]+[Ctrl]+[-]	Reduces the font size
[⇧]+[Ctrl]+[Page ↑]/[Page ↓]	Goes to the previous/next tab
[F11]	Activates or deactivates full screen mode

Table 6 Keyboard Shortcuts in gnome-terminal

grub

In the Linux boot loader GRUB, you can use the cursor keys to select an operating system or a Linux variant and then start it by pressing [Enter]. GRUB also provides the option of interactively changing the parameters of a menu item or running custom commands. Table 7 summarizes the most important shortcuts for this. The table refers to the current GRUB version 2.

Keyboard Shortcut	Function
Esc	Exits graphics mode and activates text mode.
C	Starts the command mode for the interactive execution of GRUB commands. When entering commands, file names can be completed with a Tab as in the shell.
E	Starts the editor for the selected menu item.
P	Enables the interactive GRUB functions by entering a password. This is only necessary if GRUB is protected by a password.
Ctrl+X or F10	Starts the menu item previously changed with E.

Table 7 Keyboard Shortcuts for Interactive Control of GRUB 2

info

info starts the online help system of the same name. To navigate in the help text, you can use the keyboard shortcuts summarized in Table 8. You can also read info texts more conveniently via the pinfo command from the package of the same name, with the emacs editor or in the Gnome and KDE help systems.

Keyboard Shortcut	Function
Spacebar	Scrolls text down
Backspace	Scrolls text up
B, E	Jumps to the beginning/end of the info unit (*beginning/end*)
Tab	Moves the cursor to the next cross-reference
Enter	Tracks a cross-reference to another info unit
N	Displays the next info unit in the same hierarchy level (*next*)
P	Displays the previous info unit in the same hierarchy level (*previous*)
U	Jumps *up* one hierarchy level
L	Jumps back to the last text displayed (*last*)
H	Displays detailed operating instructions (*help*)
?	Displays an overview of commands
Q	Exits info (*quit*)

Table 8 info Keyboard Shortcuts

joe

joe is a simple editor whose keyboard shortcuts are based on the WordStar word processor (see Table 9). The editor can also be started under the name jmacs or jpico. Other keyboard shortcuts that are compatible with emacs or Pico then apply.

Keyboard Shortcut	Function
`Ctrl`+`K`, `H`	Shows/hides the help window
`Ctrl`+`K`, `E`	Loads a new file
`Ctrl`+`K`, `D`	Saves the file (optionally under a new name)
`Ctrl`+`Y`	Deletes a line
`Ctrl`+`⇧`+`-`	Undoes the deletion
`Ctrl`+`C`	Exits joe (with confirmation prompt to save)

Table 9 Keyboard Shortcuts in joe

konsole

KDE fans usually run shell commands in the konsole program. Most keyboard shortcuts are passed directly to the shell by this program. However, there are also some konsole-specific keyboard shortcuts, which are summarized in Table 10.

Keyboard Shortcut	Function
`⇧`+`Ctrl`+`A`	Marks the tab for activity
`⇧`+`Ctrl`+`C`	Copies the selected text to the clipboard
`⇧`+`Ctrl`+`F`	Searches for a text in the terminal outputs
`F3`	Repeats the backward search
`⇧`+`F3`	Repeats the forward search
`⇧`+`Ctrl`+`I`	Marks the tab for longer periods of inactivity
`⇧`+`Ctrl`+`N`	Opens a new terminal window
`⇧`+`Ctrl`+`Q`	Closes the window
`⇧`+`Ctrl`+`T`	Opens a new terminal tab
`⇧`+`Ctrl`+`V`	Pastes the contents of the clipboard
`⇧`+`Ctrl`+`W`	Closes the tab

Table 10 Keyboard Shortcuts in konsole

Keyboard Shortcut	Function
⇧ + Ctrl + +	Enlarges the font
⇧ + Ctrl + -	Reduces the font size
⇧ + Ctrl + ← / →	Goes to the previous/next tab

Table 10 Keyboard Shortcuts in konsole (Cont.)

less

The less command displays texts. While the program is running, you can use the cursor keys to scroll through the text, search for texts, start the editor set by the $EDITOR environment variable, and so on (see Table 11). less is also used to display man help texts.

Keyboard Shortcut	Function
Cursor keys	Scrolls the text up or down
Pos1, End	Jumps to the beginning or end of the text
G, ⇧ + G	Jumps to the beginning or end of the text
/ pattern Enter	Searches forwards
? pattern Enter	Searches backwards
N	Repeats the forward search (*next*)
⇧ + N	Repeats the backward search
V	Starts the editor set by $EDITOR or $VISUAL
Q	Quits less (*quit*)
H	Displays a help text with additional shortcuts

Table 11 Keyboard Shortcuts for less

man

The man command displays the documentation for important commands, configuration files, C functions, and so on. To enable simple scrolling through the help text, you can use the less command. For this reason, the same shortcuts apply within man as for less (see Table 11).

mutt

mutt is an email client for text mode. The program is well suited for reading and reply-ing to emails stored locally on the computer. Died-in-the-wool Linux fans even prefer mutt in graphical user interfaces because the program can be controlled particularly efficiently using the keyboard. Table 12 only contains the most important shortcuts.

Keyboard Shortcut	Function
Cursor keys	Moves the cursor through the mail list
Enter	Displays the first lines of the selected email
Spacebar	Scrolls through the text of an email
T	Deletes the email
I	Switches from the email view back to the inbox
Y	Displays the next email
M	Composes a new email using the editor set by $EDITOR or $VISUAL
R	Replies to the email
?	Displays a help text with all keyboard shortcuts

Table 12 Keyboard Shortcuts for mutt

nano

nano is a minimalist editor that is particularly suitable for beginners. The most import-ant keyboard shortcuts (see also Table 13) are always displayed in the two bottom text lines.

Keyboard Shortcut	Function
Ctrl + A	Moves the cursor to the beginning of the line
Ctrl + D	Deletes a character
Ctrl + E	Moves the cursor to the end of the line
Ctrl + H	Deletes one character backward
Ctrl + ^	Sets a marker point
Ctrl + K	Deletes the current line or the selected text

Table 13 Keyboard Shortcuts for nano

Keyboard Shortcut	Function
Ctrl + U	Reinserts the most recently deleted text
Ctrl + R	Inserts a text file into the text
Ctrl + O	Saves the file
Ctrl + X	Exits the editor

Table 13 Keyboard Shortcuts for nano (Cont.)

screen

You can use the screen command to run multiple terminal sessions in parallel (*multiplex*). screen is controlled by keyboard shortcuts that all start with Ctrl + A.

Keyboard Shortcut	Function
Ctrl + A, ?	Displays the online help
Ctrl + A, *	Lists all sessions
Ctrl + A, 0 through 9	Switches between the first and tenth session
Ctrl + A, C	Creates a new session (*create*)
Ctrl + A, D	*Detaches* the current session from screen (*detach*)
Ctrl + A, ⇧ + H	Activates/deactivates logging
Ctrl + A, K	Ends the active session
Ctrl + A, N	Switches to the next session
Ctrl + A, \	Ends all sessions and screen

Table 14 Keyboard Shortcuts in screen

Text console

If you work directly in text consoles, that is, not in KDE, Gnome, or another desktop system, and also not via SSH, then some special keyboard shortcuts apply. Table 15 summarizes the most important shortcuts.

Keyboard Shortcut	Function
Ctrl + Alt + Fn	Switches from graphics mode to text console *n*
Alt + F1	Switches back to graphics mode

Table 15 Keyboard Shortcuts in Text Consoles

Keyboard Shortcut	Function
Alt + Fn	Switches from a text console to text console *n*
Alt + → /+ ←	Switches to the previous/next text console
⇧ + ↑ / ↓	Scrolls forward/backward page by page
Ctrl + Alt + Del	Terminates Linux via shutdown (caution!)

Table 15 Keyboard Shortcuts in Text Consoles (Cont.)

vi/vim

The vi editor is a veteran in the history of Unix. Almost all Linux distributions install the vi-compatible vim program by default, which you can start via both vi and vim.

vi provides a similar number of functions to emacs but is even more difficult to learn to use. Table 16 summarizes the basic commands.

Keyboard Shortcut	Function
I	Activates the insert mode.
A	Activates the insert mode. The text input starts at the next character.
Esc	Activates the standard mode or cancels the command input.

Commands in standard mode

D, W	Deletes a word.
D, D	Deletes the current line.
n D, D	Deletes *n* lines.
P	Inserts the last deleted text after the cursor position.
⇧ + P	Inserts the most recently deleted text before the cursor position.
.	Repeats the last command.
U	Undoes the last change (*undo*).
⇧ + U	Undoes all changes in the current line.
Ctrl + R	Undoes undo (*redo*, as of Vim 7).
: w	Saves the file.
: q	Exits vim.

Table 16 Basic Commands

Keyboard Shortcut	Function
`:` q!	Exits vim even if there are unsaved files.
Commands in insert mode	
`Ctrl`+`O` *command*	Runs the command without leaving the insert mode.

Table 16 Basic Commands (Cont.)

The main fundamental difference from other editors is that vi distinguishes between different modes.

► **Insert mode**

To enter text, you must switch to insert mode using `I` (*insert*) or `A` (*append*). vim then displays the text -- INSERT -- in the bottom line on the far left. In insert mode, you can enter text, move the cursor, and delete individual characters (`Del` and `Backspace`). The difference between `I` and `A` is that with `I`, the input starts at the current cursor position, but with `A`, input starts at the character behind it.

► **Complex command mode**

However, most commands are entered in complex command mode, which can be activated using `:`. You may have to exit the insert mode beforehand by pressing `Esc`.

Table 17 provides an overview of the most important keyboard shortcuts for deleting text. Table 18 lists useful commands for moving the cursor. Table 19, Table 20, and Table 21 deal with how to copy, select, and edit text.

Keyboard Shortcut	Function
Keyboard Shortcuts in insert mode	
`Del`, `Backspace`	The usual meaning
Commands in standard mode	
`X`	Deletes the character at the cursor position or the selected text
`⇧`+`X`	Deletes the character before the cursor
`D`, `D`	Deletes the current line
`D` *cursor command*	Deletes the text according to the command for cursor movement (see Table 18) Examples: `D`, `$` deletes everything to the end of the line. `D`, `B` deletes the previous word `D`, `W` deletes the next word

Table 17 Deleting Text

Keyboard Shortcut	Function
Cursor keys	Standard meaning
[H]/[L]	Moves the cursor to the left/right
[J]/[K]	Moves the cursor down/up
[⇧]+[H]/[⇧]+[L]	Moves the cursor to the beginning or end of the current page
[⇧]+[M]	Moves the cursor to the center of the current page
[B]/[W]	Moves the cursor one word to the left/right
[E]	Moves the cursor to the end of the word
[G], [E]	Moves the cursor to the beginning of the word
[,]	Moves the cursor to the beginning of the current/next block
[{], [}]	Moves the cursor to the beginning of the current/next paragraph
[^], [$]	Moves the cursor to the beginning or end of the line
[⇧]+[G]	Moves the cursor to the end of the line
[G], [G]	Moves the cursor to the beginning of the line
n [⇧]+[G]	Moves the cursor to line n
n [｜]	Moves the cursor to column n
[%]	Moves the cursor to the corresponding parenthesis character ()[]{}

Table 18 Keyboard Shortcuts for Cursor Movement in Standard Mode

Keyboard Shortcut	Function
[V]	(De)activates the character selection mode
[⇧]+[V]	(De)activates the line selection mode
[Ctrl]+[V]	(De)activates the block selection mode
[A], [W]	Extends the selection by one word
[A], [S]	Extends the selection by one block
[A], [P]	Extends the selection by one paragraph
[A], [B]	Extends the selection by one () level
[A], [⇧]+[B]	Extends the selection by one {} level

Table 19 Selecting Text

Keyboard Shortcut	Function
G , V	Selects the most recently selected text again
O	Toggles the cursor position between the start and end of the selection

Table 19 Selecting Text (Cont.)

Keyboard Shortcut	Function
Y	Copies the selected text to the copy register
Y , Y	Copies the current line to the copy register
Y *cursor command*	Copies the text captured by the cursor movement Example: Y , } copies text to the end of the paragraph

Table 20 Copying Text to the Copy Register

Keyboard Shortcut	Function
X	Deletes the selected text
Y	Copies the selected text to the copy register
~	Changes between uppercase and lowercase
Y	Merges the selected lines into one long line
G , Q	Performs a line break (for continuous text)
> , <	Indents or outdents the text by one tab position
=	Indents the text according to the current indent mode
!sort	Sorts the lines using the external sort command

Table 21 Editing Selected Text

Searching and Replacing

In the default mode, / *searchtext* Enter moves the cursor to the text being searched for. N repeats the search, and ⇧ + N repeats the search backwards. To search backward from the outset, you must start the search using ? *search term*. Table 22 summarizes the most important special characters for searching for patterns.

Character	Meaning
.	Any character.
^ $	Start of line/end of line.
\< \>	Start of word/end of word.
[a-e]	A character between *a* and *e*.
\s, \t	A space or a tab character.
\(\)	Summarizes a search pattern as a group.
\=	The search term must occur 0 times or once.
*	The search term may occur any number of times (even 0 times).
\+	The search term must occur at least once.

Table 22 Special Characters in the Search Term

vim distinguishes between uppercase and lowercase during the search. If you don't want this, you must initiate the search pattern with /c (applies only to this search) or run : set ignorecase (applies to all subsequent searches).

Use : set incsearch to activate the incremental search: vim moves the cursor to the first matching location as soon as you enter the search text using /*searchterm*. Enter ends the search, and Esc cancels it. After the search, all matches in the text remain highlighted until you perform a new search or run : nohlsearch.

To replace all occurrences of the text *abc* with *efg* without prompting, you need to run : %s/*abc*/*efg*/g in standard mode. ' ' then takes you back to the start of the search. Table 23 summarizes some variants of the search-and-replace command.

Keyboard Shortcut	Function
: %s/*abc*/*efg*/g	Replaces all occurrences of *abc* with *efg* without query
: %s/*abc*/*efg*/gc	Replaces all occurrences of *abc* with *efg* with query
: %s/*abc*/*efg*/gci	Replaces without considering uppercase or lowercase

Table 23 Searching and Replacing

The Author

 **Michael Kofler** is a programmer and one of the most successful and versatile computing authors in the German-speaking world. His current topics include AI, Linux, Docker, Git, hacking and security, Raspberry Pi, and the programming languages Swift, Java, Python, and Kotlin. Michael also teaches at the Joanneum University of Applied Sciences in Kapfenberg, Austria.

- Your complete, cross-distribution, professional guide to Linux, for beginners and advanced users

- Get detailed instructions for installation, configuration, and administration, on both desktop and server

- Set up security, virtualization, and more

Michael Kofler

Linux

The Comprehensive Guide

Beginner or expert, professional or hobbyist, this is the Linux guide you need! Install Linux and walk through the basics: working in the terminal, handling files and directories, using Bash, and more. Then get into the nitty-gritty details of configuring your system and server, from compiling kernel modules to using tools like Apache, Postfix, and Samba. With information on backups, firewalls, virtualization, and more, you'll learn everything there is to know about Linux!

1178 pages, pub. 05/2024
E-Book: $54.99 | **Print:** $59.95 | **Bundle:** $69.99

www.rheinwerk-computing.com/5779

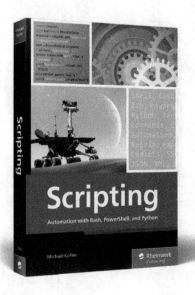

- Learn to work with scripting languages such as Bash, PowerShell, and Python

- Get to know your scripting toolbox: cmdlets, regular expressions, filters, pipes, and REST APIs

- Automate key tasks, including backups, database updates, image processing, and web scraping

Michael Kofler

Scripting

Automation with Bash, PowerShell, and Python

Developers and admins, it's time to simplify your workday. With this practical guide, use scripting to solve tedious IT problems with less effort and fewer lines of code! Learn about popular scripting languages: Bash, PowerShell, and Python. Master important techniques such as working with Linux, cmdlets, regular expressions, JSON, SSH, Git, and more. Use scripts to automate different scenarios, from backups and image processing to virtual machine management. Discover what's possible with only 10 lines of code!

472 pages, pub. 02/2024
E-Book: $44.99 | **Print:** $49.95 | **Bundle:** $59.99
www.rheinwerk-computing.com/5851